Assessing COVID-19 and Other Pandemics and Epidemics using Computational Modelling and Data Analysis

Subhendu Kumar Pani • Sujata Dash
Wellington P. dos Santos
Syed Ahmad Chan Bukhari • Francesco Flammini
Editors

Assessing COVID-19 and Other Pandemics and Epidemics using Computational Modelling and Data Analysis

 Springer

Editors
Subhendu Kumar Pani
School of Computer Science and Engg.
Orissa Engineering College
Bhubaneswar-Khordha, Odisha, India

Wellington P. dos Santos
Department of Biomedical Engineering
Federal University of Pernambuco
Recife, Pernambuco, Brazil

Francesco Flammini 🄳
Innovation, Design and Engineering
Mälardalen University
Västerås, Sweden

Sujata Dash
Department of Computer Science
Maharaja Sriram Chandra Bhanja Deo
University (Erstwhile North Orissa
University)
Takatpur, Baripada, Odisha, India

Syed Ahmad Chan Bukhari
Comp Science, Mathematics and Science
St. John's University
Jamaica, NY, USA

ISBN 978-3-030-79755-3   ISBN 978-3-030-79753-9   (eBook)
https://doi.org/10.1007/978-3-030-79753-9

This Springer imprint is published by the registered company Springer Nature Switzerland AG
The registered company address is: Gewerbestrasse 11, 6330 Cham, Switzerland

Preface

Overview

This book covers the topic of COVID-19 and Other Pandemics and Epidemics data analytics using computational modelling is an emerging field of research at the intersection of information science and computer science. Biomedical and health informatics is an emerging field of research at the intersection of information science, computer science, and healthcare. Pandemics and epidemics usher into a new era that will bring tremendous opportunities and challenges due to the availability of a huge amount of medical data. The COVID-19 outburst is among the most important tragedies the world has ever faced. It has already killed hundreds of thousands of people; millions of people are infected, billions are under lodown, and this is costing trillions of USDs to the world economy. The aim of pandemics and epidemics research is to ensure high-quality, efficient healthcare as well as better treatment and quality of life by efficiently analysing the abundant medical and healthcare data, including a patient's data, electronic health records (EHRs), and lifestyle. Earlier, it was common requirement to have a domain expert develop a model for biomedical or healthcare; however, recent advancements in representation learning algorithms allow to automatically learn the pattern and representation of the given data for the development of such model. Medical image mining, a novel research area, due to its large number of medical images, is increasingly generated and stored digitally. These images are mainly in the form of computed tomography (CT), X-ray, nuclear medicine imaging (PET, SPECT), magnetic resonance imaging (MRI), and ultrasound. Patients' biomedical images can be digitised using data mining techniques and may help in answering several important and critical questions related to healthcare. Image mining in medicine can help uncover new relationships between different data and reveal new useful information that can be helpful for scientists in treating their patients.

The book will play a vital role in improvising human life to a great extent. All the researchers and practitioners will be highly benefited by those who are working in the field of biomedical, health informatics, and artificial intelligence. This book would be a good collection of state-of-the-art approaches for data mining–based medical and health-related applications.

Objective

The objective of this book is to present innovative solutions utilising informatics to deal with various issues related to the COVID-19 outbreak, including health data analytics, information exchange, knowledge sharing, Internet of Things (IoT)-based solutions, and the implementation, assessment, adoption, and management of healthcare informatics solutions. Hence, this book informs the audience about the recent findings/results concerning a wide variety of COVID-19 and other pandemics and epidemics using computational modeling and data analysis. It will be very beneficial for new researchers and practitioners working in the field to quily know the best performing methods. They would be able to compare different approaches and can carry forward their research in the most important area which has direct impact on betterment of human life and health.

Organisation

This book, *Assessing COVID-19 and Other Pandemics and Epidemics Using Computational Modelling and Data Analysis*, consists of 19 edited chapters. A brief summary of the chapters is given below:

Chapter "Artificial Intelligence (AI) and Big Data Analytics for COVID-19 Pandemic" has summarised state-of-the-art applications of AI and big data analytics in combating the COVID-19 crisis up to now. Construction of a vast cyber infrastructure has also been recommended to stimulate global collaborations for future research on AI and big data analytics in the field of predictive, diagnostic, and therapeutic techniques against COVID-19 and other similar pandemics in the future. Public transportation is not safe during the COVID-19 pandemic even after lodown, because it will be very hard to maintain social distancing in public transport. So, to resolve this problem, chapter "COVID-19 TravelCover: Post-Lodown Smart Transportation Management System" has come up with an idea of making an intelligent application to schedule the timings of transportation, avoiding over occupancy of public transport, providing them the shortest route to reach their desired destination, providing them proper guidelines, and also providing them information of the nearest hospitals for any emergency. Chapter "Diverse Techniques Applied for Effective Diagnosis of COVID-19" intends to provide detailed information on numerous comprehensive techniques that are applied for the diagnosis of COVID-19. Machine learning techniques that can help diagnose COVID19 using X-ray or CT scan images of patients are highlighted in chapter "A Review on Detection of COVID-19 Patients Using Deep Learning Techniques".

Chapter "Internet of Health Things (IoHT) for COVID-19" intends to provide holistic information on how the concept of the Internet of Health Things (IoHT) could help in the management of COVID 19 diseases. Chapter "Diagnosis for COVID-19" presents a systematic review of the origin of the coronavirus; its types, mode of transmission, and symptoms; current developments in diagnosing and testing; and vaccine trials. Chapter "IoT in Combating COVID-19 Pandemics: Lessons for Developing Countries" highlights the role that digital technologies such as Internet of Things (IoT) play in mitigating the effect of the outbreak when combined with technologies such as artificial intelligence (AI) and cloud, which offers a wider application during this unfortunate crisis. Chapter "Machine Learning Approaches for COVID-19 Pandemic" intends to provide detailed information on the usage of machine learning and its significance in recognising existing drugs that show potential for the management of COVID-19 patients. Chapter "Smart Sensing for COVID-19 Pandemic" intends to provide detailed information on the function of smart technology in reducing the spread of COVID-19 with precise emphasis on improvement in the field of drone, machine learning, artificial intelligence, sensor technology, and mask. Chapter "eHealth, mHealth, and Telemedicine for COVID-19 Pandemic" intends to provide detailed information on the application of e-health, m-health, and telemedicine as a pre-emptive measure to increase clinical care. Chapter "Prediction of Care for Patients in a COVID-19 Pandemic Situation Based on Hematological Parameters" analyses intelligent classifiers that are able to make hospitalisation predictions considering three possible scenarios: regular ward, semi-intensive care unit, and intensive care unit, corresponding to mild (non-critical), moderate, and serious cases. Chapter "Bioinformatics in Diagnosis of COVID-19" focusses on the various bioinformatics and computational tools developed and used in COVID-19 diagnosis. Chapter "COVID-19 Detection Using Discrete Particle Swarm Optimization Clustering with Image Processing" aims to extract and assess the corona virus disease (COVID-19) induced pneumonia infection in lung using X-ray images. Chapter"LSTM-CNN Deep Learning–Based Hybrid System for Real-Time COVID-19 Data Analysis and Prediction Using Twitter Data" has proposed a LSTM-CNN Deep Learning Based Hybrid Single Window System for real-time COVID-19 data analysis and prediction using Twitter data. Chapter "An Intelligent Tool to Support Diagnosis of Covid-19 by Texture Analysis of Computerized Tomography X-ray Images and Machine Learning" has proposed an automatic system for COVID-19 diagnosis using machine learning techniques and CT X-ray images named IKONOS-CT. Chapter "Analysis of Blochain-Baed COVID-19 Data" provides a strong reason for using blochain for data validation with the case study of Mipasa –- an open data platform to support COVID-19 response (powered by the IBM Blochain Platform and the IBM Cloud and provided by a blochain-based platform called HECERA). Chapter "Intelligent Systems for Dengue, Chikungunya and Zika Temporal and Spatio-Temporal Forecasting: A Contribution and a Brief Review" presents a literature review to identify methods of predicting cases of arboviruses, as well as the prediction of breeding sites. Chapter "Machine Learning Approaches for Temporal and Spatio-Temporal Covid-19 Forecasting: A Brief Review and a Contribution" throws insight on the use of artificial intelligence and models which were designed using machine-learning algorithms with features for temporal and spatio-temporal investigation and prediction of cases of COVID-19. Chapter "Image Reconstruction for COVID-19 Using Multifrequency Electrical Impedance Tomography"presents the basic characteristics of a wireless, low-cost, and portable MfEIT system and definitions and modelling of the two-dimensional D-bar method for image reconstruction.

Target Audiences

The book's target audience are researchers in various fields, including computer science, medical informatics, healthcare IOT, artificial intelligence, machine learning, image processing, and clinical big data analytics. In addition, as the book aims to contain the literature review on each of the topics, chronological development of different topics, as well as updated information, any advanced course on this topic can use it as a reference book/ textbook with selected chapters. Graduate students, researchers, and professionals interested in machine learning, IoT, expert systems healthcare, information retrieval, big data analytics, and various machine-learning areas will benefit by reading the book.

Bhubaneswar, India   Subhendu Kumar Pani
Baripada, India   Sujata Dash
Recife, Brazil   Wellington P. dos Santos
Jamaica, NY, USA   Syed Ahmad Chan Bukhari
Västerås, Sweden   Francesco Flammini

Acknowledgements

The editors would like to anowledge the help of all the people involved in this project and, more specifically, thank the reviewers who took part in the review process. Without their support, this book would not have become a reality.

First, the editors would like to thank each one of the authors for their time, contribution, and understanding during the preparation of the book.

Second, the editors wish to anowledge the valuable contributions of the reviewers regarding the improvement of quality, coherence, and content presentation of chapters.

Last but not the least, the editors wish to anowledge the love, understanding, and support of their family members during the preparation of the book.

Bhubaneswar, Odisha, India	Subhendu Kumar Pani
Baripada, India	Sujata Dash
Recife, Brazil	Wellington P. dos Santos
NY, USA	Syed Ahmad Chan Bukhari
Mälardalen University, Sweden	Francesco Flammini

Contents

Contributors

Mayowa J. Adeniyi Department of Physiology, Edo State University Uzairue, Iyamho, Edo State, Nigeria

Charles Oluwaseun Adetunji Applied Microbiology, Biotechnology and Nanotechnology Laboratory, Department of Microbiology, Edo University Iyamho, Auchi, Edo State, Nigeria

Olorunsola Adeyomoye Department of Physiology, University of Medical Sciences, Ondo City, Nigeria

Aisha Aldosery Institute for Risk and Disaster Reduction, University College London, UCL-IRDR, London, UK

Enoch Alex Department of Human Physiology, Ahmadu Bello University Zaria, Kaduna State, Nigeria

Tercio Ambrizzi Department of Atmospheric Sciences, IAG, University of São Paulo, USP, São Paulo, Brazil

Saanya Aroura SRM University Delhi-NCR, Rajiv Gandhi Education City, Sonepat, Haryana, India

Bishvajit Bakshi Centre for Management of Health Services, Indian Institute of Management-Ahmedabad, Ahmedabad, Gujarat, India

Jonathan Bandeira Graduate Program in Computer Engineering, Polytechnique School of the University of Pernambuco, Recife, Brazil

Selma Basibuyuk Bogaziçi University, Institute of Environmental Sciences, Istanbul, Turkey

Pedro Bertemes-Filho Department of Electrical Engineering, Universidade do Estado de Santa Catarina, Joinville, Santa Catarina, Brazil

Sitanath Biswas Department of Computer Science, Maharaja Sriram Chandra Bhanja Deo University (Erstwhile North Orissa University), Wellington dos Santos Takatpur, Baripada, Odisha, India

Iuri Valério Graciano Borges Department of Atmospheric Sciences, IAG, University of São Paulo, USP, São Paulo, Brazil

Luiza Campos Department of Civil Environmental and Geomatic Engineering, University College London, UCL, London, UK

Ayobami Dare Department of Physiology, School of Laboratory Medicine and Medical Sciences, College of Health Sciences, Westville Campus, University of KwaZulu-Natal, Durban, South Africa

Sujata Dash Department of Computer Science, Maharaja Sriram Chandra Bhanja Deo, University (Erstwhile North Orissa University), Wellington dos Santos Takatpur, Baripada, Odisha, India

Ana Clara Gomes da Silva Department of Biomedical Engineering, Federal University of Pernambuco, Recife, Pernambuco, Brazil

Cecilia Cordeiro da Silva Center for Informatics, Federal University of Pernambuco, CIn-UFPE, Recife, Brazil

Abel Guilhermino da Silva Filho Center for Informatics, Federal University of Pernambuco, CIn-UFPE, Recife, Brazil

Lucas Job Brito de Araújo Center for Informatics, Federal University of Pernambuco, CIn-UFPE, Recife, Brazil

Valter Augusto de Freitas Barbosa Academic Unit of Serra Talhada, Rural Federal University of Pernambuco, Serra Talhada, Brazil

Clarisse Lins de Lima Graduate Program in Computer Engineering, Polytechnique School of the University of Pernambuco, Recife, Brazil

Maíra Araújo de Santana Graduate Program in Computer Engineering, Polytechnique School of the University of Pernambuco, Recife, Brazil

Ricardo Emmanuel de Souza Department of Biomedical Engineering, Federal University of Pernambuco, Recife, Pernambuco, Brazil

Samuel Barbosa Jatobá de Souza Center for Informatics, Federal University of Pernambuco, CIn-UFPE, Recife, Brazil

Wellington P. dos Santos Department of Biomedical Engineering, Federal University of Pernambuco, Recife, Pernambuco, Brazil

Livia Dutra Department of Atmospheric Sciences, IAG, University of São Paulo, USP, São Paulo, Brazil

Flávio Secco Fonseca Graduate Program in Computer Engineering, Polytechnique School of the University of Pernambuco, Recife, Brazil

Larisa Garipova K.G. Razumovsky Moscow State University of Technologies and Management (the First Cossack University), Moscow, Russia

Juliana Carneiro Gomes Graduate Program in Computer Engineering, Polytechnique School of the University of Pernambuco, Recife, Brazil

Barnini Goswami Department of Computer Science Engineering, Krishna Engineering College, Ghaziabad, UP, India

Archana Gupta SRM University Delhi-NCR, Rajiv Gandhi Education City, Sonepat, Haryana, India

Kajal Gupta Department of Computer Science Engineering, Krishna Engineering College, Ghaziabad, UP, India

Olga Isabekova K.G. Razumovsky Moscow State University of Technologies and Management (the First Cossack University), Moscow, Russia

Arush Jain Department of Information Technology, G. L. Bajaj Institute of Technology and Management, Greater Noida, UP, India

Kate Jones Centre for Biodiversity and Environment Research, Department of Genetics, Evolution and Environment, University College London, UCL, London, UK

Luiz Antônio Albuquerque Júnior Center for Informatics, Federal University of Pernambuco, CIn-UFPE, Recife, Brazil

Tadepalli Sarada Kiranmayee Department of Computer Science, University of Manitoba, Winnipeg, MB, Canada

Natalia Koriagina E. A. Vagner Perm State Medical University, Perm, Permskaja Oblast, Russia

Patty Kostkova Department of Atmospheric Sciences, IAG, University of São Paulo, USP, São Paulo, Brazil

K. N. Krishnamurthy Department of Agricultural Statistics, Applied Mathematics and Computer Science, University of Agricultural Sciences, Bengaluru, Karnataka, India

Antônio Ravely T. Lima Department of Biomedical Engineering, Federal University of Pernambuco, Recife, Pernambuco, Brazil

Babita Majhi Department of CSIT, Guru Ghasidas Vishwavidyalaya, Central University, Bilaspur, Chhatisgarh, India

Ritanjali Majhi School of Management, NIT Surathkaul, Surathkaul, Karnataka, India

Shreya Majumdar Department of Computer Science Engineering, Krishna Engineering College, Ghaziabad, UP, India

David William Cordeiro Marcondes Department of Electrical Engineering, Universidade do Estado de Santa Catarina, Joinville, Santa Catarina, Brazil

Gabriel Souza Marques Center for Informatics, Federal University of Pernambuco, CIn-UFPE, Recife, Brazil

Aras Ismael Masood Information Technology Department, Technical College of Informatics, Sulaimani Polytechnic University, Sulaymaniyah, Iraq

Tiago Lima Massoni Department Systems and Computing, Federal University of Campina Grande, Campina Grande, Brazil

Giselle Machado Magalhães Moreno Department of Atmospheric Sciences, IAG, University of São Paulo, USP, São Paulo, Brazil

Kamalakanta Muduli Department of Mechanical Engineering, Papua New Guinea University of Technology, Lae, Papua New Guinea

Anwar Musah Institute for Risk and Disaster Reduction, University College London, UCL-IRDR, London, UK

Olugbemi Tope Olaniyan Laboratory for Reproductive Biology and Developmental Programming, Department of Physiology, Edo University Iyamho, Iyamho, Nigeria

Tássia D. Muniz S. Oliveira Department of Biomedical Engineering, Federal University of Pernambuco, Recife, Pernambuco, Brazil

Anand Bhushan Pandey Department of Information Technology, G. L. Bajaj Institute of Technology and Management, Greater Noida, UP, India

Pramit Pandit Department of Agricultural Statistics, Bidhan Chandra Krishi Viswavidyalaya, Mohanpur, West Bengal, India

Oyekola Peter Department of Mechanical Engineering, Papua New Guinea University of Technology, Lae, Papua New Guinea

Ekaterina Petukhova K.G. Razumovsky Moscow State University of Technologies and Management (the First Cossack University), Moscow, Russia

Anjali Priyadarshini SRM University Delhi-NCR, Rajiv Gandhi Education City, Sonepat, Haryana, India

Hari Mohan Rai Department of Electronics and Communication Engineering, Krishna Engineering College, Ghaziabad, UP, India

Adimuthu Ramasamy Department of Business Studies, Papua New Guinea University of Technology, Lae, Papua New Guinea

Maksim Rebezov Prokhorov General Physics Institute, Russian Academy of Sciences, Moscow, Russia

K.G. Razumovsky Moscow State University of Technologies and Management (the First Cossa University), Moscow, Russia

Flaviano Palmeira Santos Department of Biomedical Engineering, Federal University of Pernambuco, Recife, Pernambuco, Brazil

Mohammad Ali Shariati K.G. Razumovsky Moscow State University of Technologies and Management (the First Cossack University), Moscow, Russia

Sanjana Sharma SRM University Delhi-NCR, Rajiv Gandhi Education City, Sonepat, Haryana, India

Eduardo Luiz Silva Center for Informatics, Federal University of Pernambuco, CIn-UFPE, Recife, Brazil

Arun Kumar Singh Department of Information Technology, G. L. Bajaj Institute of Technology and Management, Greater Noida, UP, India

Suchismita Swain Department of Mechanical Engineering, Temple City Institute of Technology and Engineering, Khordha, Odisha, India

Rahul Thangeda School of Management, NIT Warangal, Warangal, Telengana, India

Ruppa K. Thulasiram Department of Computer Science, University of Manitoba, Winnipeg, MB, Canada

Arianne Sarmento Torcate Graduate Program in Computer Engineering, Polytechnique School of the University of Pernambuco, Recife, Brazil

Ashish Tripathi Department of Information Technology, G. L. Bajaj Institute of Technology and Management, Greater Noida, UP, India

Merve Tunali Bogaziçi University, Institute of Environmental Sciences, Istanbul, Turkey

Vaibhav Tyagi Department of Information Technology, G. L. Bajaj Institute of Technology and Management, Greater Noida, UP, India

Bhimavarapu Usharani Department of Computer Science and Engineering, Koneru Lakshmaiah Education Foundation, Vaddeswarm, AP, India

Mêuser Jorge Silva Valença Graduate Program in Computer Engineering, Polytechnique School of the University of Pernambuco, Recife, Brazil

Prem Chand Vashist Department of Information Technology, G. L. Bajaj Institute of Technology and Management, Greater Noida, UP, India

Julia Grasiela Busarello Wolff Department of Electrical Engineering, Universidade do Estado de Santa Catarina, Joinville, Santa Catarina, Brazil

Orhan Yenigün Bogaziçi University, Institute of Environmental Sciences, Istanbul, Turkey

Abbreviations

ACE-2	Angiotensin Converting Enzyme 2
AE	Autoencoders
AI	Artificial Intelligence
AIC	Akaike Information Criteria
ANFIS	Adaptive Neuro Fuzzy Inference System
ANN	Artificial Neural Network
AR	Augmented Reality
ARDS	Acute Respiratory Distress Syndrome
ARIMA	Auto-Regressive Integrated Moving Average
BCNN	Bayesian Convolutional Neural Networks
BDPSO	Blended Discrete Particle Swarm Optimization
Bi-LSTM	Bidirectional LSTM
CCG	Contract Creation Graph
CET	Central European Time Zone
CIG	Contract Innovation Graph
CIoMT	Cognitive Internet of Medical Things
CLAHE	Contrast Limited Adaptive Histogram Equalization
CNN	Convolution Neural Network
Co-TNTs	Cobalt-Functionalized TiO_2 Nanotubes
COVID-19	Corona Virus Disease 2019
CovidSIM	Covid Simulation Tool
COVNet	COVID-19 Detection Neural Network
CREST	Cas 13-Based Rugged Equitable Scalable Testing
CRISPR	Clustered Regularly Interspaced Short Palindromic Repeats
CSV	Comma-Separated Values
CT Scan	Computed Tomography Scan
CT	Computerized Tomography
CXR	Chest X-ray
DBN	Deep Belief Networks
DCNN	Deep Convolutional Neural Network
DETECTR	DNA Endonuclease Targeted CRISPR Trans Reporter

DL	Deep Learning
EBI	European Bioinformatics Institute
ECDC	European Centre for Disease Prevention and Control
ELISA	Enzyme-Linked Immunosorbent Assay
ES	Exponential Smoothing
FN	False Negatives
FP	False Positives
GAN	Generative Adversarial Networks
Gis	Global Information Systems
Glm	Generalized Logistic Growth Model
Gru	Gated Recurrent Units
EHR	Electronic Health Record
HIPAA	Health Insurance Portability and Accountability
HM	Hidden Markov
HOMA-IR	Homeostasis Model Assessment Insulin Resistance
ID	Identity
IEEE	Institute of Electrical and Electronics Engineers
IoHT	Internet of Health Things
IoT	Internet of Things
ISO	International Standard Organization
LASSO	Least Absolute Shrinkage and Selection Operator
LMIA	Loop-Mediated Isothermal Amplification
LSTM	Long Short Term Memory
MERS	Middle East Respiratory Syndrome
MFG	Money Flow Graph
mHealth	Mobile Health
ML	Machine Learning
MS	Mass-Spectrometry
NARNN	Nonlinear Autoregressive Neural Network
NGO	Non-governmental Organization
NGS	Next Generation Sequencing
NLP	Natural Language Processing
NP	Nasopharyngeal
OP	Oropharyngeal
PCR	Polymerase Chain Reaction
PIR	Protein Information Resource
PPE	Personal Protective Equipment
PSO	Particle Swarm Optimization
qPCR	Quantitative Polymerase Chain Reaction
RBD	Receptor Binding Domain
RF	Random Forest
RFID	Radio Frequency Identification
RNN	Recurrent Neural Network
rRT-PCR	Real Time Reverse Transcription Polymerase Chain Reaction
RT-PCR	Real-Time Polymerase Chain Reaction

RT-PCR	Reverse Transcriptase Polymerase Chain Reaction
RT-qPCR	Real-Time Quantitative Polymerase Chain Reaction
SARS	Severe Acute Respiratory Syndrome
SARS-CoV	Severe Acute Respiratory Syndrome Corona Virus
SARS-CoV-2	Severe Acute Respiratory Syndrome Coronavirus-2
SCM	Supply Chain Management
SEIR	Susceptible, Exposed, Infectious, Recovered
SIB	Swiss Institute of Bioinformatics
SIR	Susceptible Infected Recovery
SVM	Support Vector Machine
TB	Tuberculosis
TMB	Tetramethyl Benzidine
TN	True Negatives
TP	True Positives
URLs	Uniform Resource Locator
USSD	Unstructured Supplementary Service Data
UTR	Untranslated Region
UTXOs	Unspent Transaction Outputs
VADR	Viral Annotation Define R
VAE	Variational Auto Encoder
WHO	World Health Organization

Editors' Biography

Subhendu Kumar Pani received his PhD from Utkal University, Odisha, India, in the year 2013. He is working as professor and principal of Krupajal Computer Academy, Odisha, India. He has more than 17 years of teaching and research experience. His research interests include data mining, big data analysis, web data analytics, fuzzy decision making, and computational intelligence. He is the recipient of five researcher awards. In addition to research, he has guided two PhD students and 31 MTech students. Dr Pani has published 51 international journal papers (25 Scopus index). His professional activities include roles as book series editor (CRC Press, Apple Academic Press, and Wiley-Scrivener), associate editor, editorial board member, and/or reviewer of various international journals. He is associate of a number of conference societies. Dr Pani has more than 150 international publications, 5 authored books, 15 edited and upcoming books, and 20 book chapters to his credit. He is a fellow of SSARSC and life member of IE, ISTE, ISCA, OBA, OMS, SMIACSIT, SMUACEE, and CSI.

Sujata Dash is currently working as an associate professor of computer science at North Orissa University in the Department of Computer Science, Baripada, India. She has been consistently delivering her services in teaching and guiding students for more than two and a half decades. She is a recipient of Titular Fellowship from the Association of Commonwealth Universities, UK. Sujata has worked as a visiting professor in the Computer Science Department at the University of Manitoba, Canada. She has published more than 170 technical papers in international journals/proceedings at international conferences; edited book chapters for reputed publishers like Springer, Elsevier, IEEE, and IGI Global USA; obtained 10 national and international patents; and has published many text books, monographs, and edited books. She is a member of international professional associations like ACM, IRSS, CSI, IMS, OITS, OMS, IACSIT, IST, and IEEE and is a reviewer of around 15 international journals which include World Scientific, Bioinformatics, Springer, IEEE ACCESS, Inderscience, and Science Direct publications. In addition, she is a member of the editorial board of around 10 international journals. Sujata has visited many countries and delivered keynotes, invited speech, and chaired many special sessions at international conferences in India and abroad. Her current research interest includes machine learning, data mining, big data analytics, bioinformatics, soft computing, and intelligent agents.

Wellington P. dos Santos is associate professor in the Department of Biomedical Engineering at Federal University of Pernambuco (UFPE), Recife, Brazil. He obtained his PhD in electrical engineering from the Federal University of Campina Grande (UFCG), Campina Grande, and Master of Electrical Engineering and Bachelor of Electronic and Electrical Engineering from UFPE, Recife, Brazil. His main research interests are: diagnostic support systems, digital epidemiology, applied neuroscience, serious games in health, and artificial intelligence applied to health.

Syed Ahmad Chan Bukhari is an assistant professor and director of healthcare informatics at St. John's University, New York. He received his PhD in computer science from the University of New Brunswi, Canada, and then went on to complete his postdoctoral fellowship at Yale School of Medicine, where he worked with Stanford University, Centre of Expanded Data Annotation and Retrieval (CEDAR), to develop data submission pipelines to improve scientific experimental reproducibility. His current research efforts are concentrated on addressing several core problems in healthcare informatics and data science. He mainly focuses on devising techniques to semantically confederate heterogeneous biomedical data and develop AI-based predictive models for clinical outcomes. These techniques also alleviate many data access–related challenges faced by healthcare providers. Dr Bukhari is a senior IEEE member and a distinguished ACM speaker who serves as an editorial board member of multiple scientific journals. In September 2019, he was awarded with the IEEE Technological Innovation Award. His research work has been published in top-tier journals and pied up by various scientific blogs and international media.

Francesco Flammini got his master (2003) and doctoral (2006) degrees in computer engineering from the University of Naples Federico II, Italy. He is currently a professor of computer science at Mälardalen University (Sweden). He has been an associate professor leading the Cyber-Physical Systems environment at Linnaeus University (Sweden). He has worked for 15 years in private and public companies, including Ansaldo STS (now Hitachi Rail) and IPZS (Italian State Mint and Polygraphic Institute), leading international projects addressing intelligent transportation and infrastructure security. He is an IEEE senior member and an ACM distinguished speaker.

Artificial Intelligence (AI) and Big Data Analytics for the COVID-19 Pandemic

Pramit Pandit, K. N. Krishnamurthy, and Bishvajit Bakshi

1 Introduction

On March 11, 2020, the World Health Organization (WHO) has proclaimed the Coronavirus Disease 2019 (COVID-19) outbreak as a pandemic. This deadly viral disease has triggered a global public health emergency, affecting nearly 229 countries, areas, or territories and resulting in over 137 million confirmed cases and over 2.95 million deaths worldwide by the middle of April, 2021 [57]. Figures 1 and 2 depict the global patterns of COVID-19 confirmed and death cases, respectively. Even though Wuhan city of China is regarded as the epicentre of this fatal disease [58, 60], it has significantly affected other developed as well as developing countries including the USA, India, Italy, France, Spain, and so on [47, 49]. It has presented humanity with an unprecedented challenge, forcing half of the planet to shut down in order to avert its spread by maintaining social distancing.

COVID-19 is caused by one of the recent and novel strains of coronavirus: Severe Acute Respiratory Syndrome Coronavirus-2 (SARS-CoV-2). It belongs to the Nidovirales (order) and Coronaviridae (family) according to the classification systems, with a genome size of around 30 kb, enveloped in single-stranded positive-sense RNA [32]. It is transmitted by inhalation or contact with infected droplets or

P. Pandit
Department of Agricultural Statistics, Bidhan Chandra Krishi Viswavidyalaya, Mohanpur, West Bengal, India

K. N. Krishnamurthy
Department of Agricultural Statistics, Applied Mathematics and Computer Science, University of Agricultural Sciences, Bengaluru, Karnataka, India

B. Bakshi (✉)
Centre for Management of Health Services, Indian Institute of Management-Ahmedabad, Ahmedabad, Gujarat, India

© The Author(s), under exclusive license to Springer Nature Switzerland AG 2022
S. K. Pani et al. (eds.), *Assessing COVID-19 and Other Pandemics and Epidemics using Computational Modelling and Data Analysis*,
https://doi.org/10.1007/978-3-030-79753-9_1

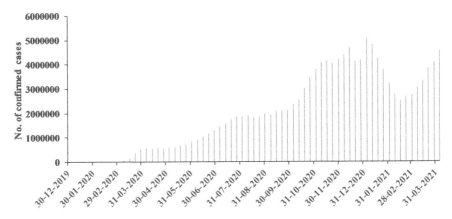

Fig. 1 The global weekly pattern of COVID-19 confirmed cases

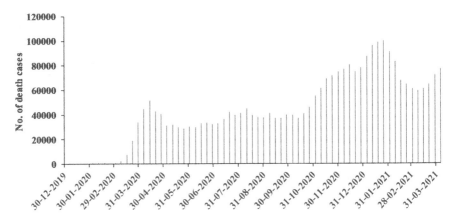

Fig. 2 The global weekly pattern of COVID-19 death cases

fomites and has a 3–14 days' incubation period [23]. However, the overall fatality rate is remarkably higher in the case of elderly people with compromised health conditions [18, 64]. Because of the seriousness of the ongoing situation, it is critical to estimate and comprehend the disease transmission dynamics in the early stages of the outbreak. Different characteristics of the COVID-19 incidents are illustrated in Fig. 3.

Throughout the world, numerous scientists, clinicians, and policymakers are constantly looking for new ways to track and manage the disease transmission, which in turn necessitates the development of advanced methods of computation. Artificial intelligence (AI) has played a vital role in this regard to combat the COVID-19 pandemic by providing improved techniques for pattern recognition, prediction, forecasting, and optimisation. Raza [43] has carried out an extensive meta-analysis to point out the areas of COVID-19 research, where AI techniques

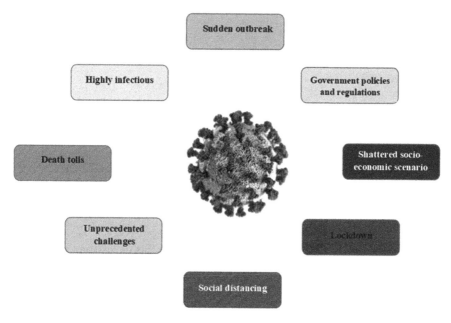

Fig. 3 Different characteristics of the COVID-19 incidents

have prominently been applied and where the research gaps are required to be bridged.

In addition, the sudden influx in the number of cases and health data has resulted in a valuable pool of data and knowledge [51, 56]. Consequently, there is a pressing need to store such a vast volume of data to conduct research on the virus, pandemic, and countermeasures. Big data is a cutting-edge technology that can digitally store such a vast volume of data. It aids computational analysis by revealing hidden patterns, trends, correlations, and differences [62, 63]. Big data analytics can be used to perform clinical trials as well as to speed up the treatment process by analysing the patients' records, thanks to its comprehensive data capturing capabilities [22].

We have explored different indexing databases (PubMed, Web of Science, Google Scholar, and Scopus) and preprint (bioRxiv, medRxiv, and arXiv) servers to carry out our search for relevant literatures. We have utilised 'AI', 'big data analytics', and 'COVID-19 research' as keywords. The number of publications on AI and big data analytics in COVID-19 research available on these platforms are graphically presented in Fig. 4. The highest publication count is observed in the case of Google Scholar as it indexes almost all the journals, books, conference proceedings, preprints, etc.

With this backdrop, we have made an attempt in this chapter to provide research communities with an insight into the ways AI and big data analytics can be phenomenal to combat the COVID-19 pandemic. The rest of the chapter is structured as follows. AI and big data analytics for the COVID-19 pandemic are

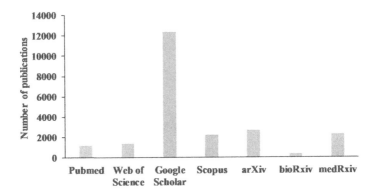

Fig. 4 The number of publications on AI and big data analytics in COVID-19 research

discussed in Sects. 2 and 3, respectively. Section 4 presents the existing challenges and the way forward. Section 5, finally, concludes the chapter.

2 AI for COVID-19 Pandemic

To battle against this novel global pandemic, scientists in the medical industry are thriving for new techniques to monitor the infected patients at different time points, to identify the best clinical trials, to develop an effective vaccine, and to track the disease spread. Recent studies have provided a clear indication that the application of AI can be a game-changer in the health sector because of its faster processing capacity, improved efficiency, and better scale-up [17]. Different AI applications for COVID-19 are presented in Fig. 5.

2.1 Screening and Treatment

Early screening and proper treatment are critical in managing the COVID-19 pandemic. The reverse transcription-polymerase chain reaction (RT-PCR) detection technique is currently the benchmark method for classifying respiratory viruses. This method detects antibodies in response to infection using a nasopharyngeal swab or sputum sample. Nevertheless, the researchers are still making efforts to improve this approach as well as other options [16, 20]. These methods, on the other hand, besides being typically expensive and time-consuming, have a low true-positive rate and necessitate specialised equipment. Furthermore, most of the countries are running short on test kits due to budgetary and technological constraints. As a result, the standard method is unsuitable for quick detection and monitoring. Using smart devices in conjunction with AI frameworks can be

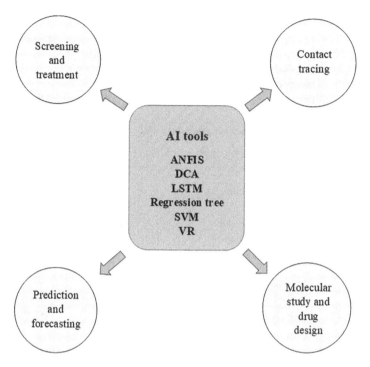

Fig. 5 AI for COVID-19

a convenient and low-cost strategy for COVID-19 identification [41]. COVID-19 can also be diagnosed using radiological images such as X-rays and CT scans [10, 31, 33, 36, 48]. Ozturk et al. [35] have designed an auxiliary tool to increase the accuracy of COVID-19 diagnosis model based on a deep learning algorithm. Utilising raw chest X-ray images of 127 infected patients, their model has achieved a remarkable degree of accuracy; binary class with 98.08% and multi-class with 87.02%. However, the development of automated AI-based models to detect, diagnose and predict COVID-19 infections for a larger population still remains an open research problem.

2.2   Contact Tracing

If a person is diagnosed with COVID-19, the next vital step is contact tracing in order to prevent the disease from spreading further. Contact tracing is basically a method of identifying and managing individuals, who have recently been exposed to a COVID-19 patient. In general, the process identifies the infected person for a 14-day follow-up after the exposure. This method, if properly implemented, has the potential to break the transmission chain of the novel corona virus and

Table 1 Contact tracing applications

Application	Location tracking mechanism	Country	Launch period
The Alipay Health Code	GPS, Global System for Mobile Communications (GSM), and transaction history of credit cards	China	February, 2020
CoronApp	GPS	Colombia	March, 2020
TraceTogether	Bluetooth	Singapore	March, 2020
Aarogya Setu	Bluetooth and location-generated social graph	India	April, 2020
COVIDSafe	Bluetooth	Australia	April, 2020
TraceCovid	Bluetooth	UAE	April, 2020
Smittestopp	Bluetooth and GSM	Norway	April, 2020
Mask.ir	GSM	Iran	May, 2020
MyTrace	Bluetooth and Google	Malaysia	May, 2020
Immuni	Bluetooth and Google	Italy	June, 2020
Corona-Warn-App	Bluetooth and Google	Germany	June, 2020
StopCovid	Bluetooth	France	June, 2020
SwissCovid	Bluetooth and Google	Switzerland	June, 2020
NHS Covid-19	Bluetooth	UK	September, 2020

suppress the outbreak by increasing the likelihood of adequate controls. Various infected countries have developed different digital contact tracing mechanisms with smartphone applications using various technologies such as Bluetooth, global positioning system (GPS), system physical address, etc. In comparison to a non-digital system, the digital contact tracing system can operate in almost real time and at a much faster rate. Almost all of these digital apps are designed to collect individual personal data, which will be analysed by the AI tools in order to track down a person, who is susceptible to the novel virus due to his/her recent contact chain.

Lalmuanawma et al. [30] have provided painstaking efforts in listing such AI-based contact tracing applications. Some of the most popular contact tracing applications are listed in Table 1. According to the studies [3, 12, 50], more than 30 countries have successfully implemented automated contact tracing using centralised, decentralised, or a combination of both approaches in order to reduce effort as well as to improve the efficiency of conventional healthcare diagnosis processes.

2.3 Prediction and Forecasting

Prediction and forecasting of the COVID-19 outbreak will help the government, public health agencies and policy-planners to handle the pandemic in a much

Table 2 AI models for prediction and forecasting of the COVID-19 pandemic

Authors	AI method	Area of application
Al-Qaness et al. [7]	Adaptive Neuro-Fuzzy Inference System (ANFIS)	Forecasting of the confirmed cases of COVID-19 in China
Chakraborty and Ghosh [14]	Hybrid Wavelet-ARIMA model and regression tree	Real-time forecasting and risk assessment of COVID-19 cases
Chimmula and Zhang [15]	Deep Learning using Long Short-Term Memory (LSTM) network	Prediction of the end point of the pandemic outbreak in Canada
Fong et al. [21]	Hybridised deep learning and fuzzy rule induction method	Forecasting in early epidemic composite Monte Carlo simulation utilising incomplete or limited data
Pokkuluri and Nedunuri [38]	Non-linear hybrid cellular automata classifier	Prediction of affected, recovered, and death cases
Yang et al. [61]	Modified SIER-LSTM model	Prediction of epidemic peaks and sizes

smarter way. Models traditionally used to track and forecast COVID-19 outbreak are epidemiological models such as SIR (susceptible, infected, and removed), SEIR (susceptible, exposed, infected, and recovered), SIRD (susceptible, infected, recovered, and dead), extended SIR model incorporating time-varying quarantine protocols such as government-level macro isolation policies and community-level micro inspection measures, ARIMA (autoregressive integrated moving average) model [19, 39], etc.

Different AI-based models, which provide freedom from the various stringent assumptions [25], are also employed on the available outbreak and demographic data to predict its spread [44]. Table 2 briefs a few AI-based prediction models reported in the COVID-19 literature. However, with increased data availability, the performance of these models is likely to be improved.

2.4 *Molecular Study and Drug Design*

Molecular biology and AI are well recognised as interdisciplinary fields of research [34, 52]. In order to solve complex biological questions as well as to analyse and interpret results that are unsolved in traditional laboratory methods, modern molecular biology requires the help of sophisticated software. For molecular design, simulation, modelling, and drug development, machine learning methods, especially deep learning, have been used [24, 65]. After the COVID-19 outbreak, scientists and healthcare experts all over the world have been urging for a viable option to discover drugs and vaccines for the SARS-CoV-2, and AI technology has been proven to be an exciting path in this regard [8, 26, 27]. These can help to fight

Table 3 AI in the molecular study and drug design for SARS-CoV-2

Authors	AI method	Area of application
Senior et al. [45]	Google DeepMind	Prediction of protein structures of SARS-CoV-2
Alimadadi et al. [6]	Deep learning	Vaccine design by utilising predicted protein structures
Beck et al. [11]	Deep learning-based drug-target interaction model	Comparison of antiviral drugs to combat COVID-19
Calvelo et al. [13]	Virtual reality (VR)	Understanding of the three-dimensional structure of SARS-CoV-2
Kumar et al. [29]	DCA (DMax Chemistry Assistant)	Navigation of the available chemical space of potential small molecule inhibitors of SARS-CoV-2
Kowalewski and Ray [28]	Support vector machine (SVM)	Prediction of novel drugs for SARS-CoV-2 using machine learning from a >10 million chemical space

the emerging COVID-19 pandemic by assisting in the identification of leads for therapies and vaccinations, proposing new inhibitors, exploring structural effects on genetic variation in viruses, and so on [9]. Table 3 summarises the application of AI in molecular study and drug design.

3 Big Data Analytics for COVID-19 Pandemic

The era of big data has arrived as a consequence of rapid growth in data volume and technological advancement [53]. A huge amount of patient data such as medical reports, X-ray images, lists of physicians and nurses, travel history, etc. have necessitated the application of big data analytics in the COVID-19 scenario. As the number of COVID-19 patients is skyrocketing day by day, big data analytics can play a pivotal role in analysing these data sets and extracting patterns. On the line of Pham et al. [37], big data analytics for COVID-19 are graphically represented in Fig. 6.

Alamo et al. [5] have carried out a comprehensive review on open-data resources for monitoring, modelling, and forecasting the COVID-19 pandemic. Tables 4, 5 and 6 provide open-data institutions, open-source communities, and regional data resources for the COVID-19, respectively.

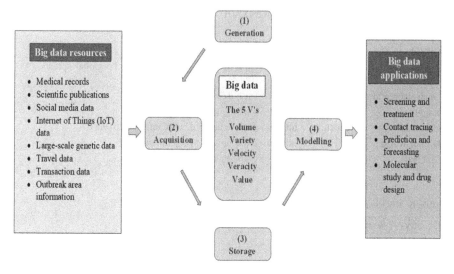

Fig. 6 Big data analytics for the COVID-19 (adapted from Pham et al. [37] licensed under HYPERLINK https://creativecommons.org/licenses/by/4.0/ CC BY 4.0)

Table 4 Open-data institutions for the COVID-19 (adapted from Alamo et al. [5])

Institution	URL
WHO	https://www.who.int/
Johns Hopkins University	https://coronavirus.jhu.edu/
European Union	https://data.europa.eu/euodp/en/data/
University of Oxford	https://www.bsg.ox.ac.uk/news/coronavirus-research-blavatnik-school
Africa Centers for Disease Control and Prevention	https://africacdc.org/
Google	https://google.com/covid19-map/
Organization for Economic Co-operation and Development	https://www.oecd.org
The Institute for Health Metrics and Evaluation	http://www.healthdata.org/
Institute for Data Valorization	https://ivado.ca/en/
New England Complex Systems Institute	https://necsi.edu/
Medical Research Council Center for Global Infectious Disease Analysis	https://www.imperial.ac.uk/mrc-global-infectious-disease-analysis/covid-19/
The World Bank Group	https://data.worldbank.org/
EuroMOMO	https://www.euromomo.eu/
United States National Institutes of Health	https://www.nih.gov/health-information/coronavirus
United States National Institute of Standards and Technology	https://randr19.nist.gov/

Table 5 Open-source communities for the COVID-19 (adapted from Alamo et al. [5])

Community	URL
GitHub	https://github.com/open-covid-19/data
Zindi	https://zindi.africa/competitions/predict-the-global-spread-ofcovid-19/data
Open Source Initiative	https://opensource.org/node/1062
Kaggle	https://www.kaggle.com/tags/covid19
Harvard Dataverse Repository	https://dataverse.harvard.edu/dataverse/2019ncov
Drupal World	https://www.drupal.org/project/covid_19
Red Hat	https://www.redhat.com/

Table 6 Regional COVID-19 data resources (adapted from Alamo et al. [5])

Country	Source	URL
Argentina	Ministry of Health	https://www.argentina.gob.ar/coronavirus/informe-diario
Australia	Australian Health Department	https://www.health.gov.au/news/health-alerts/novelcoronavirus-2019-ncov-health-alert/coronaviruscovid-19-current-situation-and-case-numbers
China	China National Health Commission	http://en.nhc.gov.cn/DailyBriefing.html
France	Public France Health System	https://www.santepubliquefrance.fr
Germany	Robert Koch Institute	https://www.rki.de/EN/Home/homepage_node.html
India	Government of India	https://www.mygov.in/covid-19/
Italy	Italian Civil Protection Department	http://www.protezionecivile.gov.it/attivita-rischi/rischio-sanitario/emergenze/coronavirus
Spain	Ministry of Health	https://www.mscbs.gob.es/profesionales/saludPublica/ccayes/alertasActual/nCov-China/home.htm
USA	Centers for Disease Control and Prevention	https://www.cdc.gov/

3.1 Screening and Treatment

Qin et al. [40] have employed social media search indexes (SMSI) to find new suspicious or confirmed cases of COVID-19. For the purpose of identification, SMSI compiles a list of keywords that surface on the social networking sites, including symptoms like cough, fever, etc. The prime benefit of using keywords as a strategy is that it allows health officials to outline suitable responses in order to construct the necessary warning system [2]. However, the reliability and accuracy of these posts still remain the major bottlenecks of this approach.

3.2   Contact Tracing

Another role of big data analytics is contact tracing, which is of paramount importance for healthcare organisations and government policymakers to effectively control the COVID-19 pandemic [37]. It makes use of posts with meta-data, social media tags, information of metro smartcards, vehicle logs, credit card transactions, etc. Although not much precise, identifying important features from location meta-data to verify that a person is at a particular location at a particular time can provide a tracking system, which can monitor people even if they do not have any tracking equipment with them [2]. Zhou et al. [66] have considered leveraging big data for spatial analysis methods and GIS technology, which has the potential to facilitate the acquisition and integration of heterogeneous data from various health data resources. In addition, big data can ensure that the information required for determining the transmission rate on a large scale is readily available. However, data architecture is critical in a big data environment to ensure merging, sharing, and analysis of the data.

3.3   Prediction and Forecasting

Data scientists and researchers are employing new big data-driven approaches to monitor COVID-19 cases in real time. As coronavirus spreads faster than the appearance of its symptoms, people need to be warned with high speed and effectiveness. The development of such an early warning system at a large scale is possible by augmenting big data and AI in a single framework. In several countries, big data analytics are being harnessed to further speed up the monitoring of infectious populations. For instance, the Chinese government has used big data analytics and AI in the form of millions of surveillance cameras, drones, and facial recognition technologies to monitor the movement of its citizens, to decide whether people are properly abiding by the quarantine rules, etc. [54]. Taiwan's healthcare system has recognised the strength of technology, and their swift response to the spread of coronavirus has been instrumental in slowing down the infection rate. Taiwan has set a great example in the human healthcare sector by implementing a sound action plan using big data [55]. However, dealing with the increasing number of scientific publications and the newly generated huge amount of data still remains a daunting task for the researchers.

3.4   Molecular Study and Drug Design

Big data can be very crucial in gaining insights for drug discovery. Within this short span of time, few attempts have already been made to develop a suitable vaccine by using big data analytics. Ahmed et al. [4] have utilised the GISAID database (www.gisaid.org/CoV2020/) to extract amino acid residues with a view to find potential targets for developing future vaccines against the COVID-19 pandemic. Abdelmageed et al. [1] have obtained a massive data set from the National Centre of Biotechnology Information (NCBI) to facilitate the vaccine production. Following a two-step procedure, different peptides are tried for the purpose of new vaccine development. In the first step, analysis of the whole genome of SARS-CoV-2 is carried out in a comparative genomic approach so that a potential antigenic target can be identified. Then, using an Artemis Comparative Tool, the human coronavirus reference sequence is analysed to encode the four major structural proteins (envelope protein, nucleocapsid protein, membrane protein, and spike) of coronavirus.

4   Existing Challenges and the Way Forward

Despite the fact that AI and big data analytics have shown great potentials in the fight against the COVID-19 pandemic, there are still issues that must be resolved in the near future. As the number of confirmed cases (infected and dead) is increasing considerably, various strategies such as social distancing, lockdown, large-scale screening, and testing have already been taken. During the COVID-19 pandemic, officials have asked people to share their personal details such as medical reports, GPS location, travel records, etc. so that AI as well as big data analytics can be employed to keep an eye on the situation. People, on the other hand, are often hesitant to share such details due to privacy concerns. Blockchain technology, federated learning, and incentive mechanisms can be used to resolve this issue. Government organisations have also to play a pivotal role in this regard to harmonise the approaches executed by different entities. Seoul Metropolitan Government has already implemented 'AI monitoring calling system', a framework for automatic health check-up [46]. A collaboration between the Zhejiang Provincial Government and Alibaba DAMO Academy has also been evident to develop an AI platform for automatic COVID-19 testing and analysis [59].

In spite of the sheer power of AI and big data analytics, augmentation of AI and big data is still in its infancy in the case of combatting the COVID-19 pandemic. The process of utilising big data obtained from the social media platforms is a challenging task as the filtering of relevant patterns from the cumulative noise is still quite complicated. In addition, when it comes to diagnosis and prognosis, AI needs to be more sensitive [42]. However, the prime challenge often encountered by the big data platforms and applications is the lack of standard datasets. Even though several AI-based big data analytics have been employed recently, the findings remain inconclusive, as in most of the cases, neither different algorithms are attempted in the same data set nor the sample sizes used in the studies are comparable [54]. Furthermore, clinical data should be systematically managed and processed for the sake of easy and secure accessibility. Hence, the construction of a vast cyber-

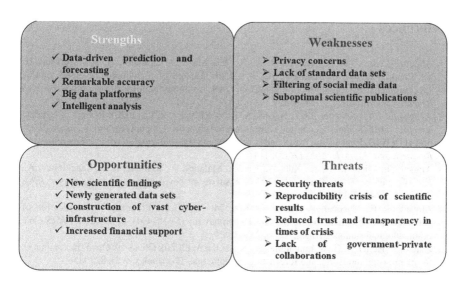

Fig. 7 SWOT analysis of AI and big data analytics in combatting COVID-19

infrastructure to stimulate global collaborations is a prerequisite for research on AI and big data analytics in the field of predictive, diagnostic, and therapeutic techniques against COVID-19 and other similar pandemics in the future. Figure 7 illustrates the SWOT (Strengths, Weaknesses, Opportunities, Threats) analysis of AI and big data analytics in combatting COVID-19.

5 Conclusion

The outbreak of the COVID-19 pandemic has touched almost every corner of the world. This lethal disease has shattered the social and economic securities of people to a great extent. In this chapter, we have summarised the state-of-the-art applications of AI and big data analytics in combating the COVID-19 crisis up to now. Various big data components have been found in literature, where AI has offered to play significant roles. The advanced analytical tools have not only helped in recognising the signs and symptoms of the disease, but also in forecasting the outbreak. In addition, AI-based big data analytics have achieved notable success in contact tracing, molecular study, and drug design. However, data privacy concerns and lack of standard data sets are identified as the major impediments restricting the success of AI and big data analytics in this regard. Hence, a centralised collection of all the data related to COVID-19 is necessitated to unearth the potential of AI-based big data analytics in fighting the pandemic.

References

1. Abdelmageed, M. I., Abdelmoneim, A. H., Mustafa, M. I., Elfadol, N. M., Murshed, N. S., Shantier, S. W., & Makhawi, A. M. (2020). Design of a multiepitope-based peptide vaccine against the E protein of human COVID-19: An immunoinformatics approach. *BioMed Research International, 26*, 286–297.
2. Agbehadji, I. E., Awuzie, B. O., Ngowi, A. B., & Millham, R. C. (2020). Review of big data analytics, artificial intelligence and nature-inspired computing models towards accurate detection of COVID-19 pandemic cases and contact tracing. *International Journal of Environmental Research and Public Health, 17*(15), 1–16.
3. Ahmed, N., Michelin, R. A., Xue, W., Ruj, S., Malaney, R., Kanhere, S. S., Seneviratne, A., Hu, W., Janicke, H., & Jha, S. K. (2020). A survey of covid-19 contact tracing apps. *IEEE Access, 8*, 134577–134601.
4. Ahmed, S. F., Quadeer, A. A., & McKay, M. R. (2020). Preliminary identification of potential vaccine targets for the COVID-19 coronavirus (SARS-CoV-2) based on SARS-CoV immunological studies. *Viruses, 12*(3), 254–268.
5. Alamo, T., Reina, D. G., Mammarella, M., & Abella, A. (2020). Covid-19: Open-data resources for monitoring, modeling, and forecasting the epidemic. *Electronics, 9*(5), 827–856.
6. Alimadadi, A., Aryal, S., Manandhar, I., Munroe, P. B., Joe, B., & Cheng, X. (2020). Artificial intelligence and machine learning to fight COVID-19. *Physiological Genomics, 52*(4), 200–202.
7. Al-Qaness, M. A., Ewees, A. A., Fan, H., & Abd El Aziz, M. (2020). Optimization method for forecasting confirmed cases of COVID-19 in China. *Journal of Clinical Medicine, 9*(3), 674–688.
8. Alsharif, M. H., Alsharif, Y. H., Albreem, M. A., Jahid, A., Solyman, A. A. A., Yahya, K., Alomari, O. A., & Hossain, M. S. (2020). Application of machine intelligence technology in the detection of vaccines and medicines for SARS-CoV-2. *European Review for Medical and Pharmacological Sciences, 24*(22), 11977–11981.
9. Amaro, R. E., & Mulholland, A. J. (2020). A community letter regarding sharing biomolecular simulation data for COVID-19. *Journal of Chemical Information and Modeling, 60*(6), 2653–2656.
10. Apostolopoulos, I. D., & Mpesiana, T. A. (2020). COVID-19: Automatic detection from x-ray images utilizing transfer learning with convolutional neural networks. *Physical and Engineering Sciences in Medicine, 43*(2), 635–640.
11. Beck, B. R., Shin, B., Choi, Y., Park, S., & Kang, K. (2020). Predicting commercially available antiviral drugs that may act on the novel coronavirus (SARS-CoV-2) through a drug-target interaction deep learning model. *Computational and Structural Biotechnology Journal, 18*, 784–790.
12. Braithwaite, I., Callender, T., Bullock, M., & Aldridge, R. W. (2020). Automated and partly automated contact tracing: A systematic review to inform the control of COVID-19. *The Lancet Digital Health, 2*(11), 607–621.
13. Calvelo, M., Piñeiro, Á., & Garcia-Fandino, R. (2020). An immersive journey to the molecular structure of SARS-CoV-2: Virtual reality in COVID-19. *Computational and Structural Biotechnology Journal, 18*, 2621–2628.
14. Chakraborty, T., & Ghosh, I. (2020). Real-time forecasts and risk assessment of novel coronavirus (COVID-19) cases: A data-driven analysis. *Chaos, Solitons and Fractals, 135*, 1–10.
15. Chimmula, V. K. R., & Zhang, L. (2020). Time series forecasting of COVID-19 transmission in Canada using LSTM networks. *Chaos, Solitons and Fractals, 135*, 1–6.
16. Corman, V. M., Landt, O., Kaiser, M., Molenkamp, R., Meijer, A., Chu, D. K., Bleicker, T., Brünink, S., Schneider, J., Schmidt, M. L., & Mulders, D. G. (2020). Detection of 2019 novel coronavirus (2019-nCoV) by real-time RT-PCR. *Euro Surveillance, 25*(3), 1–8.

17. Davenport, T., & Kalakota, R. (2019). The potential for artificial intelligence in healthcare. *Future Healthcare Journal, 6*(2), 94–98.
18. Dowd, J. B., Andriano, L., Brazel, D. M., Rotondi, V., Block, P., Ding, X., Liu, Y., & Mills, M. C. (2020). Demographic science aids in understanding the spread and fatality rates of COVID-19. *Proceedings of the National Academy of Sciences, 117*(18), 9696–9698.
19. Fanelli, D., & Piazza, F. (2020). Analysis and forecast of COVID-19 spreading in China, Italy and France. *Chaos, Solitons and Fractals, 134*, 1–5.
20. Fomsgaard, A. S., & Rosenstierne, M. W. (2020). An alternative workflow for molecular detection of SARS-CoV-2–escape from the NA extraction kit-shortage, Copenhagen, Denmark, March 2020. *Euro Surveillance, 25*(14), 1–14.
21. Fong, S. J., Li, G., Dey, N., Crespo, R. G., & Herrera-Viedma, E. (2020). Composite Monte Carlo decision making under high uncertainty of novel coronavirus epidemic using hybridized deep learning and fuzzy rule induction. *Applied Soft Computing, 93*, 106–119.
22. Haleem, A., Javaid, M., Khan, I. H., & Vaishya, R. (2020). Significant applications of big data in COVID-19 pandemic. *Indian Journal of Orthopaedics, 54*, 526–528.
23. Huang, C., Wang, Y., Li, X., Ren, L., Zhao, J., Hu, Y., & Cao, B. (2020). Clinical features of patients infected with 2019 novel coronavirus in Wuhan, China. *Lancet, 395*(10223), 497–506.
24. Jing, Y., Bian, Y., Hu, Z., Wang, L., & Xie, X. Q. S. (2018). Deep learning for drug design: An artificial intelligence paradigm for drug discovery in the big data era. *The AAPS Journal, 20*(3), 1–10.
25. Jordan, M. I., & Mitchell, T. M. (2015). Machine learning: Trends, perspectives, and prospects. *Science, 349*(6245), 255–260.
26. Kaushal, K., Sarma, P., Rana, S. V., Medhi, B., & Naithani, M. (2020). Emerging role of artificial intelligence in therapeutics for COVID-19: A systematic review. *Journal of Biomolecular Structure & Dynamics, 39*, 1–16.
27. Kaushik, A. C., & Raj, U. (2020). AI-driven drug discovery: A boon against COVID-19? *AI Open, 1*, 1–4.
28. Kowalewski, J., & Ray, A. (2020). Predicting novel drugs for SARS-CoV-2 using machine learning from a >10 million chemical space. *Heliyon, 6*(8), 1–14.
29. Kumar, A., Loharch, S., Kumar, S., Ringe, R. P., & Parkesh, R. (2021). Exploiting cheminformatic and machine learning to navigate the available chemical space of potential small molecule inhibitors of SARS-CoV-2. *Computational and Structural Biotechnology Journal, 19*, 424–438.
30. Lalmuanawma, S., Hussain, J., & Chhakchhuak, L. (2020). Applications of machine learning and artificial intelligence for Covid-19 (SARS-CoV-2) pandemic: A review. *Chaos, Solitons and Fractals, 110*, 1–10.
31. Lee, E. Y., Ng, M. Y., & Khong, P. L. (2020). COVID-19 pneumonia: What has CT taught us? *The Lancet Infectious Diseases, 20*(4), 384–385.
32. Li, Q., Guan, X., Wu, P., Wang, X., Zhou, L., Tong, Y., Ren, R., Leung, K. S., Lau, E. H., Wong, J. Y., & Xing, X. (2020). Early transmission dynamics in Wuhan, China, of novel coronavirus–infected pneumonia. *The New England Journal of Medicine, 382*, 1199–1207.
33. Litjens, G., Kooi, T., Bejnordi, B. E., Setio, A. A. A., Ciompi, F., Ghafoorian, M., Van Der Laak, J. A., Van Ginneken, B., & Sánchez, C. I. (2017). A survey on deep learning in medical image analysis. *Medical Image Analysis, 42*, 60–88.
34. Olson, A. J. (2018). Perspectives on structural molecular biology visualization: From past to present. *Journal of Molecular Biology, 430*(21), 3997–4012.
35. Ozturk, T., Talo, M., Yildirim, E. A., Baloglu, U. B., Yildirim, O., & Acharya, U. R. (2020). Automated detection of COVID-19 cases using deep neural networks with X-ray images. *Computers in Biology and Medicine, 121*, 1–11.
36. Panwar, H., Gupta, P. K., Siddiqui, M. K., Morales-Menendez, R., & Singh, V. (2020). Application of deep learning for fast detection of COVID-19 in X-rays using nCOVnet. *Chaos, Solitons and Fractals, 138*, 1–8.

37. Pham, Q. V., Nguyen, D. C., Huynh-The, T., Hwang, W. J., & Pathirana, P. N. (2020). Artificial intelligence (AI) and big data for coronavirus (COVID-19) pandemic: A survey on the state-of-the-arts. *IEEE Access, 4,* 1–19.

38. Pokkuluri, K. S., & Nedunuri, S. U. D. (2020). A novel cellular automata classifier for covid-19 prediction. *Journal of Health Science, 10*(1), 34–38.

39. Prem, K., Liu, Y., Russell, T. W., Kucharski, A. J., Eggo, R. M., Davies, N., Flasche, S., Clifford, S., Pearson, C. A., Munday, J. D., & Abbott, S. (2020). The effect of control strategies to reduce social mixing on outcomes of the COVID-19 epidemic in Wuhan, China: A modelling study. *The Lancet Public Health, 5*(5), 261–270.

40. Qin, L., Sun, Q., Wang, Y., Wu, K. F., Chen, M., Shia, B. C., & Wu, S. Y. (2020). Prediction of number of cases of 2019 novel coronavirus (COVID-19) using social media search index. *International Journal of Environmental Research and Public Health, 17*(7), 1–14.

41. Rao, A. S. S., & Vazquez, J. A. (2020). Identification of COVID-19 can be quicker through artificial intelligence framework using a mobile phone–based survey when cities and towns are under quarantine. *Infection Control and Hospital Epidemiology, 41*(7), 826–830.

42. Rawat, W., & Wang, Z. (2017). Deep convolutional neural networks for image classification: A comprehensive review. *Neural Computation, 29*(9), 2352–2449.

43. Raza, K. (2020). Artificial intelligence against COVID-19: A meta-analysis of current research. In *Big data analytics and artificial intelligence against COVID-19: Innovation vision and approach* (pp. 165–176). Singapore: Springer.

44. Santosh, K. C. (2020). AI-driven tools for coronavirus outbreak: Need of active learning and cross-population train/test models on multitudinal/multimodal data. *Journal of Medical Systems, 44*(5), 1–5.

45. Senior, A. W., Evans, R., Jumper, J., Kirkpatrick, J., Sifre, L., Green, T., Qin, C., Žídek, A., Nelson, A. W., Bridgland, A., & Penedones, H. (2020). Improved protein structure prediction using potentials from deep learning. *Nature, 577*(7792), 706–710.

46. Seoul Metropolitan Government. (2020). *Seoul introduces the COVID-19 AI monitoring call system.* http://english.seoul.go.kr/seoul-introduces-the-covid-19-%E3%80%8Cai-monitoring-call-system%E3%80%8D/

47. Shams, S. A., Haleem, A., & Javaid, M. (2020). Analyzing COVID-19 pandemic for unequal distribution of tests, identified cases, deaths, and fatality rates in the top 18 countries. *Diabetes and Metabolic Syndrome: Clinical Research and Reviews, 14*(5), 953–961.

48. Shen, D., Wu, G., & Suk, H. I. (2017). Deep learning in medical image analysis. *Annual Review of Biomedical Engineering, 19,* 221–248.

49. Singh, A. K., & Misra, A. (2020). Impact of COVID-19 and comorbidities on health and economics: Focus on developing countries and India. *Diabetes and Metabolic Syndrome: Clinical Research and Reviews, 14*(6), 1625–1630.

50. Skoll, D., Miller, J. C., & Saxon, L. A. (2020). COVID-19 testing and infection surveillance: Is a combined digital contact tracing and mass testing solution feasible in the United States? *Cardiovascular Digital Health Journal, 1*(3), 149–159.

51. Tang, W., Liao, H., Marley, G., Wang, Z., Cheng, W., Wu, D., & Yu, R. (2020). The changing patterns of coronavirus disease 2019 (COVID-19) in China: A tempogeographic analysis of the severe acute respiratory syndrome coronavirus 2 epidemic. *Clinical Infectious Diseases, 71*(15), 818–824.

52. Thackeray, S. J., & Hampton, S. E. (2020). The case for research integration, from genomics to remote sensing, to understand biodiversity change and functional dynamics in the world's lakes. *Global Change Biology, 26*(6), 3230–3240.

53. Tsai, C. W., Lai, C. F., Chao, H. C., & Vasilakos, A. V. (2015). Big data analytics: A survey. *Journal of Big Data, 2*(1), 1–32.

54. Verma, S., & Gazara, R. K. (2021). Big data analytics for understanding and fighting COVID-19. In *Computational intelligence methods in COVID-19: Surveillance, prevention, prediction and diagnosis* (pp. 333–348). Singapore: Springer.

55. Wang, C. J., Ng, C. Y., & Brook, R. H. (2020). Response to COVID-19 in Taiwan: Big data analytics, new technology, and proactive testing. *JAMA, 323*(14), 1341–1342.

56. Wang, Y., Kung, L., & Byrd, T. A. (2018). Big data analytics: Understanding its capabilities and potential benefits for healthcare organizations. *Technological Forecasting and Social Change, 126*, 3–13.
57. WHO. (2021). *WHO Coronavirus (COVID-19) Dashboard.* https://covid19.who.int/
58. Wu, J. T., Leung, K., Bushman, M., Kishore, N., Niehus, R., de Salazar, P. M., Cowling, B. J., Lipsitch, M., & Leung, G. M. (2020). Estimating clinical severity of COVID-19 from the transmission dynamics in Wuhan, China. *Nature Medicine, 26*(4), 506–510.
59. Xiong, J., Yan, J., Fu, K., Wang, K., & He, Y. (2021). Innovation in an authoritarian society: China during the pandemic crisis. *The Journal of Business Strategy, 10*, 2020–2023.
60. Yang, H., Bin, P., & He, A. J. (2020). Opinions from the epicenter: An online survey of university students in Wuhan amidst the COVID-19 outbreak1. *Journal of Chinese Governance, 5*(2), 234–248.
61. Yang, Z., Zeng, Z., Wang, K., Wong, S. S., Liang, W., Zanin, M., Liu, P., Cao, X., Gao, Z., Mai, Z., & Liang, J. (2020). Modified SEIR and AI prediction of the epidemics trend of COVID-19 in China under public health interventions. *Journal of Thoracic Disease, 12*(3), 165–175.
62. Yaqoob, I., Hashem, I. A. T., Gani, A., Mokhtar, S., Ahmed, E., Anuar, N. B., & Vasilakos, A. V. (2016). Big data: From beginning to future. *International Journal of Information Management, 36*(6), 1231–1247.
63. Yassine, A., Singh, S., Hossain, M. S., & Muhammad, G. (2019). IoT big data analytics for smart homes with fog and cloud computing. *Future Generation Computer Systems, 91*, 563–573.
64. Zaki, N., Alashwal, H., & Ibrahim, S. (2020). Association of hypertension, diabetes, stroke, cancer, kidney disease, and high-cholesterol with COVID-19 disease severity and fatality: A systematic review. *Diabetes and Metabolic Syndrome: Clinical Research and Reviews, 14*(5), 1133–1142.
65. Zhang, L., Tan, J., Han, D., & Zhu, H. (2017). From machine learning to deep learning: Progress in machine intelligence for rational drug discovery. *Drug Discovery Today, 22*(11), 1680–1685.
66. Zhou, C., Su, F., Pei, T., Zhang, A., Du, Y., Luo, B., Cao, Z., Wang, J., Yuan, W., Zhu, Y., & Song, C. (2020). COVID-19: Challenges to GIS with big data. *Geography and Sustainability, 1*(1), 77–87.

COVID-19 TravelCover: Post-Lockdown Smart Transportation Management System

Hari Mohan Rai, Barnini Goswami, Shreya Majumdar, and Kajal Gupta

1 Introduction

The outbreak of coronavirus pandemic, 2019 (COVID-19), has created a global health crisis that has had a deep impact on the way we perceive our world and our everyday lives. COVID-19 is highly contagious and mortality also very high, this makes the disease all the more dangerous [5, 14, 21]. One of the major prevention is to maintain social distancing, which would smash the chain of the expansion of the disease. Lockdown has been implemented to implement this idea but we can't keep lockdown in a country for a very long time, otherwise the country's economy will drastically decline. Post lockdown many things will change around us. And after the daily routine resumes post lockdown, the public transportation system will play a crucial role, as it is most commonly used by people, and to prevent the further spread of COVID-19 cases, social distancing needs to be maintained at all the public places [12, 22]. So there has to be proper management for public transportation which can allow its use, without further producing more COVID-19 cases.

To solve this problem we have been implementing a solution, by building an intelligent application to schedule the timings of various transports, avoiding the over occupancy of buses/railway stations, etc. [16].

This work describes a unique approach to solve the issue of maintaining social distance while traveling due to the COVID-19 outbreak and also to make public transportation function like it used to function early. Our application will also check

H. M. Rai (✉)
Department of Electronics and Communication Engineering, Krishna Engineering College, Ghaziabad, UP, India

B. Goswami · S. Majumdar · K. Gupta
Department of Computer Science Engineering, Krishna Engineering College, Ghaziabad, UP, India

the user's authenticity and accountability by asking for their Aadhaar Card (12-digit individual unique identification number issued by Government of India, used widely for identification and address proof) details to enter while they log in to our application. The purpose of travel will be kept track of for future references and maintaining a priority queue of allocation of the ticket. This app will assign the shortest possible routes between source and destination, after the checking of the route availability, while ticket allocation, to provide ample amount of space inside the bus for people to maintain the social distance criteria. There is also a ticket validation system on the basis of masked and unmasked images of the user. So, this app will avoid congestion, which will help in maintain social distancing.

1.1 Problem Statement

COVID-19 is a lethal virus that spreads via direct contact with infected items or people. Therefore, it is extremely important to maintain social distance and wear mask to slow down or stop the transmission of this virus. Almost every nation in the world has imposed a full lockdown for the same reason, but this cannot be sustained for an extended period of time. As a result of the lockdown, several industries will be virtually closed, with daily wage employees being the most impacted. The GDP growth of many nations has slowed, and in some cases has even gone negative; similarly, the GDP growth of India has been also affected, as can be seen from the most current data. Worldwide corona cases are increasing but here we are specifically talking about India as our research is particularly based on India's data. Here are some characteristics of the current scenario of India with respect to COVID-19 till first November 2020 [7].

- Total corona cases 81,84,082
- 91.54% Cured/discharged/migrated (74,91,513)
- 6.97% Active cases (5,70,458)
- 1.49% Deaths (1,22,111)
- Total COVID-19 confirmed cases = Cured/discharged/migrated + Active cases + Deaths

Analyzing the data trends over the past few months, according to the report published on Government of India COVID-19 dashboard [7], we observe the following:

- The surge of new coronavirus cases started somewhere in the month of April, with 129 new cases in Delhi and 74 new ones in Tamil Nadu had come up due to the Nizamuddin Markaz congregation.
- In May, the growth rate of new infections encountered on daily basis was drastically increasing due to which the initial 21-day lockdown period had to be extended a few times.

- Finally, on June 08, 2020, the 75-day lockdown ended and the unlocking process, divided into several stages, commenced.

Also, unforeseen employment issues became a threat to the survival of daily wagers. Thousands of migrant workers were seen hurdling at Anand Vihar bus terminal and railway station of New Delhi, wanting to go back to their villages due to a shortage of food and resources in the city. This in turn led to a dramatic rise in COVID-19 cases in the capital. Incidents like these compelled us to think of a way of managing such traffic on roads and compelled people to follow the guidelines in public places.

Post lockdown many things have changed around us. As of now, we can see that the lockdown has been resumed in India, and day by day cases are getting increased. People are going out and maybe getting in touch with infected people, which is increasing the infection throughout the country. Thus, the public transportation system will play a crucial role, as it is most commonly used by people, and to prevent the further spread of COVID-19 cases, social distancing needs to be maintained at all the public places. So there has to be proper management for public transportation which can allow its use, without further producing more COVID-19 cases.

1.2 Related Works

This section includes methods that formulated the problem, addressed a central or related problem, used a similar methodology as our work to a similar problem, and also how our work is inspired by their work.

By clearly describing previous work, we can better describe the current limitations and the need for a new methodology. It also allows demonstrating our area of work and will help others to relate our current work to other scientific areas. We collected, analyzed, and coded the author-assigned keywords of other research works to start a discussion on the topic "Intelligent post lockdown transportation management system." Presently, how to maintain social distancing while traveling is a boiling area of research. Managing traffic, safer routes, and also telling the public about the shortest route to reach out to their destination have been used in designing many applications in the past [25]. Some of those existing approaches from the past are discussed in this section.

Existing work in the field motivated our design idea to tackle the needs of a safer traveling environment for the public. A handful of solutions have been suggested pertaining to the issue in discussion.

Transit app was designed for aggregating and mapping real-time public transit data and crowdsourcing user data to determine the actual location of buses and trains. This app was first released for iPhone users, then it was launched for the android users as well. It offers upcoming departure times for all nearby transit lines and alerts for various types of transportation where available, including both bus and train. If it is found that public transit is not cooperating, people can easily request an Uber or grab the closest bicycle.

Whether you ride the train, subway/underground/tube, bus, light rail, ferry or metro, use bikes, ride-sharing like Uber, getting the best urban mobility information is critical [17].

Moovit [6] guides you from point A to B most easily and efficiently. Get train and bus times, maps, and live arrival times with ease so you can plan your trip with confidence. Find critical alerts and service disruptions for your favorite lines. Get step-by-step directions of optimal route bus, train, metro, bike, or a combination of them [13].

GoTo is the first ever Real-time Bus Booking Platform for Intercity Travels (Harsh Vardhan [9]). It is focused on saving your precious time which is generally wasted waiting for buses at the boarding point, with no clue of bus arrival. Though booking operators claim that you can track the bus, in most cases you are unable to get its actual expected time of arrival.

All the abovementioned applications help people to book tickets, reach their destination in time, provide them with a scheduled timetable, and notify them of numerous modes of transportation. There are several such applications like Chalo, UTS, NextBus, etc. So, after exploring all the previous works we have come up with an application which in addition to existing features provides new and innovative ways to implement social distancing in public transport [2, 3, 20]. These features include priority-based ticket allocation, validating the ticked through masked images, etc.

"Data Privacy" is one of the most important concerns in any data collection module. It has been observed that many of the existing applications are asking for way too much information from people for providing more efficient services, which is not ethical. The Supreme Court of India has recognized the right to privacy as a fundamental right under Article 21 of the Constitution as a part of the right to "life" and "personal liberty" [4].

H. R. Schmidtke [18] presented and reviewed services based on location tracking for COVID-19 contact tracking. In his observation, it clearly states how vulnerable a user can get if he/she shares a large amount of his/her personal data with such COVID-19-related applications. For instance, they have talked about the Cambridge Analytica Case [8].

Kumar et al. [11] proposed a new digital twin-centric method for the prediction of the intention of the driver and also for avoiding the overcrowded traffic paths.

H. R. Schmidtke [19] studied the approaches used for the verification method by the various smart transportation systems. In his study, he also discussed about the intelligent environment aspects that authenticate local safety and security techniques.

After exploring several design options, we eventually decided to develop a website which will have features from several of our brainstormed and existing ideas named COVID-19 TravelCover. By integrating the features of distinct applications together in one platform, COVID-19 TravelCover offers a more holistic and effective solution than the existing applications by providing them with a variety of tailor-made features as per user's requirements. Another highlight is that the currently available applications fail to provide a solution to the COVID-19.

But in our work, we have made sure that no such information is asked from the user which can pose data privacy concerns to our users, but ensuring the authenticity of the users at the same time (by asking for Aadhaar Card details while registration, used as a legally acceptable document for unique identification in India) [10]. Also, we have verified the perspective of an intelligent environment by making all the processes digital and by minimizing the physical contact during the overall process.

1.3 Scope and Objective

The goals of the proposed work are as follows:

1. It will provide the shortest route for the traveler's desired destination thus saving time.
2. This app will be used to make public transportation possible while maintaining social distancing. It will help in avoiding occupancy not only on the roads but also inside public transport.
3. We have tried to make it user friendly so many people can easily use it.
4. It helps to detect on a mass scale whether people are wearing a mask or not. In the case of defaulters, they will not be allowed to take up the journey. This would implement strict rule implementation through our application.
5. It will keep a track of the travel of healthcare officials for analysis by maintaining a priority.
6. Payments will be online so no requirement for handling physical currency is present.
7. For future scope, we want to collaborate with the government to install cameras in all buses and through those cameras; using OpenCV libraries, we can implement mask detection failing which the person will be charged with penalty accordingly.
8. We are also planning to implement a better ticket validation mechanism such that before boarding particular public transportation we will be required to scan a QR code which will click a photo of the user then and there, thus increasing efficiency and reliability of our application.
9. For further research purposes, we can align our application with appropriate cloud storages to handle the data efficiently and expand the functionality of our application.

1.4   Novelty

In our application, we are using a fully customized and extremely simple to understand algorithm to allot the shortest available route to the user to take up their journey. Our application will check the user's authenticity and accountability by asking for their Aadhaar card (identity proof) while they sign up for our app. This application will provide services based on the preference of people's urgency for the purpose of their travel. There is a whole different step which the user is required to go through in order to validate the allotted ticket based on the mask detection mechanism, failing which the user will not be allowed to take up the journey. We strive to maintain social distancing not only on the roads but also inside the public transports as well. The efficient allocation of available transport resources ensures less congestion on roads. Through our application we will allow only 50% of actual occupancy allocation in a mode of particular transport, thus providing ample space inside the transports as well.

1.5   Scientific Contribution

1. A very simple and hassle-free, shortest path allocating algorithm that can be used by anyone, even those new to the development domain.
2. Through this application, we provide a methodology that can be used to implement priority-based ticket allocation, thus allowing health officials to reach the needy on time.
3. We have tried to introduce an idea of 50% occupancy of transports which can considerably reduce the risk of people getting affected with coronavirus.
4. As we are able to maintain a track record of travel history, we can analyze the data to track down people visiting red zone areas (areas having a heavy number of cases within a short period).
5. We have seen that people failing to generate masked images will not receive validation for their tickets. Thus, we have introduced a way for stricter implementation of the safety rules.
6. *Impact on the Environment.*
   The design and implementation of our application would require a greater number of vehicles to be available as we have planned 50% occupancy of the buses in order to implement social distancing inside the bus itself. But at the same time, we are implementing the efficient utilization of our resources. When we have no option at that time we should ensure minimal wastage of resources. Analyzing the pattern of demand based on the area, the buses can be scheduled accordingly so that the minimum number of buses can be implemented to cater to people's needs. On a small level, we are also saving paper by making payments and issue of tickets online, thus saving trees.

The organization of this chapter is arranged as Section 2 briefs about the proposed methodology employed in this work including COVID-19 TravelCover Architecture and Software Designing. Section 3 explains the Algorithm and working of the proposed work consisting of Route Allocation, fare calculation, Unique Ticket Number Generation, Ticket Validation through mask detection, Security features, Imposing guidelines, and Customer-first approach. Section 4 briefs the discussion over the result and outcome, and the final Sect. 5 provides concluding remarks and future work of this chapter.

2 Proposed Methodology

2.1 COVID-19 TravelCover Architecture

In data flow architecture, we have presented the flow diagram of the proposed work with respect to the input and output flow of the data in different phases of the application. The flowchart shown in Fig. 1 depicts the data flow diagram of our COVID-19 TravelCover application.

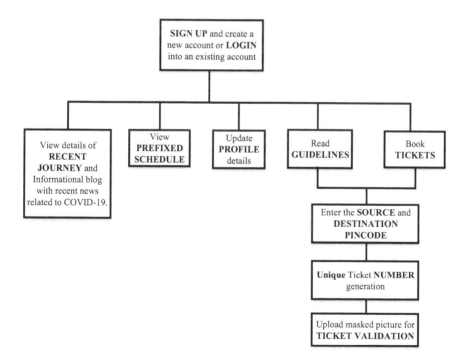

Fig. 1 The flow architecture of proposed COVID-19

The application required some standard details while using it for the first time, such as name, email id along with Aadhaar card number for checking the authenticity of the user, from the SIGN-UP section. Then the users will also be required to enter their home and "usual" destination address details so that these options are readily available for the user while booking tickets. The personal details of the user will be available in the profile section which he/she can change as and when required. While booking tickets users will have to either enter or choose from available options the required source and destination pin codes after carefully viewing the prefixed timetable and deciding an appropriate time. They will also have to go through guidelines compulsorily before navigating to the "Book Tickets" options. After submitting the details, the user will receive a unique ticket number which subject to validation will complete the process of booking the tickets. For ticket validation, the user will be required to upload a masked image of themselves from their corresponding device, if the image fulfills the criteria then will give a validation signal which has to be shown while boarding the vehicle.

2.2 Software Designing

In this section, the structure of the software application is explained elaborately which has been aided with a flowchart representing all the pages of our application. Figure 2 presents the block diagram depicting the overall software designing steps and processes.

(a) **Index.html** is the first page of our website ("welcome page"). It will direct a user either to "registration.html" (registration page) or login.html (login page).

(b) **Registration.html** page a user will fill in their required details and the details will get stored in the backend database register.java (/register). After this, it will direct the user to the next step in the registration process.

(c) **Registration2.html**, the user will fill in the other required details remained and the details will get stored in the backend database register2.java (/register2). If the user is already registered, then the user can directly go to the "LOGIN PAGE".

(d) **Login.html**–It is where his/her credentials are already there in the database login.java (/login Action). Then the next page after the registration page will be home.html where a user can store his/her HOME ADDRESS so that it will get stored in the database Home.java (/source), with the help of which user need not enter their home address every time they want to book tickets. Similarly, the destination.html page will ask all the usual destination details where a user usually goes and the details will get stored in destination.java (/destination). Then comes the Dashboard page (dashboard.jsp) where the user's recent journey will be displayed. Dashboard directs to the other pages:

(e) **Guidelines.jsp**, here the user can find the few most important guidelines for the prevention from COVID-19.

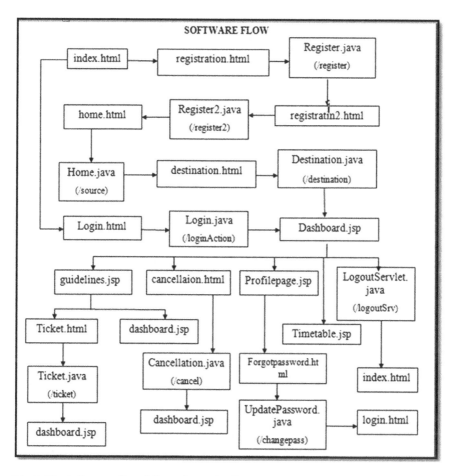

Fig. 2 Block diagram of consisting software details

(f) **Cancellation.html** from this page the user can cancel his/her booked ticket, and it will then update in the database cancellation.java (/cancel) and then redirects to the dashboard.jsp.

(g) **Profilepage.jsp**, where people can see the information they have entered during the registration process.

(h) **Timetable.jsp,** here the user will find the timetable which will make it easy for the user to see timings and book tickets according to their convenience.

(i) **LogoutServlet.java** (/logoutSev), here the user can log out their account, then will get directed to the index.html page. From the guidelines.jsp, the user can get back to the dashboard or to the Ticket.html where the users will book their tickets and the data will get stored into the database through Ticket.java (/ticket) and then directs again to dashboard.jsp.

(j) **Profilepage.jsp** (PROFILE PAGE) contains a forgot password option which redirects to the Forgotpassword.html (FORGOT PASSWORD PAGE) where the user will have the provision to update his/her password and the password will get updated in the database through Updatepassword.java (/changepass) and then this page will redirect to the login.html.

3 Algorithm and Working

3.1 Route Allocation

This section explains the process or steps for determining the shortest distance and fare calculation for each route. Whenever we want to go from one place to another, we might have more than one possible path to choose from, to take up the journey. In such a case, we all prefer going through the path which involves the least time to reach the destination, as it would save valuable time. In our application also, we tend to improve customer satisfaction by allotting the shortest path possible as it would save the customer's time and money. Presently, we are using a very simple to use and understand algorithm for shortest route allocation as it complies with the simplicity of our project and is easily modifiable. This algorithm also uses the concept of priority while allotting the tickets. This is implemented in a way that if two users are booking tickets at the same time then, the user with higher associated priority will be allotted the ticket first. The users are required to enter the purpose of travel while booking the tickets. Now using pattern matching, the algorithm will look for words like "doctor", "health worker", "medicine", etc. in the purpose entered by the user to associate a greater priority with that user.

The steps to find the shortest route and how to decide the priority has been explained through the example. Let us suppose we are considering a small district, comprising of nine PIN codes. Also, four possible routes are connecting these pin codes among themselves. Let the pin codes be numbered as **1, 2, 3, and so on till 9,** and the routes are named as **R1, R2, R3, and R4** as shown below. Now, the task is to find the shortest route possible that can be allocated to the user. Destination page will look for the address of where the customer wants to go most often may be on a daily basis (if any), therefore his usual destination like office address, college address, hospital address, etc. Table 1 shows the route number and vehicle stops details for finding the shortest distance based on Fig. 3.

Table 1 contains the route details, the first column represents the particular route name and the rest of the columns consist of the bus stops in order of their traversal during the journey. Let us assume a person "A" wants to go from pin code "1" to pin code "5". Now on referring to the above figures it is clear that all the four routes can take A to his destination. The number of stops between the source and destination in corresponding routes is as follows:

Table 1 Vehicle route and corresponding stop details

ROUTE NAME	S1	S2	S3	S4	S5	S6	S7	S8	S9
R1	1	2	3	4	5	6	0	0	0
R2	1	2	8	3	4	5	6	0	0
R3	1	2	8	7	4	5	6	0	0
R4	1	2	3	8	7	4	5	6	0

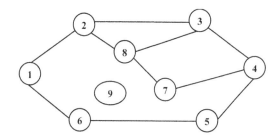

Fig. 3 Shortest route calculation steps using random users

R1–4 stops
R2–5 stops
R3–5 stops
R4–6 stops

From the above data, it is clear that "R1" will be the best and shortest route possible between the specified source and destination. Thus, route R1 will be allotted to "A."

3.2 Fare Calculation

While finding the shortest route we will be keeping a track of the number of stops between the source and destination along a particular route. This same data will be re-utilized by us in calculating the fare incurred on the user. We have decided on a standard rate for traversing between two stops, that is, 10 rupees per stop, and based on this the final price will be displayed to the user. For instance, let us consider the above case again where "A" wants to go from pin code "1" to "5." We have already found that the "R1" route will be allotted to "A" so the fare corresponding to that will be (Price per stop) * (Number of Stops to be traversed) = 4X10 = 40/−. Thus, "A" will have to pay 40 rupees to take up the journey.

3.3 Unique Ticket Number Generation

In order to avoid reusing the same ticket twice or sharing the ticket by users, our application will generate a unique ticket number to uniquely identify each journey. The uniqueness of every ticket is ensured by using a combination of source and destination pin code, alphabets, and a series of random numbers generated using the random() function. The pattern of each ticket number has been designed in a way such that it can directly indicate the source, destination, and username of the user. For example, the ticket number niniTNBF1T50.387 indicates the user name "nini", the source pin code as "1," and the destination pin code as "5".

3.4 Ticket Validation Through Mask Detection

Every ticket generated has to be validated based on the provided details and mask detection technique. This extra step in the process of booking tickets has been introduced to ensure that the user is complying with the guidelines given by the government. The validation has been implemented using a flask app. For the validation, the user will be required to upload their recent photo from their device, which will be verified by IBM visual recognition either as "Masked" or "Without mask." This page has to be shown by the user while boarding the transport, failing which the user will not be allowed to enter the bus.

We have planned an alternative and more effective method to impose the guidelines as well. If we are able to install cameras in every public transport, then we can use OpenCV (Open Source Computer Vision Library), an open-source computer vision and machine learning software library of python, to distinguish between people who are not wearing and those who are not wearing a mask properly, so that required action can be taken against such people.

3.5 Security Features

Our application ensures security in all respects either for user authentication or for ticket generation in the following ways:

(a) While signing up the user is required to enter their "AAdhaar Card Number" for authentication. Aadhaar is an initiative by the Unique Identification Authority of India laid down by "Targeted Delivery of Financial and Other Subsidies, Benefits and Services Act, 2016 (Aadhaar Act 2016)".

(b) As discussed in the previous section, every journey will be associated with a unique ticket number to ensure that there is no fraudulent use of the ticket generated.

(c) We constantly urge the user to keep changing their passwords and choose strong passwords with special characters and length greater than eight characters to reduce the risk of getting hacked.

3.6 Imposing Guidelines

(a) We have designed our application in such a way that the users can get the option for booking tickets at the end of the guidelines page. This will force the user to go through the guidelines at least once before going to book the tickets. We know that some people will just scroll through the points without reading but people are becoming self-aware nowadays so even if some people keep the points in mind, it will prove beneficial to society.

(b) The most powerful feature of our application is that even when the ticket is generated for a particular user, it has to be validated to allow the user to take up the journey. This validation will be based on image recognition and only properly masked images will only get validated.

(c) We have planned to reduce the bus occupancy by 50%. This will allow the passengers to maintain social distancing even inside the bus. We are also planning to impose a penalty on people who will not be abiding by the rules specified by the government.

(d) As we have seen already that route allocation will be based on the shortest path, this will reduce the time spent in public transport by the user thus reducing exposure and eventually reducing the risk of infection.

(e) We are also promoting digital methods of payment in order to ensure a contactless money transaction.

3.7 Customer-First Approach

Our application has been built keeping the customer as a priority. As we have discussed earlier, that shortest path will be allocated to the user which will reduce the time of travel, will be pocket friendly, and owing to minimum exposure will reduce the risk of people getting affected with COVID-19. The ticket allocation is done on a priority basis, giving doctors and health workers the highest priority, thus ensuring their availability at required places on time.

Also, we would ensure 50% occupancy of transports in order to provide our users with ample space to follow social distancing. This feature will result in more number of buses to be scheduled at regular intervals of time, but the lives of the users cannot be negotiated with. Lastly, the application has been made extremely user friendly, thus making booking, canceling tickets, and updating information easier.

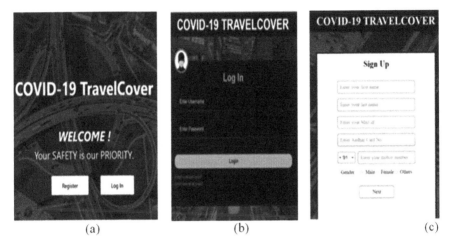

(a)   (b)   (c)

Fig. 4 User Interface of COVID-19 TravelCover: (**a**) Welcome Page, (**b**) Login page, (**c**) Sign Up page

3.8   User Interface (UI)

In this section, we have explained and presented each page of the application along with a brief description. All the main pages of and interface used in the application such as HOME page, Login Page, Ticket Booking, etc. are visualized with their steps to be used. Figure 4 shows the user interface of (a) Cover page, (b) Login page, and (c) Registration page, respectively.

Cover Page It is the welcome or cover page of the application which will guide the user to follow the initial steps.

Login Page Those who already have an account here on our website can LOG IN directly by entering the username and password. If a person has forgotten his/her password, then they can click on the Forgot your Password option, or if a person does not have an account can click on the Don't have an account option.

Registration Page If a person does not have an account can directly go to the registration page. On this SIGN-UP page, the user needs to enter First Name, Last Name, Email Address, Mobile Number, and Gender. Click on the Next button to complete his/her registration form. On the next page, you need to enter Username, confirm the email, enter the password of your choice, and re-enter the password.

Home Address In this home address page, the user can save the address of theirs so they can easily choose home rather than entering the address of their home every time. In this box, the user needs to fill up their Username, Address of their home, Choose the State, Pin Code, and then click on the SUBMIT button.

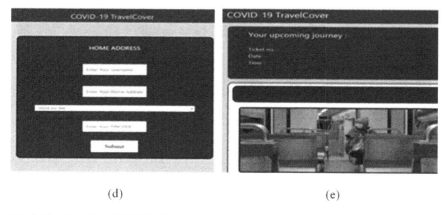

(d)   (e)

Fig. 5 User Interface of COVID-19 TravelCover: (**d**) Home Address Page, (**e**) Dashboard page

Dashboard An informative page about the user's upcoming journey and this page also has a blog. This page also has a Cancel Your Ticket button (for the user to cancel his/her ticket). And from now, on every page, there will be a hamburger menu to toggle between pages. The home address page and Dashboard page visualization have been shown in Fig. 5.

Guidelines This page provides users with a set of general instructions that are essential to be followed in public places and transport. We have followed guidelines issued by the Ministry of Health & Family Welfare-Government of India for domestic travel [15].

Time Table On this page, the timetable shown, which is dynamic, changes as per the user's details. The changes in the time of the buses will get updated as soon as the database gets updated. UI of Guidelines page and Time table page has been shown in Fig. 6.

Ticket-Booking This box appears for the booking of the ticket; here, the user needs to:

(a) Enter the source pin code (from where he/she will start his/her journey).
(b) Enter the destination pin code (where he/she will end his/her journey).
(c) Enter the purpose for which the user wants to move from one place to another (this purpose entry will help to define the priority of one's work).
(d) Enter the date (day, month, and year) and time.

After the completion of the box, the user needs to click on the Proceed button, Fig. 7 visualizes the ticket booking page and cancelation page, respectively.

Cancelation Page
1. After clicking on the "Cancel your ticket" button on the dashboard page, this cancelation page will appear, the user needs to enter his/her ticket number, then will click on the "Cancel Request" button.

(f)

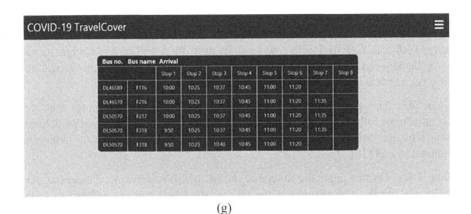

(g)

(h)

Fig. 6 User Interface of COVID-19 TravelCover: (**f**) Guidelines page, (**g**) Time Table page, (**h**) blog page

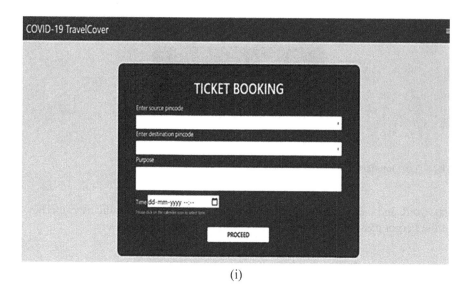

(i)

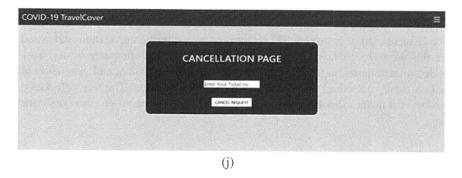

(j)

Fig. 7 User Interface of COVID-19 TravelCover: **(i)** Ticket Booking Page, **(j)** Cancelation page

2. As soon as the user clicks on the "Cancel Request" button, a prompt pops up where the user needs to enter his/her password and then needs to click on the OK button.
3. After clicking on the previous prompt's OK button, another prompt will pop up which will ask the user for his/her cancelation's confirmation.
4. As soon as the user confirms his/her cancelation, another prompt will appear with the confirmation of the user's ticket's cancelation.

Mask Detection System This system is made to detect the user is wearing a mask or not. The user needs to upload his/her picture and will get detected by our app.

Forgot Password Here on this page the user can change their password, if they forgot it by entering specified details such as New Password, Confirm New

Fig. 8 User Interface of COVID-19 TravelCover Forgot Password Page

Password, and then click on the SUBMIT button. Figure 8 shows the user interface of the Forgot password page and Mask detection system, respectively.

4 Result and Discussion

The proposed method for the smart transport management system has been tested on a real-time database collected through the application "COVID-19 TravelCover". We have randomly collected 100 real-time data using the application and based on this we have verified the system validity. The system efficiency of our work is determined based on the dynamic fare calculation, validation, and generation of tickets (mask detection) and dynamic time table generation. The fare calculation system and ticket validation system have been explained based on some real-time data.

4.1 Fare Calculation

We have explained the fare calculation method with the help of an example and taken a few real-time data from the system for briefing purposes. Let us take an example and check whether the price generated is correct or not, where the price per stop is 10/- rupees.

We have used a very simple and user-friendly technique to calculate the fare of individual users. After receiving the inputs from the user for source and destination, respectively, we will refer to the "route details" table (shown above in the second picture). First, we will find the routes which consist of both the source pin code and destination pin code as its stops and create an array at the backend with a list of such routes present in the table. Then we will calculate the number of stops between source and destination in each of these routes and will find the minimum value among them. This minimum value will represent the number of stops between the source and the destination. Thus, now we just have to multiply 10 with the

Table 2 Real-time users' ticket details and fare calculation table

UNAME	S_PIN	D_PIN	DTIME	PRICE	TNO	PRIORITY
Nini	1	5	2020-07-07 T12:57	50	niniTNBF1T50.387	1
Kaju	1	8	2020-07-04 T12:49	30	kajuTNBF1T80.139	

minimum value received and that will be the price incurred upon the user to take up the journey.

Table 2 provides the individual user's ticket details consisting of username (UNMAE), source PIN(S_PIN), Destination PIN(D_PIN), Date and Time (DTIME), Price Calculated(PRICE), Ticket Number(TNO), and priority(**PRIORITY**). The bus or vehicle route details are charted in Table 2, where we have considered four routes (R1 to R4) and nine bus stops represented by S1 to S9. Based on Tables 1 and 2, we have explained two cases for automatically calculating the fare of the ticket of every individual user.

Case 1 For username "nini", we can see in Table 2 that the price calculated is Rs 50/- between pin code "1" and "5," which has been allocated according to "R1." Since all four routes consist of the stops 1 and 5, hence based on shortest routes determination it has opted route 1 which has only 5 stops difference, hence the price calculated is 50.

Case 2 For username "kaju", the price calculated is 30/- between pin codes "1" and "8", which has been allocated according to "R2", because route R2 provides the shortest path between stops 1 and 8.

4.2 Validation of Tickets

As mentioned earlier in the chapter, the ticket validation system is implemented through a python-based flask app. We have used the "Visual Recognition" service provided by IBM for this purpose. We have used supervised learning methodology to initially train the model to make it reliable. The model is trained to differentiate between masked and unmasked images using a large dataset of 100 images among which 50 images were masked and the rest of them were unmasked ones. The visual recognition service has allowed us to design a customized layout to implement the machine learning model. Only after a rigorous test of the trained model, the model has been incorporated within the project.

After a unique ticket generation is done for a particular journey, the user will see a button called "Click to validate", clicking on which the user will be redirected to the validation page. On the validation page, the user is required to upload an image by browsing through their device. After uploading the image as soon as the "Predict" button will be clicked, the assessment result will be displayed below the image uploaded by the user. Only if the user is able to show this validation page to

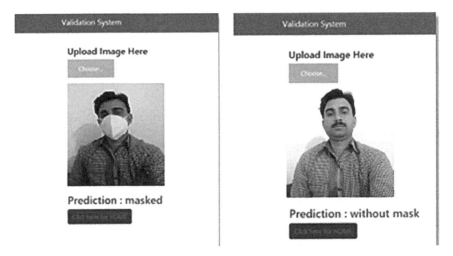

Fig. 9 Ticket validation using mask detection, masked (left) and unmasked image (right)

the regulation authority at the time of boarding the bus, will he/she be allowed to take up their journey? Figure 9 shows that the flask app is able to detect the masked and unmasked images pretty well.

4.3 Security Features

Our application ensures security as well as authenticity in a way that every different journey will be associated with a unique ticket number. A unique ticket number will not only help us in distinguishing different users and keeping a track of them but also would avoid the risk of double-spending (i.e. using the same ticket to travel twice), thus enhancing the authenticity of our application. Figure 10 shows the ticket details table of our database and we can see in the "TNO" column (marked with a red circle) that each and every ticket number is unique.

4.4 Priority

One of the topmost concerns that the world is facing nowadays is safety related to health. In such a scenario, the demand for health officials is quite high as compared to the actual health workers available. Thus, efficient time management has become very necessary for health workers. Through our application, we put forward an effort to help the health officials. This will be ensured with the help of maintaining priority. The health workers and patients will be given the highest priority (=1, according to

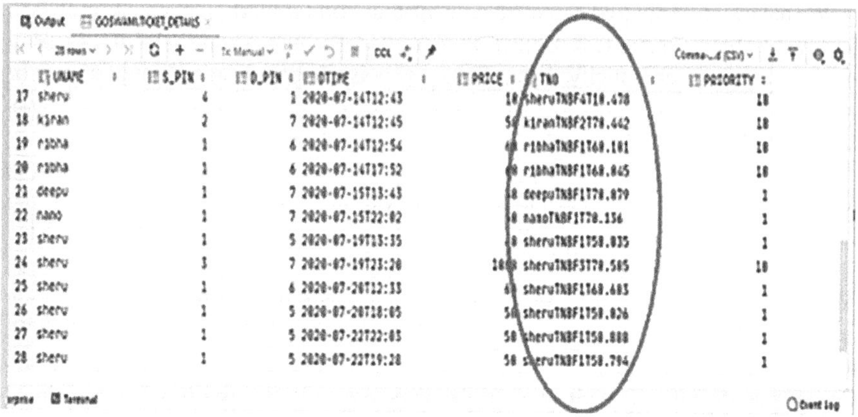

	UNAME	S_PIN	D_PIN	DTIME	PRICE	TNO	PRIORITY
17	sheru	4	1	2020-07-14T12:43	10	sheruTNBF4T10.478	10
18	kiran	2	7	2020-07-14T12:45	50	kiranTNBF2T70.442	10
19	ribha	1	6	2020-07-14T12:54	60	ribhaTNBF1T60.101	10
20	ribha	1	6	2020-07-14T17:52	60	ribhaTNBF1T60.045	10
21	deepu	1	7	2020-07-15T13:43	50	deepuTNBF1T70.079	1
22	nano	1	7	2020-07-15T22:02	50	nanoTNBF1T70.136	1
23	sheru	1	5	2020-07-19T13:35	60	sheruTNBF1T50.035	1
24	sheru	3	7	2020-07-19T23:20	1000	sheruTNBF3T70.505	10
25	sheru	1	6	2020-07-20T12:33	60	sheruTNBF1T60.603	1
26	sheru	1	5	2020-07-20T18:05	50	sheruTNBF1T50.026	1
27	sheru	1	5	2020-07-22T22:03	50	sheruTNBF1T50.008	1
28	sheru	1	5	2020-07-22T19:20	50	sheruTNBF1T50.794	1

Fig. 10 COVID-19 TravelCover ticket details table with unique ticket number

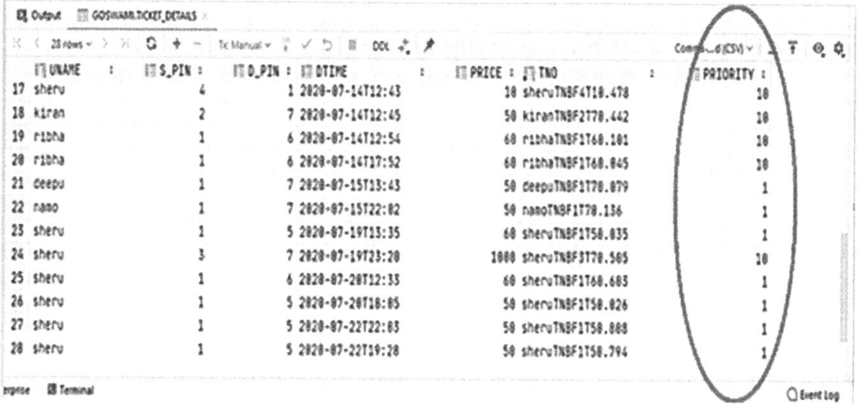

	UNAME	S_PIN	D_PIN	DTIME	PRICE	TNO	PRIORITY
17	sheru	4	1	2020-07-14T12:43	10	sheruTNBF4T10.478	10
18	kiran	2	7	2020-07-14T12:45	50	kiranTNBF2T70.442	10
19	ribha	1	6	2020-07-14T12:54	60	ribhaTNBF1T60.101	10
20	ribha	1	6	2020-07-14T17:52	60	ribhaTNBF1T60.045	10
21	deepu	1	7	2020-07-15T13:43	50	deepuTNBF1T70.079	1
22	nano	1	7	2020-07-15T22:02	50	nanoTNBF1T70.136	1
23	sheru	1	5	2020-07-19T13:35	60	sheruTNBF1T50.035	1
24	sheru	3	7	2020-07-19T23:20	1000	sheruTNBF3T70.505	10
25	sheru	1	6	2020-07-20T12:33	60	sheruTNBF1T60.603	1
26	sheru	1	5	2020-07-20T18:05	50	sheruTNBF1T50.026	1
27	sheru	1	5	2020-07-22T22:03	50	sheruTNBF1T50.008	1
28	sheru	1	5	2020-07-22T19:20	50	sheruTNBF1T50.794	1

Fig. 11 COVID-19 TravelCover ticket details database with user's priority

our application). Figure 11 shows the priority of the users during ticket allocation itself where each user will be associated with some priority of the purpose entered by them, based on which the sequence of allocation will be determined.

We performed a small survey with our friends, family, and other acquaintances. In that survey, we asked each individual a few questions related to our application. Some of the questions were, "Are you satisfied with the management of transportation prevalent nowadays in terms of safety?", "Will you prefer an application which will allow you to book buses based on your source and destination, online?", "What do you think of the idea of ticket allocation on the basis of priority?". After carefully analyzing the results we found that around 60% of the people were willing to associate with an application as ours, thus indicating the need and success of our application.

We have also compared our work with various similar work or applications based on Social Distancing, Mask Detection, Location Record, Shortest Route, Priority Assignment, and Simplicity. From Table 3, it is observed that no application or work includes all the features except our proposed COVID-19 TravelCover, whereas almost all the application are user friendly and simple to use.

MOOVIT[6] is capable of finding the shortest path as well as location tracking, whereas TRANSIT [17] and GoTo (Harsh Vardhan [9]) only able to record the location. SOCIAL DISTANCING APP [24] and Coready [24] are developed for maintaining the social distancing very efficiently along with location tracking. Among all applications, MASK DETECTOR [1] is only capable of detecting the face mask. But our proposed application has all the features like social distancing, mask detection, location tracking, priority assignment, and also user friendly. Hence, our developed application is far superior as compared to all applications presented in Table 3.

5   Conclusion and Future Scope

This chapter has provided a dependable solution to tackle the growth of infections of COVID-19 and would facilitate proper public transportation management post lockdown. Once the lockdown ends, there needs to be elaborate planning to manage all the people using public transport as offices, factories, etc. will start functioning. So, to ensure everyone's well-being and safety, we are introducing an intelligent post-lockdown management system for public transport which will help in avoiding over occupancy of public transport and will help people to maintain social distancing which will lower their risk of getting infected by COVID-19. In our application, we will have ticket allocation based on the urgency of the purpose of travel; this will help the doctors and other health workers get to their destination in time as in tough times of the COVID-19 pandemic, these people hold the highest priority. Through our ticket booking system, we will try to reduce the transit time for users as much as possible thus saving time and comparatively reducing the risk of the individual getting infected. We will also strive to avoid over-occupancy of the public vehicles and ensure social distancing by allowing the commuters to board only the allowed seats as available seats would be arranged in a pattern to facilitate social distancing. All the money transactions would be done through e-currency, thus reducing the chances of infection being spread through physical currency. We also have a strong ticket validation system aligning with the basic guidelines to allow only people with precautions to board the bus.

In the future scope, we will implement the neural network, reinforcement learning, on this platform to automatically categorize the type of passengers.

Table 3 Comparison of the proposed COVID-19 TravelCover with the state-of-art methods

Applications	Parameters					
	Social distancing	Mask detection	Location record	Shortest route	Priority assignment	Simplicity
MOOVIT [6]	NA	NA	√	√	NA	√
TRANSIT [17]	NA	NA	√	NA	NA	√
GoTo [9]	NA	NA	√	NA	NA	√
SOCIAL DISTANCING APP [23]	√	NA	√	NA	NA	√
MASK DETECTOR [1]	NA	√	NA	NA	NA	√
Coready [24]	√	NA	√	NA	NA	√
COVID-19 TravelCover	√	√	√	√	√	√

Acknowledgments We would like to acknowledge the "IBM Hack Challenge 2020" for providing the platform and cloud services and also for providing a problem statement for implementing this idea. This project is selected and implemented under the mentor's supervision in the hackathon and we strive to come up with better ways of practical implementation of our idea, thus promising a broader future scope.

References

1. Aniket sindhu. (2020). *Mask detector: Mask detection from camera/photos – Apps on Google Play*. https://play.google.com/store/apps/details?id= com.aniket.maskdetector& hl=en_US&gl=US
2. Cats, O., Koutsopoulos, H. N., Burghout, W., & Toledo, T. (2011). Effect of real-time transit information on dynamic path choice of passengers. *Transportation Research Record: Journal of the Transportation Research Board, 2217*(1), 46–54. https://doi.org/10.3141/2217-06.
3. Cats, O., & Loutos, G. (2016). Real-time bus arrival information system: An empirical evaluation. *Journal of Intelligent Transportation Systems: Technology, Planning, and Operations, 20*(2), 138–151. https://doi.org/10.1080/15472450.2015.1011638.
4. Data Protection Laws. (2018). *Data protected India | insights | Linklaters*. Linklaters. https://www.linklaters.com/en/insights/data-protected/data-protected%2D%2D-india
5. Di Girolamo, N., & Meursinge Reynders, R. (2020). Characteristics of scientific articles on COVID-19 published during the initial 3 months of the pandemic. *Scientometrics, 125*(1), 795–812. https://doi.org/10.1007/s11192-020-03632-0.
6. Erez, N., Bick, R., & Evron, Y. (2012). *Moovit*. https://moovit.com/
7. Government of India. (2020). *#IndiaFightsCorona COVID-19*. COVID-19 Dashbaord. https://www.mygov.in/covid-19/
8. Graham-Harrison, E., & Cadwalladr, C. (2018). Revealed: 50 million Facebook profiles harvested for Cambridge Analytica in major data breach. *The Guardian*, 1–5. https://www.theguardian.com/news/2018/mar/17/cambridge-analytica-facebook-influence-us-election
9. Sharma, H. V., Sagar, A., Kaushik, D., Sharma, N., Sagar, R., Rajput, R., & Sharma, M. (2020). *Home – GoTo Bus*. https://www.gotobus.in/
10. India, G. of. (2020). *Unique Identification Authority of India*. 21287. https://uidai.gov.in/
11. Kumar, S. A. P., Madhumathi, R., Chelliah, P. R., Tao, L., & Wang, S. (2018). A novel digital twin-centric approach for driver intention prediction and traffic congestion avoidance. *Journal of Reliable Intelligent Environments, 4*(4), 199–209. https://doi.org/10.1007/s40860-018-0069-y.
12. Majumdar, A., Malviya, N., & Alok, S. (2020). An overview on Covid-19 outbreak: Epidemic to pandemic. *International Journal of Pharmaceutical Sciences and Research, 11*(5), 1958–1968. https://doi.org/10.13040/IJPSR.0975-8232.11(5).1958-68.
13. Matheus, R., Janssen, M., & Maheshwari, D. (2018). *Data science empowering the public: Data-driven dashboards for transparent and accountable decision-making in smart cities*. Government Information Quarterly, February, 0–1. https://doi.org/10.1016/j.giq.2018.01.006.
14. Mohan, M., & Mishra, S. (2020). India's Response to the COVID-19 Pandemic: A Frontal Assault on the "Historically Dispossessed." *International Journal of Health Services*, 002073142096843. https://doi.org/10.1177/0020731420968438.
15. NCDC. (2020). *COVID19 Guidelines*. National Centre for Disease Control (NCDC). https://ncdc.gov.in/index1.php?lang=1&level=1&sublinkid=703&lid=550
16. Nguyen, Q. T., Ba, N., & Phan, T. (2015). Scheduling problem for bus rapid transit routes. *Advances in Intelligent Systems and Computing, 358*, 1–415. https://doi.org/10.1007/978-3-319-17996-4.

17. Sam Vermette, G. C. (2012). *Transit: Bus & Subway Times.* https://play.google.com/store/apps/details?id=com.thetransitapp.droid&hl=en
18. Schmidtke, H. R. (2020). Location-aware systems or location-based services: A survey with applications to CoViD-19 contact tracking. *Journal of Reliable Intelligent Environments.*https://doi.org/10.1007/s40860-020-00111-4.
19. Schmidtke, H. R. (2018). A survey on verification strategies for intelligent transportation systems. *Journal of Reliable Intelligent Environments, 4*(4), 211–224. https://doi.org/10.1007/s40860-018-0070-5.
20. Shalaby, A., & Farhan, A. (2004). Prediction model of bus arrival and departure times using AVL and APC data. *Public Transportation, 7*(1), 41–61.
21. Singh, H. P., Khullar, V., & Sharma, M. (2020). Estimating the impact of Covid-19 outbreak on high-risk age group population in India. *Augmented Human Research, 5*(1), 18. https://doi.org/10.1007/s41133-020-00037-9.
22. Singhal, T. (2020). A review of coronavirus disease-2019 (COVID-19). In *Indian Journal of Pediatrics* (Vol. 87, issue 4, pp. 281–286). Springer. https://doi.org/10.1007/s12098-020-03263-6.
23. Syook. (2020). *The social distancing app – apps on Google Play.* Syook.Com. https://play.google.com/store/apps/details?id=com.syook.socialdistancing&hl=en&gl=US
24. Trushal Sardhara. (2020, August). *Social Distancing: Here's an App for That – IEEE Spectrum.* IEEE Spectrum. https://spectrum.ieee.org/news-from-around-ieee/the-institute/ieee-products-services/social-distancing-heres-an-app-for-that
25. Yu, B., Lam, W. H. K., & Tam, M. L. (2011). Bus arrival time prediction at bus stop with multiple routes. *Transportation Research Part C: Emerging Technologies, 19*(6), 1157–1170. https://doi.org/10.1016/j.trc.2011.01.003.

Diverse Techniques Applied for Effective Diagnosis of COVID-19

Charles Oluwaseun Adetunji (ID), **Olugbemi Tope Olaniyan,**
Olorunsola Adeyomoye, Ayobami Dare, Mayowa J. Adeniyi, Enoch Alex,
Maksim Rebezov, Natalia Koriagina, and Mohammad Ali Shariati

1 Introduction

Coronaviruses are single-stranded RNA viruses with genome size ranging from 26 to 32 KB [7]. They have spikes that project from their surfaces. In humans, they cause respiratory tract infection which can be mild or severe. Mild infection could show symptoms such as common cold, while the severe form could cause severe acute respiratory syndrome (SARS), Middle East Respiratory Syndrome (MERS), and COVID-19 infection [52]. SARS-CoV-1 was the first strain of the SARS coronavirus identified to be responsible for the outbreak that occurred between the years 2002 and 2004 [9, 49]. In December 2019, another strain of SARS coronavirus

C. O. Adetunji (✉)
Applied Microbiology, Biotechnology and Nanotechnology Laboratory, Department of Microbiology, Edo University Iyamho, Auchi, Edo State, Nigeria
e-mail: adetunji.charles@edouniversity.edu.ng

O. T. Olaniyan
Laboratory for Reproductive Biology and Developmental Programming, Department of Physiology, Edo University Iyamho, Iyamho, Nigeria

O. Adeyomoye
Department of Physiology, University of Medical Sciences, Ondo City, Nigeria

A. Dare
Department of Physiology, School of Laboratory Medicine and Medical Sciences, College of Health Sciences, Westville Campus, University of KwaZulu-Natal, Durban, South Africa

M. J. Adeniyi
Department of Physiology, Edo State University Uzairue, Iyamho, Edo State, Nigeria

S. K. Pani et al. (eds.), *Assessing COVID-19 and Other Pandemics and Epidemics using Computational Modelling and Data Analysis*,
https://doi.org/10.1007/978-3-030-79753-9_3

broke out in the City of Wuhan in China. This strain was named SAR-CoV-2 which causes Coronavirus disease 2019 (COVID-19) [22–24].

Some of the diagnostic techniques that could be applied for quick identification of SARS-CoV-2 include reverse transcription polymerase chain (RT-PCR), real-time quantitative polymerase chain reaction (RT-qPCR) [22], an enzyme-linked immunosorbent assay (ELISA), IITF, lateral flow immunoassay, and clustered regularly interspaced short palindromic repeats (CRISPRs) [50]. In addition, imaging techniques such as the computerized tomography (CT) and X-rays could also be utilized to diagnose the disease [23, 24]. All the methods highlighted above could be affected by inadequate sample volume, inappropriate collection of samples, inaccuracy of the technique, inappropriate window for the sample collection, and contamination [26, 36]. There is a need for the development a rapid and cheaper, way to diagnose the disease. Furthermore, there is need to develop home-based kits that can be used by individuals to know their status and immediately self-isolate if confirmed positive. Even though vaccines are now available that can prevent infestation of this dangerous virus, the role of diagnosis of the disease in prevention of the spread of the virus cannot be overemphasized. Moreover, the application of molecular and advanced biotechnological techniques has been documented to play a crucial role for rapid detection of COVID 19 diseases [1, 4, 20].

Therefore, this chapter intends to provide a detailed information on the application on some techniques that could be applied for adequate and quick detection of COVID 19 diseases.

2 General Overview on COVID-19

The term "severe acute respiratory syndrome coronavirus 2" used in depicting the virus clearly indicates the lungs are the major hubs affected by the virus. For COVID-19, symptoms appear between 1 and 14 days following exposure to the virus. It is worth noting that 33% of people infected usually do not present with noticeable features. The virus spreads through contact, infected surfaces,

E. Alex
Department of Human Physiology, Ahmadu Bello University Zaria, Kaduna State, Nigeria

M. Rebezov
Prokhorov General Physics Institute, Russian Academy of Sciences, Moscow, Russia

K.G. Razumovsky Moscow State University of Technologies and Management (the First Cossack University), Moscow, Russia

M. A. Shariati
K.G. Razumovsky Moscow State University of Technologies and Management (the First Cossack University), Moscow, Russia

N. Koriagina
E. A. Vagner Perm State Medical University, Perm, Permskaja Oblast, Russia

aerosols, and respiratory droplets. Even though, symptoms appear variable, most of the widely reported symptoms include pneumonia, shortness of breath, cough, anosmia, sore throat, and many other symptoms [38]. Pathophysiologic mechanism involves modulation of renin-angiotensin-aldosterone system [35]. Specifically, the virus invades the host cells through angiotensin converting enzyme II, an enzyme that is present in large quantity in pneumocytes [51]. Hence therapeutic goals antagonizing angiotensin II activity might prove helpful [15]. As expected, innate immunity, specifically inflammatory response occurs during viral invasion. However, in COVID-19, there appears to be an association between COVID-19 severity and the levels of inflammatory cytokines such as monocyte chemoattractant protein-1, gamma-interferon, interleukin-IB, and many more. This implies that the barrage of diseases characterizing the condition is attributable to severe inflammatory responses [43].

3   COVID-19 and Mental Health

A Chinese study showed that 96.2% of COVID-19 inpatients experienced post-traumatic stress symptoms [6]. Even in healthcare providers, a study reported that the prevalence of anxiety and depression stands at 45% and 51%, respectively [22–24]. A study conducted in Ibadan, Nigeria, on the health implication of COVID-19-induced lockdown on adult residents revealed that sleeplessness correlated positively with anxiety and depression [2]. In Spain, an elevated level of anxiety and stress was orchestrated by COVID-19 lockdown measures in young children [46]. About one quarter of older people with depression may be adversely affected by COVID-19 measures, most especially, social distancing measures.

4   COVID-19 Diagnosis and Management

Identification of COVID-19 has been done using real-time polymerase chain reaction (TaqPath COVID-19 kit), transcription-mediated amplification, reverse transcription loop-mediated isothermal amplification, and nasopharyngeal swab [3]. Serology test which detects serum or plasma antibodies is also available [45]. Non-invasive techniques such as chest computerized tomography, computerized tomography, and X-rays are also beneficial as far as examination of the pulmonary structure is concerned.

One of the ways to prevent COVID-19 is through social distancing (at least 6 feet). Other preventive strategies include regular hand washing, avoiding unnecessary movement, social gathering restrictions, and use of face mask. Face mask was found very effective as a preventive measure [17]. In addition, vaccines are now available to induce an artificial active immunity against the virus [5]. However, it has been observed that many COVID-19 patients have benefited from numerous anti-

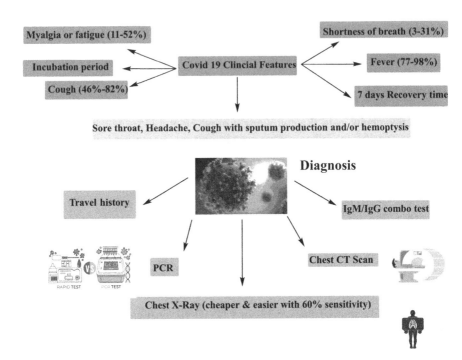

Fig. 1 Several techniques that could be applied for the identification of COVID-19

COVID-19 drugs such as vasodilators, antiviral drugs, corticosteroids, and many more. Figure 1 illustrates diverse COVID-19 clinical features, symptoms, and the various types of techniques applied for the detection of COVID-19.

5 Different Comprehensive Techniques for Rapid Detection of COVID-19

Prompt diagnosis of COVID-19 is vital to prevent the spread of the disease. Diagnosis of COVID-19 involves using different techniques to identify the occurrence of the virus using various samples. Antibodies play a momentous function in the detection of infectious pathogens. Once there is an exposure to SARS-CoV-2, the body's immune system produces antibodies (human IgG) which can be detected between the 7th and 14th days after infection and may even persist after the clearance of the virus [25]. Enzyme-linked immunosorbent assay (ELISA) detects the presence of human IgG against SARS-CoV-2 in the serum or plasma. As described in [33], patient samples and standards are placed in microplate wells containing specific bound antigens in the well surface. An enzyme conjugate is added and allowed to incubate for 60 mins at room temperature in order to form antigen-antibody complex. The mixture is washed 4-5 times using wash buffer to remove unbound materials. A Tetramethylbenzidine (TMB) substrate is added and the mixture is incubated for another 20 mins; this causes changes in color of the mixture. The intensity of the color corresponds to the quantity of the human IgG in the sample. A stop is added and the absorbance read at 450 nm wavelength using the microplate reader. The concentration of the antibodies can be obtained by using absorbance of the sample and the standard. The real-time reverse transcription polymerase chain (rRT-PCR) can be used for qualitative detection of SARS-CoV-2 nucleic acid [41]. Samples from oropharyngeal and nasopharyngeal swabs, sputum, and aspirate from lower respiratory tract are used. The RNA from the virus is made undergo reverse transcription to cDNA which is amplified using a thermal cycler. During amplification, probes bound to template are cut off using Taq enzyme. The polymerase chain reaction (PCR) instrument will draw a real-time amplifier curve based on the signal received, and this will either show the occurrence or absence of novel SARS-CoV-2. The CRISPR-Cas systems allow for easy diagnosis of COVID-19 in less than 40 mins. This method has advantage over rRT-PCR because of its ability to simultaneously perform reverse transcription and thermocycling amplification using loop-mediated replication for SARS-CoV-2 RNA [53]. The Cas systems detect the viral sequences when the presence of the virus is confirmed. The system is rapid, sensitive, and the reporting format is user's friendly. Some of the other diagnostic techniques that has been used to check for the presence of SARS-CoV-2 in samples include indirect immunofluorescence assay, lateral flow assay, X-rays, and the use of computerized tomography.

6 Performance of Several Laboratory Diagnostic Evaluations and Platforms

Diagnosis of COVID-19 involves confirmation of coronavirus using molecular technique. Also test that detects serum antibodies against the virus (serological tests) is also available. With respect to molecular test, the challenge is the time required for the result to be available. For serological tests, the issue is the sensitivity of the technique. Yet rapidity and sensitivity of diagnosis are crucial in the prevention of disease progression and community transmission.

As part of measures to unravel the challenge, Biotechnology Companies are working very hard to ensure rapid techniques are developed. An example of this is a new multiplex real-time polymerase chain reaction developed by Thermo Fisher Scientific called TaqPath COVID-19 kit. The kit contains both controls and assays that are required for the detection of ribonucleic acid from coronavirus-2019 virus.

Many studies on COVID-19 diagnosis identified RT-PCR as a consensus technique for COVID-19 validation. Sensitivity of this technique is dependent on specimen type. For instance, analysis of sputum specimens yielded a sensitivity of 69%, while that of nasopharyngeal swabs yielded 63% [22–24]. The time the specimen is collected is very crucial. Samples collected earlier before disease onset or later may yield a false negative result. Evidence exists to prove that sensitivity of the test using sputum specimen is dependent on the timing of sample collection. Moreover, it has been demonstrated that COVID-19 normally shows the greatest viral load during the first week after the onset of symptom and decline thereafter [22–24].

7 Alternative Methods for the SARS-CoV-2 Detection

The application of molecular analytical test has become highly relevant in the identification of COVID-19 disease. The development of these tests is based on proper understanding of protein and genetic make-up of SARs-CoV-2 that causes infection. RNA is an ideal biomolecule that promotes the nucleic acid specificity via base pair complementarity, as reported by RNA interference or modification study [31]. These properties of RNA and other similar biomolecules have transformed molecular diagnosis by promoting a quick and precise detection of nucleic acid for infectious pathogens such as SARs-CoV-2. Among the relevant molecular techniques used in COVID-19 identification are CRISPR-based high sensitivity enzymatic test, RNA aptamer, DETECTR, and next-generation gene sequencing.

8 CRISPR-Based Techniques

A special member of the CRISPR-associated (Cas) proteins, Cas13, can be used to identify the viral molecules of COVID-19 under the influence of a predesigned guide CRISPR RNA (crRNA), which in turn initiates trans-cleavage of reporter-coupled single-strand nucleic acids by Cas13 to produce fluorescent signals or readable colorimetric signals on a lateral flow strip. Using this method, crRNA sequences are specifically selected to reduce relevant sequence related to the genomes of other human respiratory viruses [8, 32]. Thus, this test can constantly detect SARS-CoV-2 target sequences and present the test outcome in few minutes [8, 54, 55].

CRISPR-associated (Cas) is a gene editing technology. In the diagnosis of COVID-19, many CRISP-based methods are available as potential options. All-in-One Dual CRISPR Cas 12a, unarguably, is a very accurate and highly sensitive technique. It has been designed for identification of COVID-19 virus [19]. One of the advantages of AIOD-CRISPR could applied in point-of-care screening [12]. CRISP Cas 12-based detection developed by Broughton and colleagues is very rapid and takes less than 40 minutes compared to other isothermal nucleic acid-oriented techniques. It has the potential of being used to diagnose COVID-19 from RNA samples of COVID-19 patients [10].

9 DNA Endonuclease Targeted CRISPR Trans Reporter (DETECTR)

Another technique that can be utilized for the diagnosis of COVID-19 is called DNA endonuclease targeted CRISPR trans reporter (DETECTR) assay [8] with accuracy level comparable to real-time polymerase chain reaction with an advantage of thermocycling avoidance. Besides this, it is also less complex, easy, and portable.

10 CAS 13-Based Rugged Equitable Scalable Testing (CREST)

A very sensitive, easy, and portable method called Cas 13-based Rugged equitable scalable testing (CREST) was introduced by Rauch et al. [44]. It is a cheap technique that makes use of available protein.

10.1 Amplification-Free Assay

Amplification-free assay is a mobile phone-oriented diagnostic method which detects COVID-19 virus in patient's nasal swab [16]. It is a portable and affordable technique.

10.2 Specific High Enzymatic Reporter Unlocking

This is a new diagnostic method based on real-time reverse transcriptase developed by Jasper Fuk-woo and has been shown to be more specific and sensitive than RdRp-p2 approach. This is a lateral flow-based assay which was developed by Zheng's laboratory with promising roles in timing and cost economy and highly sensitive pathogen detection [32]. It has been discovered that with the aid of CRISPR, it is possible to detect COVID-19. This is because the approach is more sensitive with a higher level of specificity when compared to real-time polymerase chain reaction using metagenomics [16].

10.3　Post Analysis Phase

It has been discovered that reporting, verification, interpretation, and documentation of COVID-19 results take place during the post analysis phase. This phase is as important as the previous two phases.

10.4　RNA Aptamers

Aptamers comprise of nuclei acid oligonucleotides or small peptide molecules with high specificity for binding specific target molecules, thereby resulting to their increased sensitivity and precise detection. As a result of their high reproducibility and purity as well as increase stability and reversibility, with high presence of target molecules, aptamers are considered as a new diagnostic device [37]. For adequate diagnosis of COVID-19, aptamers are used together with RT-PCR, or ELISA and are used as aptamer-linked immune sorbent assay, and cantilever-based aptasensors. The increased sensitivity of aptasensors as well as quick and easy diagnostic process is highly important in point-of-care diagnosis of SARs-CoV-2. This alternative test has been applied to diagnose Norovirus strains and bovine viral diarrhea efficiently relative to PCR test [14, 39]. With COVID-19 infestation, aptamers that recognize the viral protein has been used quick identification of SARs-CoV-2 and may act as an antiviral agent in treating COVID-19 disease [11, 57].

10.5　Next-Generation Sequencing (NGS)

NGS techniques such as (Explify®) are sophisticated molecular procedures that started in 2005 and currently used in genomic research. Till date, this technique has become a gold standard in genomic sequencing and very useful in the detection of several diseases and genetic mutations (Morozova and Marra [30]. NGS DNA sequencing is majorly applied to provide adequate information on the molecular prevalence, spread, and classification of pathogens. Compared to single gene analysis done by other tests, a single NGS test can analyze large volume of gene present in the clinical sample, thus, making this technique to be highly resourceful and generally accepted as a diagnostic tool, transforming the diagnosis of pathogens. NGS technology together with other bioinformatics tools has revolutionized the study of viral parthenogenesis and diagnostics and has useful application in the current SARs-CoV-2 outbreak [28]. Thus, NGS technique has a great potential to identify unknown mutation or DNA recombination in the gene of SAR-CoV-2 within a little period of time, thereby improving the diagnosis and preventing a second or third wave as well as new outbreak of infection.

11 Molecular Diagnostic Techniques for COVID-19

COVID-19 molecular diagnosis is done using real-time polymerase chain reaction, reverse transcription loop-mediated isothermal amplification, and transcription-mediated amplification using nasopharyngeal swab. COVID-19 diagnosis consists of preliminary phase, sample analysis phase, and post-analysis phase.

11.1 Preliminary Phase

The preliminary phase consists collection of sample and sample transportation and storage. Collection of samples requires trained personnel under strict observance of standard precautionary and preventive measures. It is imperative to take note that the higher the amount of viral RNA, the higher the likelihood of COVID-19 detection. Samples required for molecular COVID-19 diagnosis are nasopharyngeal and oropharyngeal swabs, sputum, bronchial lavage, saliva, blood, and many more [56]. In Nigeria, the National Center for Disease Control stipulates that one oropharyngeal swab and one nasopharyngeal swab be collected for the test.

11.2 Analysis Phase

There are several laboratory methodologies that are available for diagnosis. The major concern of molecular diagnosis is the timing of test result. The knowledge that there exist some COVID-19 patients who practically display no symptoms of the disease (asymptomatic COVID-19 patient) altered the perception mode regarding diagnosis and highlighted the need for more effective diagnostic tools. This will help in not only early and prompt diagnosis of COVID-19 in symptomatic patients but also prediction of disease progression and prognosis. Early and prompt detection of COVID-19 is critical in order to recognize the unsuspecting population who are at risks of developing disease symptoms and early institution of treatments. As a highly contagious disease, prompt diagnosis requires alteration in traditional orientation regarding laboratory setting which mandates that tests should be done only in specific laboratory locations. Having COVID-19 screening in point-of-care or bed sites demands molecular techniques suiting mobile diagnostic laboratory model will help enormously. There are many methods including immune assays which work by detecting DNA or RNA in the sample make COVID-19 diagnosis faster.

11.3   Loop-Mediated Isothermal Amplification (LMIA)

LMIA centers on selective amplification at a constant temperature of specific nucleic acid. With the method, denaturation of initial template is not important. Hence this method offers a prompt, sensitive, and less expensive diagnosis. The procedure is economical because it does not involve the use of costly reagents and instruments. There exists modified loop-mediated isothermal amplification. One of them integrates reverse transcription with conventional loop-mediated isothermal amplification [22–24]. The procedure based on reverse transcriptase-LMIA has been showed to be relatively consistent with 88.89% sensitive when compared to real-time polymerase chain reaction technique without consuming much time [18]. LMIA works based on the principle of colorimetry which has advantage of minimizing cross contamination [27].

Numerous scientists have validated the significant LMIA and other molecular techniques in the identification and diagnosis of COVID-19 [13, 21, 29, 34, 40, 42, 47, 48].

12   Conclusion and Future Perspectives

This chapter has given a comprehensive detail on diverse facts on approaches that could be utilized for quick recognition and diagnosis of COVID-19. Relevant information was also provided on COVID-19 and mental health, COVID-19 identification and handling, and different comprehensive techniques for SARS-CoV-2 detection. Moreover, it was also established in this chapter that the application of molecular approaches has been identified as a sustainable diagnostic test that could perform a momentous function in the identification of COVID-19 disease. Typical examples of such molecular techniques highlighted in this chapter includes CRISPR-based high sensitivity enzymatic test, RNA aptamer, DETECTR, and next-generation gene sequencing. There is a need to increase the awareness about the application of next-generation sequencing most especially in the developing countries for swift identification of COVID-19 diseases.

References

1. Afzal, A. (2020). Molecular diagnostic technologies for COVID-19: Limitations and challenges. *Journal of Advanced Research, 26*, 149–115.
2. Abimbola, A. (2020). Mental health implications of lockdown during coronavirus pandemic among adult resident in Ibadan, Nigeria. *AJSW, 10*(3), 50–58. Special Issue on COVID-19.
3. Ai, T., Yang, Z., Hou, H., Zhan, C., Chen, C., Lv, W., Tao, Q., Sun, Z., & Xia, L. (2020). Correlation of chest CT and RT-PCR testing for coronavirus disease 2019 (COVID-19) in China: A report of 1014 cases. *Radiology, 296*(2), E32–E40. https://doi.org/10.1148/radiol.2020200642. Epub 2020 Feb 26. PMID: 32101510; PMCID: PMC7233399.

4. Alpdagtas, S., Ilhan, E., Uysal, E., Sengor, M., Ustundag, C. B., & Gunduz, O. (2020). Evaluation of current diagnostic methods for COVID-19. *APL bioengineering, 4*(4), 041506. https://doi.org/10.1063/5.0021554.
5. Anderson, R. M., Heesterbeek, H., Klinkenberg, D., & Hollingsworth, T. D. (2020). How will country-based mitigation measures influence the course of the COVID-19 epidemic? *Lancet, 395*(10228), 931–934.
6. Bo, H. X., Li, W., Yang, Y., Wang, Y., Zhang, Q., Cheung, T., Wu, X., & Xiang, Y. T. (2020). Posttraumatic stress symptoms and attitude toward crisis mental health services among clinically stable patients with COVID-19 in China. *Psychological Medicine*, 1–2. https://doi.org/10.1017/S0033291720000999. Epub ahead of print. PMID: 32216863; PMCID: PMC7200846.
7. Brian, D. A., & Baric, R. S. (2005). Coronavirus genome structure and replication. *Current Topics in Microbiology and Immunology, 287*, 1–30.
8. Broughton, J. P., Deng, X., Yu, G., Fasching, C. L., Servellita, V., Singh, J., Miao, X., Streithorst, J. A., Granados, A., Sotomayor-Gonzalez, A., Zorn, K., Gopez, A., Hsu, E., Gu, W., Miller, S., Pan, C. Y., Guevara, H., Wadford, D. A., Chen, J. S., & Chiu, C. Y. (2020). CRISPR-Cas12-based detection of SARS-CoV-2. *Nature Biotechnology, 38*(7), 870–874. https://doi.org/10.1038/s41587-020-0513-4. Epub 2020 Apr 16. .PMID: 32300245.
9. Chakraborty, H., & Bhattacharjya, S. (2020). Mechanistic insights of host cell fusion of SARS-CoV-1 and SARS-CoV-2 from atomic resolution structure and membrane dynamics. *Biophysical Chemistry, 265*, 106438.
10. Chen, J. S., Ma, E., Harrington, L. B., Da Costa, M., Tian, X., Palefsky, J. M., & Doudna, J. A. (2018). CRISPR-Cas12a target binding unleashes indiscriminate single-stranded DNase activity. *Science, 360*(6387), 436–439. https://doi.org/10.1126/science.aar6245. Epub 2018 Feb 15. Erratum in: Science. 2021 Feb 19;371(6531): PMID: 29449511; PMCID: PMC6628903.
11. Chen, Z., Wu, Q., Chen, J., Ni, X., & Dai, J. (2020). A DNA aptamer based method for detection of SARS-CoV-2 Nucleocapsid protein. *Virologica Sinica, 35*(3), 351–354. https://doi.org/10.1007/s12250-020-00236-z. Epub 2020 May 25. PMID: 32451881; PMCID: PMC7246297.
12. Ding, X., Yin, K., Li, Z., & Liu, C. (2020). All-in-one dual CRISPR-Cas12a (AIOD-CRISPR) assay: A case for rapid, ultrasensitive and visual detection of novel coronavirus SARS-CoV-2 and HIV virus. *Biochemistry*. https://doi.org/10.1101/2020.03.19.998724.
13. El-Tholoth, M., Bau, H. H., & Song, J. (2020). A single and two-stage, closed-tube, molecular test for the 2019 novel coronavirus (COVID-19) at home, clinic, and points of entry. *ChemRxiv : the preprint server for chemistry*. https://doi.org/10.26434/chemrxiv.11860137.
14. Escudero-Abarca, B. I., Suh, S. H., Moore, M. D., Dwivedi, H. P., & Jaykus, L.-A. (2014). Selection, characterization and application of nucleic acid aptamers for the capture and detection of human norovirus strains. *PLoS One, 9*, e106805.
15. Gurwitz, D. (2020). Angiotensin receptor blockers as tentative SARS-CoV-2 therapeutics. *Drug Development Research., 81*(5), 537–540.
16. Hou, T., Zeng, W., Yang, M., Chen, W., Ren, L., Ai, J., Wu, J., Liao, Y., Gou, X., Li, Y., Wang, X., Su, H., Gu, B., Wang, J., & Xu, T. (2020). Development and evaluation of a rapid CRISPR-based diagnostic for COVID-19. *PLoS Pathogens, 16*(8), e1008705. https://doi.org/10.1371/journal.ppat.1008705.
17. Jeremy, H., Huang, A., Li, Z., Tufekci, Z., Zdimal, V., van der Westhuizen, H.-M., von Delft, A., Price, A., Fridman, L., Tang, L.-H., Tang, V., Watson, G. L., Bax, C. E., Shaikh, R., Questier, F., Hernandez, D., Chu, L. F., Ramirez, C. M., & Rimoin, A. W. (2021). *Proceedings of the National Academy of Sciences, 118*(4), e2014564118. https://doi.org/10.1073/pnas.2014564118.
18. Hu, X., Deng, Q., Li, J., Chen, J., Wang, Z., Zhang, X., Fang, Z., Li, H., Zhao, Y., Yu, P., Li, W., Wang, X., Li, S., Zhang, L., & Hou, T. (2020). Development and clinical application of a rapid and sensitive loop-mediated isothermal amplification test for SARS-CoV-2 infection. Edited by Christina F Spiropoulou. *MSphere, 5*(4), e00808–e00820. https://doi.org/10.1128/mSphere.00808-20.

19. Jeon, Y., Choi, Y. H., Jang, Y., Yu, J., Goo, J., Lee, G., Jeong, Y. K., Lee, S. H., Kim, I. S., Kim, J. S., Jeong, C., Lee, S., & Bae, S. (2018). Direct observation of DNA target searching and cleavage by CRISPR-Cas12a. *Nature Communications, 9*(1), 2777. https://doi.org/10.1038/s41467-018-05245-x. PMID: 30018371; PMCID: PMC6050341.

20. Karthik, K., Aravindh Babu, R. P., Dhama, K., Chitra, M. A., Kalaiselvi, G., Alagesan Senthilkumar, T. M., & Raj, G. D. (2020). Biosafety concerns during the collection, transportation, and processing of COVID-19 samples for diagnosis. *Archives of Medical Research, 51*(7), 623–630. https://doi.org/10.1016/j.arcmed.2020.08.007.

21. Kashir, J., & Yaqinuddin, A. (2020). Loop Mediated Isothermal Amplification (LAMP) assays as a rapid diagnostic for COVID-19. *Medical Hypotheses, 141*, 109786. https://doi.org/10.1016/j.mehy.2020.109786.

22. Liu, R., Han, H., Liu, F., Lv, Z., Wu, K., Liu, Y., Feng, Y., & Zhu, C. (2020a). Positive rate of RT-PCR detection of SARS-CoV-2 infection in 4880 cases from one hospital in Wuhan, China, from Jan to Feb 2020. *Clinica Chimica Acta, J505*, 172–175.

23. Liu, S., Yang, L., Zhang, C., Xiang, Y. T., Liu, Z., Hu, S., & Zhang, B. (2020b). Online mental health services in China during the COVID-19 outbreak. *Lancet Psychiatry, 7*(4), e17–e18. https://doi.org/10.1016/S2215-0366(20)30077-8. Epub 2020 Feb 19. PMID: 32085841; PMCID: PMC7129099.

24. Liu, Y. C., Kuo, R. L., & Shih, S. R. (2020c). COVID-19: The first documented coronavirus pandemic in history. *Biomedical Journal, 43*(4), 328–333.

25. Long, Q. X., Liu, B. Z., Deng, H. J., Wu, G. C., Deng, K., Chen, Y. K., Liao, P., Qiu, J. F., Lin, Y., Cai, X. F., Wang, D. Q., Hu, Y., Ren, J. H., Tang, N., Xu, Y. Y., Yu, L. H., Mo, Z., Gong, F., Zhang, X. L., Tian, W. G., & Huang, A. L. (2020). Antibody responses to SARS-CoV-2 in patients with COVID-19. *Nature Medicine, 26*(6), 845–848.

26. Lu, J., Yin, Q., Li, Q., Fu, G., Hu, X., Huang, J., Chen, L., Li, Q., & Guo, Z. (2020). Clinical characteristics and factors affecting the duration of positive nucleic acid test for patients of COVID-19 in XinYu, China. *Journal of Clinical Laboratory Analysis, 34*(10), e23534.

27. Malik, Y. S., Verma, A. K., Kumar, N., Touil, N., Karthik, K., Tiwari, R., Bora, D. P., Dhama, K., Ghosh, S., Hemida, M. G., Abdel-Moneim, A. S., Bányai, K., Vlasova, A. N., Kobayashi, N., & Singh, R. K. (2019). Advances in diagnostic approaches for viral etiologies of diarrhea: From the lab to the field. *Frontiers in Microbiology, 10*, 1957. https://doi.org/10.3389/fmicb.2019.01957.

28. Massart, S., Chiumenti, M., De Jonghe, K., Glover, R., Haegeman, A., Koloniuk, I., Komínek, P., Kreuze, J., Kutnjak, D., Lotos, L., Maclot, F., Maliogka, V., Maree, H. J., Olivier, T., Olmos, A., Pooggin, M. M., Reynard, J. S., Ruiz-García, A. B., Safarova, D., Schneeberger, P. H. H., Sela, N., Turco, S., Vainio, E. J., Varallyay, E., Verdin, E., Westenberg, M., Brostaux, Y., & Candresse, T. (2019). Virus detection by high-throughput sequencing of small RNAs: Large-scale performance testing of sequence analysis strategies. *Phytopathology, 109*(3), 488–497. https://doi.org/10.1094/PHYTO-02-18-0067-R. Epub PMID: 30070618.

29. Mori, Y., Nagamine, K., Tomita, N., & Notomi, T. (2001). Detection of loop-mediated isothermal amplification reaction by turbidity derived from magnesium pyrophosphate formation. *Biochemical and Biophysical Research Communications, 289*(1), 150–154.

30. Morozova, O., & Marra, M. A. (2008). Applications of next-generation sequencing technologies in functional genomics. *Genomics, 92*(5), 255–264.

31. Morris, K. V., & Mattick, J. S. (2014). The rise of regulatory RNA. *Nature Reviews. Genetics, 15*(6), 423–437.

32. Myhrvold, C., Freije, C. A., Gootenberg, J. S., Abudayyeh, O. O., Metsky, H. C., Durbin, A. F., Kellner, M. J., Tan, A. L., Paul, L. M., Parham, L. A., Garcia, K. F., Barnes, K. G., Chak, B., Mondini, A., Nogueira, M. L., Isern, S., Michael, S. F., Lorenzana, I., Yozwiak, N. L., MacInnis, B. L., Bosch, I., Gehrke, L., Zhang, F., & Sabeti, P. C. (2018). Field-deployable viral diagnostics using CRISPR-Cas13. *Science, 360*(6387), 444–448. https://doi.org/10.1126/science.aas8836. PMID: 29700266; PMCID: PMC6197056.

33. Nagasawa, M., Yamaguchi, Y., Furuya, M., Takahashi, Y., Taki, R., Nagata, K., Suzaki, S., Kurosaki, M., & Izumi, N. (2020). Investigation of anti-SARS-CoV-2 IgG and IgM antibodies in the patients with COVID-19 by three different ELISA test kits. *SN Comprehensive Clinical Medicine*, 1–5. Advance online publication.

34. Njiru, Z. K. (2012). Loop-mediated isothermal amplification technology: Towards point of care diagnostics. edited by Philippe Büscher. *PloS Neglected Tropical Disease, 6*(6), e1572.

35. Olaniyan Olugbemi, T., Ayobami, D., Okotie, G. E., Adetunji, C. O., Oluwaseun, I. B., Bamidele, O. J., & Olugbenga, E. O. (2020). Testis and blood-testis barrier in Covid-19 infestation: Role of angiotensin converting enzyme 2 in male infertility. *Journal of Basic and Clinical Physiology and Pharmacology. DEGRUYTER, 31*(6), 1–13. https://doi.org/10.1515/jbcpp-2020-0156.

36. Tope, O. O., Adetunji, C. O., Okotie, G. E., Adeyomoye, O., Anani, O. A., & Mali, P. C. (2021). Impact of COVID-19 on assisted reproductive technologies and its multifacet influence on global bioeconomy. *Journal of Reproductive Healthcare and Medicine, 2*(Suppl_1), 92–104. https://doi.org/10.25259/JRHM_44_2020.

37. O'Sullivan, C. K. (2002). Aptasensors–the future of biosensing? *Analytical and Bioanalytical Chemistry, 372*(1), 44–48. https://doi.org/10.1007/s00216-001-1189-3. Epub 2001 Dec 13. .PMID: 11939212.

38. Page, J., Hinshaw, D., & McKay, B. (2021). In hunt for Covid-19 origin, patient zero points to second Wuhan Market – The man with the first confirmed infection of the new coronavirus told the WHO team that his parents had shopped there. *The Wall Street Journal*. Retrieved 27 Feb 2021.

39. Park, J. W., Jin Lee, S., Choi, E. J., Kim, J., Song, J. Y., & Bock Gu, M. (2014). An ultra-sensitive detection of a whole virus using dual aptamers developed by immobilization-free screening. *Biosensors & Bioelectronics, 51*, 324–329.

40. Poon, L. L., Leung, C. S., Tashiro, M., Chan, K. H., Wong, B. W., Yuen, K. Y., Guan, Y., & Peiris, J. S. (2004). Rapid detection of the severe acute respiratory syndrome (SARS) coronavirus by a loop-mediated isothermal amplification assay. *Clinical Chemistry, 50*(6), 1050–1052. https://doi.org/10.1373/clinchem.2004.032011. Epub 2004 Mar 30. PMID: 15054079; PMCID: PMC7108160.

41. Praharaj, Ira, Amita Jain, Mini Singh, Anukumar Balakrishnan, Rahul Dhodapkar, Biswajyoti Borkakoty, Munivenkatappa Ashok, Pradeep Das Das, Debasis Biswas, Usha Kalawat, Jyotirmayee Turuk, A. P. Sugunan, Shantanu Prakash, Anirudh K. Singh, Rajamani Barathidasan, Subhra Subhadra, Jyotsnamayee Sabat, M. J. Manjunath, Poonam Kanta, Nagaraja Mudhigeti, Rahul Hazarika, Hricha Mishra, Kumar Abhishek, C. Santhalembi, Manas Ranjan Dikhit, Neetu Vijay, Jitendra Narayan, Harmanmeet Kaur, Sidhartha Giri, Nivedita Gupta. (2020). Indian Journal of Medical Research. 0975–9174. https://doi.org/10.4103/0971-5916.318161

42. Pyrc, K., Milewska, A., & Potempa, J. (2011). Development of loop-mediated isothermal amplification assay for detection of human coronavirus-NL63. *Journal of Virological Methods, 175*(1), 133–136.

43. Quirch, M., Lee, J., & Rehman, S. (2020). Hazards of the cytokine storm and cytokine-targeted therapy in patients with COVID-19: Review. *Journal of Medical Internet Research, 22*(8), e20193.

44. Rauch, J. N., Valois, E., Solley, S. C., Braig, F., Lach, R. S., Baxter, N. J., Kosik, K. S., Arias, C., Acosta-Alvear, D., & Wilson, M. Z. (2020). A scalable, easy-to-deploy, protocol for Cas13-based detection of SARS-CoV-2 genetic material. *Molecular Biology*. https://doi.org/10.1101/2020.04.20.052159.

45. Salehi, S., Abedi, A., Balakrishnan, S., & Gholamrezanezhad, A. (2020). Coronavirus disease 2019 (COVID-19): A systematic review of imaging findings in 919 patients. *AJR. American Journal of Roentgenology, 215*(1), 87–93.

46. Semo, B., & Frissa, S. M. (2020). The mental health impact of the COVID-19 pandemic: Implications for sub-Saharan Africa. *Psychology Research and Behavior Management, 13*, 713–720.

47. Song, J., Liu, C., Mauk, M. G., Rankin, S. C., Lok, J. B., Greenberg, R. M., & Bau, H. H. (2017). Two-stage isothermal enzymatic amplification for concurrent multiplex molecular detection. *Clinical Chemistry, 63*(3), 714–722. https://doi.org/10.1373/clinchem.2016.263665. Epub 2017 Jan 10. PMID: 28073898; PMCID: PMC5913740.

48. Thai, H. C., Mai, Q. L., Cuong, D. V., Parida, M., Minekawa, H., Notomi, T., Hasebe, F., & Morita, K. (2004). Development and evaluation of a novel loop-mediated isothermal amplification method for rapid detection of severe acute respiratory syndrome coronavirus. *Journal of Clinical Microbiology, 42*(5), 1956–1961. https://doi.org/10.1128/jcm.42.5.1956-1961.2004.
49. Tratner, I. (2003). SRAS: 1. Le virus [SARS-CoV: 1. The virus]. *Medecine Sciences: M/S, 19*(8-9), 885–891.
50. Vandenberg, O., Martiny, D., Rochas, O., van Belkum, A., & Kozlakidis, Z. (2021). Considerations for diagnostic COVID-19 tests. *Nature Reviews Microbiology, 19*(3), 171–183.
51. Verdecchia, P., Cavallini, C., Spanevello, A., & Angeli, F. (2020). The pivotal link between ACE2 deficiency and SARS-CoV-2 infection. *European Journal of Internal Medicine., 76*, 14–20.
52. Weiss, S. R., & Navas-Martin, S. (2005). Coronavirus pathogenesis and the emerging pathogen severe acute respiratory syndrome coronavirus. *Microbiology and Molecular Biology Reviews: MMBR, 69*(4), 635–664.
53. Xiang, X., Qian, K., Zhang, Z., Lin, F., Xie, Y., Liu, Y., & Yang, Z. (2020). CRISPR-cas systems based molecular diagnostic tool for infectious diseases and emerging 2019 novel coronavirus (COVID-19) pneumonia. *Journal of Drug Targeting, 28*(7-8), 727–731.
54. Zhang, F. (2020). *A protocol for detection of COVID-19 using CRISPR diagnostics.*
55. Zhang, F., Abudayyeh, O. O., & Gootenberg, J. S. (2020). A protocol for detection of COVID-19 using CRISPR diagnostics. https://www.broadinstitute.org/files/publications/special/COVID-19%20detection%20(updated).pdf.
56. Zou, L., Ruan, F., Huang, M., Liang, L., Huang, H., Hong, Z., Yu, J., Kang, M., Song, Y., Xia, J., Guo, Q., Song, T., He, J., Yen, H. L., Peiris, M., & Wu, J. (2020). SARS-CoV-2 viral load in upper respiratory specimens of infected patients. *The New England Journal of Medicine, 382*(12), 1177–1179. https://doi.org/10.1056/NEJMc2001737. Epub 2020 Feb 19. PMID: 32074444; PMCID: PMC7121626.
57. Zou, X., Wu, J., Gu, J., Shen, L., & Mao, L. (2019). Application of aptamers in virus detection and antiviral therapy. *Frontiers in Microbiology, 10*, 1462.

A Review on Detection of COVID-19 Patients Using Deep Learning Techniques

Babita Majhi, Rahul Thangeda, and Ritanjali Majhi

1 Introduction

The world has witnessed a jolt in the year 2020.The health systems throughout the world were cautioned on January 10, 2020, when a mortality was reported in the city of Wuhan, China, because of novel corona virus [60]. Sooner the virus started to spread across the world, and it was becoming difficult for the health specialists to handle the sudden wave. This virus was named as COVID-19 (corona virus disease 2019) by the World Health Organization (WHO) and was declared as a pandemic on March 11, 2020 [64]. The pandemic has not only globally effected the health, but it has worsened up the economy as well. There were thousands of cases in countries like the USA and India on a daily basis. Governments, with no other options in hand, have started to contain people through lockdown measures. Many countries have started to follow lockdown to contain the virus. The lockdowns have helped contain the virus, but however these have weakened the economy. Hence, slowly lockdown measures were slightly open for the people with strict obligations on using mask and maintaining physical distance. The health professionals at different parts of the world are working their best to obtain drugs and vaccine for the disease.

Globally as on January 09, 2021, there are 87,589,206 confirmed cases of COVID-19 and 1,906,606 deaths as per WHO [64], and the numbers are further

B. Majhi (✉)
Department of CSIT, Guru Ghasidas Vishwavidyalaya, Central University, Bilaspur, Chhatisgarh, India

R. Thangeda
School of Management, NIT Warangal, Warangal, Telengana, India

R. Majhi
School of Management, NIT Surathkaul, Surathkaul, Karnataka, India

© The Author(s), under exclusive license to Springer Nature Switzerland AG 2022
S. K. Pani et al. (eds.), *Assessing COVID-19 and Other Pandemics and Epidemics using Computational Modelling and Data Analysis*,
https://doi.org/10.1007/978-3-030-79753-9_4

increasing throughout the world. Few of the countries are also reporting the second strain [10], and the number of confirmed cases is increasing worldwide and some of the countries have already declared lockdown as a preventive measure. At this juncture, it is important for the government to make sure a vaccine is distributed to all its citizens and at the same time keep an eye on the economy and contain the virus. Every effort has to be made to make sure the virus does not spread and further degrade the health and economy of the countries.

To maintain a proper health system and infrastructure, it is of utmost importance for the government to strategize and plan effectively. To develop right measures, one of the most important activity is to understand the probable rise of cases. This will enable the policy makers to decide on the next steps or the future course of action. Hence, estimating accurately the number of confirmed cases will be of immense help to effectively plan the infrastructure and take necessary measures. Several researchers have worked in this direction and have contributed through their studies using mathematical models to estimate COVID-19 in different regions. These models will immensely benefit it they can be used in real life. Starting from statistical time series models to machine learning models and deep learning models have been developed to estimate COVID-19 effectively. The models have shown some great sense of adaptability for the interesting results they have produced. Another important dimension to look at would be to detect the disease in early stages, as the disease intensifies it deteriorates other organs and would even lead to mortality. The sudden degradation of health of a patient with corona virus is an important concern. Added to that, availability of high end imagining technologies and regular testing methods in rural places is a big challenge. Though such are available also, it needs high degree of proficiency from the radiologists to make sure their assessment is right. Hence to avoid all such challenges and to make the health system be able to assess COVID-19 efficiently at a lower cost is important. Developing technology that would help in identifying the disease would help the government work more efficiently in containing the disease.

Hence, identifying the disease in early stages would help reduce the mortality and other serious problems. Using technologies such as artificial intelligence will help identify the disease within less time. Images of X-rays and CT scans can be used to accurately detect the spread of disease. Having such algorithms in place will need lesser human intervention and can be of immense benefit to the health sector. There has been ample contribution of using deep learning architectures to detect COVID-19. This chapter aims at reviewing articles pertaining to implementation of deep learning models and making an in-depth study to assess the various architectures, models, and inputs used. This will enable the medical practitioners to have a set of relevant information together to make sensible decisions in detecting COVID-19 and will also help policy makers develop effective strategies. Further, the next section aims at detailing the approach followed for carrying out the review and the following section elaborates various articles using deep learning algorithms to detect COVID-19. The next sections predominantly will focus on discussion of the various concepts obtained from the existing article, followed by highlighting the challenges. This chapter concludes stating the significance of implementation of deep learning to diagnose COVID-19 early with higher accuracy and at the same time highlights the significance of having more training data to appreciate the efficiency and deployment nature of the proposed models.

2 Methodology/Article Selection

There are number of articles published about COVID-19 since the pandemic has halted the world. There is good amount of research pursued in the area especially post March 2020 till the date of writing this chapter. In this chapter, the focus was to shortlist articles that had researched using images such as X-rays and CT scans to identify early the COVID-19 through the application of deep learning algorithms. There are enormous articles that have researched on implementing forecasting algorithms to assess the to-be positive cases using machine learning and deep learning. The goal of this chapter was to restrict the study to identifying and surveying articles pertaining to implementation of diagnosing COVID-19 using deep learning algorithms. A search was performed on Google scholar to identify such articles and finally a total of 60 such articles published between January 2020 and December 2020 were considered for the review. Emphasis was given to articles that studied early detection than segmentation. The following flowchart is aimed at elaborating the search procedure in details (Fig. 1).

3 Review of Literature

Artificial intelligence (AI)–based algorithms have been in use to use historical information to provide predictions for future. Development of machine learning and deep learning algorithms may be considered as integral part of AI. Over the course of time, the evolution of deep learning algorithms has gained importance due to their computing speed availability and also the data being captured. Deep learning has been excelling as one of the state-of-the-art performance in medical imaging. Medical imaging is one area where there is a dire need of such computer-aided systems which can automate certain process. This will not only save time but also will improve and benchmark the quality.

It is very important to predict clinical tasks such as to diagnose a disease for health sector. Corona virus (COVID-19) has impacted over more than 200 countries. The disease can widely spread and will be a great threat to health systems if not handled properly. COVID-19 has been a fearful disease that is threatening lives of humans. It would be a great support to the health systems if computer-aided systems can help mitigate the issues caused by COVID-19. This chapter tries to address this issue by reviewing some key articles proposed by researchers. This section is used to study in depth several articles proposed by researchers and summarize their findings.

Fig. 1 Flowchart explaining
the article search
methodology

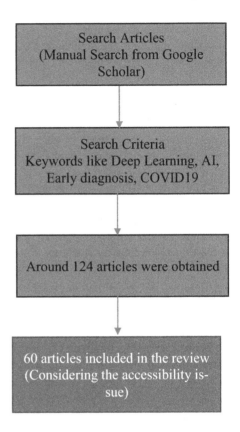

There are several articles that have used statistical, machine learning, and deep learning models to predict the number of positive cases of COVID-19 in different parts of the world [17, 31, 42, 56, 62, 70]. Arora et al. [73] have proposed deep learning–based models for predicting novel corona virus (COVID-19) positive cases from the 32 states and union territories of India. Zeroual et al. [71] developed five deep learning methods to forecast the COVID-19 spread and compared them. COVID-19 data from Italy, Spain, France, China, the USA, and Australia were used. Shastri et al. [58] predicted the future conditions of novel corona virus by proposing a deep learning–based comparative analysis of COVID-19 in India and the USA. The models have showcased some interesting results that exhibited the confidence of deploying them in real time. Due credit has to be given to all eminent researchers who have developed such efficient models in short time considering the need of the hour.

While this work is of immense use to make strategic decisions on the infrastructure and other related issues, the second most important point of research would be to identify the disease in the preliminary stages [36]. If the disease can be identified early, there is high proximity that the policy makers or the health systems will be able to make necessary preparations, and also an early detection would reduce the seriousness associated to the disease. Another important aspect is that such diagnosis will be of immense use at this juncture, as there is a huge requirement of medical staff to contain the wide spread of the disease. Hence if such facilities are available in real time, it will be easier for the medical staff to save not only lives but also time [43]. In this context, attention was given to articles that considered deep learning as an application to diagnose COVID-19. Deep learning has been one of the most useful approach with which patient's X-ray or CT scan can be studied to understand the impact of COVID-19 in the patient. The research has been promising in this area, and some of the researchers have used widely available data online to showcase the capability of their models. Brunese et al. [12] proposed three-phase approach. In the first, it was to detect the presence of pneumonia and then identify if it is COVID-19 or pneumonia and finally have localized the areas that are symptomatic of COVID-19 in X-rays. They used this deep learning model–based framework to diagnose corona virus using X-rays; 6523 chest X-rays have been used to study the effectiveness and the accuracy was 97%. The detection of COVID-19 approximately took 2.5 seconds. To develop or perform deep learning models, mostly imagining techniques such as X-rays or CT scans were used. Chaimae Ouchicha et al. [48] have proposed a deep learning model to diagnose COVID-19 from chest X-ray images. A CNN-based model CVDNET (a deep convolutional neural network) was used to classify subjects having corona virus from pneumonia related X-rays. They have used publicly available chest X-ray images. The results have been interesting, and they propose the results can be even better if there are more training data available. Rodolfo M.Pereira et al. [50] in another article have used chest X-rays with a CNN model to estimate the COVID-19 disease. They have also looked into early and late fusion techniques in the schema. TalhaBurakAlakus&IbrahimTurkoglu [4] have developed predictive deep learning models that can estimate corona virus using laboratory data. The models were able to predict with an accuracy of 86.66%. The results obtained have been tested on 500 patients and have shown some good quality results.

Heida-ri et al. [27] aimed at developing a computer-based diagnosis tool using chest X-rays to diagnoseCOVID-19. After pre-processing of the image, it is then fed into convolutional neural network (CNN) model to classify chest X-ray images. The model was developed using 8474 chest X-ray images. At 95% confidence, the model showed an accuracy of 84.5%. The significant contribution from this study is the pre-processing stage which has helped in increasing the accuracy of the mode. Hussain et al. [32] applied a 22-layer convolutional neural network architecture called as CoroDet Model to predict the COVID-19 using the X-ray and CT scan images. The proposed model for a two-class prediction was able to showcase an accuracy of 99.1%, 94.2% for a three-class prediction, and 91.2% for a four-class prediction. They assure the results are trained on one of the real-time database available, and this in turn means the training is efficient and the model performance is also better. Ahmed et al. [3] developed a deep learning platform for social distance tracking using an overhead perspective. The framework used YOLOv3 object recognition approach to identify humans in video sequences. Bounding box method is used to identify people. In addition, a tracking algorithm was used to detect individuals in video sequences such that the person who violates/crosses the social distance threshold is also being tracked.

The most important component for research using deep learning is the availability of training data sets [47]. Azemun et al. [14] have used chest radiograph images and have implemented the data on the ResNet-101 convolutional neural network architecture. The results showed a good performance. El-Rashidy et al. [19] proposed a framework that has patient layer, cloud layer, and hospital layer. Zhu, et al. [72] aimed to develop an efficient deep learning algorithm along with risk score to identify COVID-19 patients. The algorithm focused on developing a risk score based on severity of the disease.

Social distancing has proven to be a useful measure to contain COVID-19. There are few areas where individuals are noticed not following the norms of social distancing. Yang et al. [69] proposed a deep learning model that can alert individuals who are not maintaining social distancing. The authors have tested the model, and it is ready for deployment. Ghoshal and Tucker [23] have developed a Bayesian convolutional neural network (BCNN) to improve the diagnostic performance using publicly accessible chest X-rays. Amyar et al. [5] present an automatic classification segmentation tool for helping diagnosing COVID-19 using chest CT imaging. A new multitask deep learning model was proposed to identify COVID-19 patient. There were three tasks proposed by the authors and were used on different datasets. The proposed model is also tested with 1369 patients' data. The ROC showed up to 97% and this indicates the results indicated a positive performance. Xu et al. [66] studied to establish an early screening model to distinguish COVID-19. They have used CT images and the model was built using the deep learning algorithms. The model showed an accuracy of 86.7%, and the authors further suggest that this method can be used as a supplementary tool for doctors.

Afshar et al. [2] have studied the significance of using computed tomography (CT) over other diagnostic methods to detect COVID-19. In their study, a new COVID-19 CT scan dataset called as COVID-CT-MD consisting of healthy and COVID-19 subjects were developed. These data can be used to develop models that can effectively detect COVID-19. Farooq&Bazaz [21]. proposed a deep learning artificial neural network (ANN)-based algorithm for estimation COVID-19 disease. The main contribution is that unlike typical deep learning techniques, this algorithm eliminates the need train the model from scratch. On the top of it, they have developed a strategy that categorizes the model results into high risk and low risk compartments. Shan et al. [57] developed a deep learning-based segmentation method that can automatically quantify infection regions with respect to the lung. The performance of the system was evaluated by comparing the infection regions on 300 CT scans. The average dice similarity coefficient was 91.6%.

Nayak et al. [46] have proposed a deep learning assisted method to diagnose COVID-19 using X-ray images. The authors have evaluated eight CNN models and compared them for classification of COVID-19 from the normal subjects. The comparative analysis was also performed based on learning rate, batch, epochs, and the optimizers best suitable. ResNet-34 provided an accuracy of 98.33% and was found to be outperforming the rest of the models.

Narin et al. [45] in their article have stressed on the need of creating a deep learning model that can diagnose COVID-19. Five pre-trained CNN models have been proposed for the detection using chest X-rays. ResNet-50 outperformed the other four models as per their study. Soares L. P. & Soares C. P [59]. have used chest images and have fed them to a CNN model to automate the detection of COVID-19. The goal of the research article was to automate the process detecting COVID-19 using chest images.

There are good amount of annotated image datasets of X-ray images and hence a great success has been achieved while these datasets are trained using deep learning models [1]. X-rays are cheaper and easily accessible at most of the medical facilities. Hence using X-rays and developing a model would be more useful to the society. The authors have developed a model that can detect COVID-19 that resulted in an accuracy of 95.12%. Jain, R et al. [35] proposed a deep learning network to diagnose the disease after the steps of augmentation and pre-processing. In this article, chest X-rays were considered upon a CNN model and the results were compared. Civit-Masot et al. [16] analyzed the effectiveness of deep learning model based on VGG16 for the identification of pneumonia and COVID-19 using torso radiographs. The results indicate high specificity, and this helps in assessing that this model can be used for screening real time. Horry et al. [29] used a suitable convolutional neural network (CNN) model after comparing several CNN models. In their study they have used different imaging modes such as X-ray, ultrasound, and CT scan. It was concluded that ultrasound images have showcased better results when compared to CT scan and X-ray images.

Islam et al. [33] reviewed the recent deep learning techniques used for different medical imaging–based problems. They reviewed in specific to COVID-19-related literature that has been using deep learning algorithms. The article also highlighted some of the data partitioning techniques and performance measures that have been implemented by researchers. Hu et al. [30] studied a deep learning strategy for identifying and classifying COVID-19 using CT images. Weakly supervised deep learning strategy that was proposed showed a minimal requirement of manual labeling and at the same time could produce some decent results. They propose that their model can be widely deployed because of the efficiency and reduction of the manual labeling. Ouchicha et al. [48] in their article propose a CVDNet (a deep convolutional neural network) model to identify COVID-19-infected patients using X-ray images. The results show that the proposed model is working well even on a small set of data. This model can be further improved with more training data.

Ardakani et al. [7] proposed an artificial intelligence–based technique to diagnose COVID-19. Ten convolutional neural networks were used to have a comparison between them to understand which models better distinguish infection of COVID-19. The best performance among those models was obtained from ResNet-101 with an accuracy level of AUC 0.994.

Yang et al. [68] state that CT scans cannot be used to confirm COVID-19 as they do not have the ability to recognize which virus is causing viral pneumonia. And according to the experts, during pandemic if it is viral pneumonia, then it could be COVID-19. Further, the paper also talks about the pre-processing techniques applied on the image dataset of CT scan. CT scans are in sequence of slices but only few slices are taken, so this may reduce the accuracy and relevancy of the model. But expert radiologists and medical professionals are able to diagnose the disease with the help of a single slice also. Many times a single slice has enough details for correct diagnosis. This paper also throws light on the use of images gathered from research papers and those of actual CTs. The overall result of their research shows that original CT dataset when input to the deep learning model (they used ResNet-50 and DenseNet-189 models) shows better results in comparison to image dataset collected from various research papers. The availability of more number of images in dataset further increases the accuracy and other performance parameters. Asnaoui & Chawki [37] have performed a comparative study on the use of deep learning models that can detect COVID-19. They used 6087 images to evaluate the performance of the model. The results had an accuracy of close to 92.18% with Inception-ResNetV2 and 88.09% with Densnet201. Song et al. [61] have collected data comprising of 88 patients who had COVID-19, 101 patients infected with bacteria-based pneumonia and 86 healthy people from two provinces in China. The authors used deep learning–based model to accurately identify COVID-19. The results showcase that the model can identify with an AUC of 0.99. The model was also able to differentiate the COVID-19-infected and bacteria pneumonia–infected patients very accurately with an AUC of 0.95. The diagnosis of the patient would just take 30 seconds and would be of immense use to the world of health care.

Chen et al. [15] built a deep learning–based COVID-19 detection model that can work with good resolution-based CT images; 46,096 images were collected from Renmin Hospital of Wuhan University to develop the model. The model was tested on patients and has shown a per-patient accuracy of 95.24% and per image accuracy of 98.85%. The authors also claim that with the help of model they developed, the analysis time of the radiologists have come down by 65%. They conclude that this model can be a good fit in clinical activities to help radiologists take quick and timely decisions. Gozes et al. [24] propose a deep learning method that can identify and also mention the severity of COVID-19 from the CT scans. The first step in the pipeline would be to process the images and then feed into the model. The model was tested on 110 confirmed COVID-19 patients in China.

Deep learning algorithms have numerous applications in the health sector for diagnosing health-related issues. Bhattacharya et al. [11] have summarized on the research articles that have worked on developing deep learning applications for COVID-19 detection using medical image–related data. The article also discussed on some of the challenges and issues related to the implementation of the deep learning models. Sedik et al. [54] proposed a solution to detect COVID-19 using a deep learning–based model. The authors have used a CNN-based model along with Convolutional Long Short Term Memory. They have used both X-ray images and CT scan–based images as datasets to be fed to the models. The proposed models have been tested on both the datasets, and the results were found to be nearly 100% in accuracy in some cases.

Sethy et al. [55] proposed a deep feature plus support vector machine (SVM)–based method to detect the infection of COVID-19 using X-ray images. The features are extracted from a CNN and then are fed to a SVM to classify the patients. The model was compared with 13 number of CNN models that extracted features. The authors argue that when dataset is small, it would be difficult to train a deep learning model. In such cases, they suggest this model would turn out to be efficient and time saving.

Maghdid et al. [41] aimed at building a comprehensive dataset of X-rays and CT scan images from different sources. Also, a pre-trained AlexNet model was applied to the prepared data. The results showcase an accuracy of 98%.

Luz et al. [40] aimed to propose an accurate and efficient model both in terms of memory and processing time to identify COVID-19 using chest X-ray images. The authors have used EfficientNet family of deep learning methods that are known for showcasing good results; 13,569 X-ray images were used develop the model, and 231 images were used to test on the efficiency of the model. The results obtained showcase an accuracy 93.9%.

Haghanifar et al. [25] proposed a study to identify images related to pneumonia using deep CNN models from a large dataset. The authors focused on developing right features to improve the efficiency of the model. The proposed model is fully automated and is efficient mechanism to identify COVID-19.

Hemdan et al. [28] developed a new deep learning model called COVIDX-Net to help identify COVID-18 using X-ray images. The results were tested using 23 confirmed COVID positive cases. The proposed model had seven architectures of deep CNN. The model showed a f1-score of 0.89 for normal and 0.91 for COVID-19 patients.

Bai et al. [8] propose an AI model that can predict mild patients with potential threat to have the disease progressed. A sample of 133 mild COVID patient data was collected from Wuhan Pulmonary Hospital. CT scans during the admission at mild stage and scans during severe stage were considered. A logistic regression model and a deep learning model were built. The results showcase that the deep learning model outperforms the logistic regression model.

4 Discussion

The key aspect of this work is to review the existing deep learning models that can be employed to detect, segment, and predict the COVID-19 patients. The summary of the recent articles are summarized in Sect. 2 and detailed in Table 1. The study indicates that most of the articles have used X-ray and CT images to develop deep learning models. Literature review reveals that most of the articles have used X-ray images for their study, and this could be for the ease of their availability. There are not many articles that have used CT scans and X-rays together.

From Table 1, it is evident that most of the researchers have used RNN-, CNN-based models to estimate COVID-19. Also, it can be observed that the results are promising, and this indicates the possibility of using them in real life.

Table 1 Summary of some of the articles using different deep learning architecture

Authors	Year of publication	Dataset	Model used	Results
Farooq & Hafeez [22]	2020	X-rays	CovidResnet	Accuracy = 96.23 F1 score = 100
Ghoshal and Tucker [23]	2020	X-rays	Resnet-50	Entropy = 99.68
Khalifa et al. [38]	2020	pneumonia chest X-rays	Resnet-18	Accuracy = 99 F1 score = 98.97
Pathak et al. [67]	2020	CT scans	CNN	Accuracy = 93.01 Precision = 95.18
Al-antari et al. [6]	2020	X-rays	CNN	Accuracy = 97.40
Razzak et al. [53]	2020	X-rays, CT scans	Resnet-101	Accuracy = 98.75 Precision = 96.43
Ozturk et al. [49]	2020	X-rays CT scans	SAE	Accuracy = 71.92 F1 Score = 69.13
Hasan et al. [26]	2020	CT scans	CNN	Accuracy = 99.68
Moura et al. [44]	2020	X-rays	DenseNet-161	Accuracy = 99 F1 score = 99
Tartaglione et al. [18]	2020	X-rays	ResNet-18	Accuracy = 100 F1 score = 100
Bassi et al. [9]	2020	X-rays	DenseNet-201	Accuracy = 99.4 F1-score = 99.4
Ramadhan et al. [52]	2020	X-rays	COVID-Net	Accuracy = 98.4
Ouchicha et al. [48]	2020	X-rays	CVDNet	Accuracy = 97.20
Jain, G. et al. [34]	2020	CT scans	ResNet-152	Accuracy = 94.98 Precision = 91.53
Li et al. [39]	2020	CT scans	ResNet-50	AUC = 96
Bukharia et al. [13]	2020	X-rays	ResNet-50	Accuracy = 98.28 F1- score = 98.19
Farid et al. [20]	2020	CT scans	CNN	Accuracy = 94.11 F1-score = 94
Wang and Wong [63]	2020	X-rays	COVID-Net	Accuracy = 92.4 Precision = 88.9
Heidari et al. [27]	2020	X-rays	CNN	Accuracy = 98.1
Brunese et al. [12]	2020	X-rays	VCG 16	Accuracy = 97
Xu et al. [66]	2020	CT scans	ResNet	Accuracy = 86%
Rahimzadeh and Atta [51]	2020	X-rays	Concatenated CNN	Accuracy = 99.50 Precision = 35.27
Wu et al. [65]	2020	CT scans	ResNet-50	Accuracy = 76
Shan et al. [57]	2020	CT scans	VB-Net (CNN)	Dice similarity coefficient = 91.6% ± 10,0%

After reviewing several articles, it can be stated that there are two types of estimating COVID-19 using deep learning methods. First is to detect the disease by using the classification algorithms and the second is to use segmentation-based algorithms. In cases where we do not have a training set with the target variable

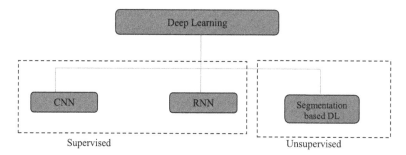

Fig. 2 Summary of the deep learning models mostly used

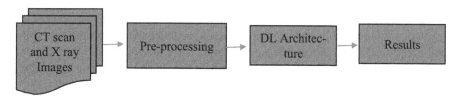

Fig. 3 Approach to detect COVID-19

defined, it is advisable to use the segmentation-based algorithms. Researchers have used both methods to essentially estimate COVID-19. Figure 2 indicates the different deep learning architectures that have been used by the researchers.

As defined in Fig. 2, it is not meant that only these algorithms have been used for estimating COVID-19. It was intended to summarize the most popularly used methods that are being used.

In most of the models that have been studied, it was noticed that X-ray and CT scan images were used for study. Pre-processing of data was performed where the data were rotated, scaled, cleaned as the need be, and then the data were fed into the deep learning models. In some cases after the modeling phase, the results were post processed. Figure 3 indicates a generic view of how the approach is built using a deep learning algorithm to estimate COVID-19.

5 Challenges

Researchers have widely contributed by proposing deep learning models once the growth and spread of COVID-19 globally was creating serious challenges. The most important challenge that would arise was the availability of data and the ground truth of the data. Though there are publicly available datasets, still there are very less datasets available. The first challenge would be to unavailability of huge public database that has X-rays and CT scans. Other challenge would be to not just have the images but the related information like their previous disease history or so. This will enable more accuracy of the model. Also, the ground truth of the data has to be true. Over the course of time, if the public databases increase, the subjects will definitely enhance the quality of the models and in future may have more accurate models.

6   Conclusion

COVID-19 diagnosis was very widely studied using deep learning methods. There are several public databases available, and this has indeed made the training convenient. The various deep learning architectures and inputs used are tabulated in Table 1. Like it was stated earlier, estimating or diagnosing COVID-19 using real-life data would be highly useful and if measured accurately would be challenging in real life. There are databases that are available. These methods would be useful in real life to estimate the disease more efficiently if there are more amount of training data available. These studies have focused on accuracy of their models but may also have to quantify the uncertainty that may arise when such models are used. Also, it is very important to have accurate ground truth to train the models more efficiently. To have trust obtained from clinicians is a must to bring technology in alignment to the computer-based medical diagnosis.

References

1. Abbas, A., Abdelsamea, M. M., & Gaber, M. M. (2020). Classification of COVID-19 in chest X-ray images using DeTraC deep convolutional neural network. *arXiv preprint arXiv:2003.13815*.
2. Afshar, P., Heidarian, S., Enshaei, N., Naderkhani, F., Rafiee, M. J., Oikonomou, A., & Mohammadi, A. (2020). COVID-CT-MD: COVID-19 computed tomography (CT) scan dataset applicable in machine learning and deep learning. *arXiv preprint arXiv:2009.14623*.
3. Ahmed, I., Ahmad, M., Rodrigues, J. J., Jeon, G., & Din, S. (2020). A deep learning-based social distance monitoring framework for COVID-19. *Sustainable Cities and Society, 65*, 102571.
4. Alakus, T. B., & Turkoglu, I. (2020). Comparison of deep learning approaches to predict COVID-19 infection. *Chaos, Solitons & Fractals, 140*, 110120.
5. Amyar, A., et al. (2020). Multi-task deep learning based CT imaging analysis for COVID-19 pneumonia: Classification and segmentation. *Computers in Biology and Medicine, 126*, 104037.
6. Al-antari, M. A., Hua, C.-H., & Lee, S. (2020). Fast deep learning computer aided diagnosis against the novel Covid-19 pandemic from digital chest x-ray images. https://doi.org/10.21203/rs.3.rs-36353/v1.
7. Ardakani, A. A., Kanafi, A. R., Acharya, U. R., Khadem, N., & Mohammadi, A. (2020). Application of deep learning technique to manage COVID-19 in routine clinical practice using CT images: Results of 10 convolutional neural networks. *Computers in Biology and Medicine, 121*, 103795.
8. Bai, X., Fang, C., Zhou, Y., Bai, S., Liu, Z., Xia, L., ... Chen, W. (2020). Predicting COVID-19 malignant progression with AI techniques. *SSRN Electronic Journal, 17*, 42.

9. Bassi, P. R. A. S., & Attux, R. (2020). A deep convolutional neural network for covid-19 detection using chest x-rays. *arXiv:2005.01578*

10. BBC. (2020). https://www.bbc.com/news/health-55312505

11. Bhattacharya, S., Maddikunta, P. K. R., Pham, Q. V., Gadekallu, T. R., Chowdhary, C. L., Alazab, M., & Piran, M. J. (2021). Deep learning and medical image processing for coronavirus (COVID-19) pandemic: A survey. *Sustainable Cities and Society, 65,* 102589.

12. Brunese, L., Mercaldo, F., Reginelli, A., & Santone, A. (2020). Explainable deep learning for pulmonary disease and coronavirus COVID-19 detection from X-rays. *Computer Methods and Programs in Biomedicine, 196,* 105608.

13. Bukhari, et al. (2020). The diagnostic evaluation of Convolutional Neural Network (CNN) for the assessment of chest X-ray of patients infected with COVID-19. https://www.medrxiv.org/content/10.1101/2020.03.26.20044610v1

14. CheAzemin, M. Z., Hassan, R., MohdTamrin, M. I., & &Md Ali, M. A. (2020). COVID-19 deep learning prediction model using publicly available radiologist-adjudicated chest X-ray images as training data: Preliminary findings. *International Journal of Biomedical Imaging.*https://doi.org/10.1155/2020/8828855.

15. Chen, J., Wu, L., Zhang, J., Zhang, L., Gong, D., Zhao, Y., . . . Yu, H. (2020). Deep learning-based model for detecting 2019 novel coronavirus pneumonia on high-resolution computed tomography. *Scientific Reports, 10*(1), 1–11.

16. Civit-Masot, J., Luna-Perejón, F., Domínguez Morales, M., & Civit, A. (2020). Deep learning system for COVID-19 diagnosis aid using X-ray pulmonary images. *Applied Sciences, 10*(13), 4640.

17. Dehesh, T., Mardani-Fard, H. A., &Dehesh, P. (2020). Forecasting of Covid-19 confirmed cases in different countries with arima models. *medRxiv.*

18. Tartaglione, E., Barbano, C. A., Berzovini, C., Calandri, M., & Grangetto, M. (2020). Unveiling covid-19 from chest x-ray with deep learning: A hurdles race with small data. *arXiv preprint arXiv:2004.05405.*

19. El-Rashidy, N., El-Sappagh, S., Islam, S. M., El-Bakry, H. M., & Abdelrazek, S. (2020). End-to-end deep learning framework for coronavirus (COVID-19) detection and monitoring. *Electronics, 9*(9), 1439.

20. Farid, A. A., Selim, G. I., Awad, H., & Khater, A. (2020). A novel approach of CT images feature analysis and prediction to screen for Corona virus disease (COVID-19). *International Journal of Scientific & Engineering Research, 11*(3) https://doi.org/10.14299/ijser.2020.03.02.

21. Farooq, J., & Bazaz, M. A. (2020). A novel adaptive deep learning model of Covid-19 with focus on mortality reduction strategies. *Chaos, Solitons & Fractals, 138,* 110148.

22. Farooq, M., & Hafeez, A. (2020). Covid-resnet: A deep learning framework for screening of covid19 from radiographs. *arXiv preprint arXiv:2003.14395.*

23. Ghoshal, B., & Tucker, A. (2020). Estimating uncertainty and interpretability in deep learning for coronavirus (COVID-19) detection. *arXiv preprint arXiv:2003.10769.*

24. Gozes, O., Frid-Adar, M., Sagie, N., Zhang, H., Ji, W., & Greenspan, H. (2020). Coronavirus detection and analysis on chest ct with deep learning. *arXiv preprint arXiv:2004.02640.*

25. Haghanifar, A., Majdabadi, M. M., & Ko, S. (2020). Covid-cxnet: Detecting covid-19 in frontal chest x-ray images using deep learning. *arXiv preprint arXiv:2006.13807.*

26. Hasan, A. M., AL-Jawad, M. M., Jalab, H. A., Shaiba, H., Ibrahim, R. W., & AL-Shamasneh, A. R. (2020). Classification of covid-19 coronavirus, pneumonia and healthy lungs in ct scans using q-deformed entropy and deep learning features. *Entropy, 22*(5), 517. https://doi.org/10.3390/e22050517.

27. Heidari, M., Mirniaharikandehei, S., Khuzani, A. Z., Danala, G., Qiu, Y., & Zheng, B. (2020). Improving the performance of CNN to predict the likelihood of COVID-19 using chest X-ray images with preprocessing algorithms. *International Journal of Medical Informatics, 144,* 104284.

28. Hemdan, E. E. D., Shouman, M. A., & Karar, M. E. (2020). Covidx-net: A framework of deep learning classifiers to diagnose covid-19 in x-ray images. *arXiv preprint arXiv:2003.11055.*

29. Horry, M. J., et al. (2020). COVID-19 detection through transfer learning using multimodal imaging data. *IEEE Access, 8,* 149808–149824. https://doi.org/10.1109/ACCESS.2020.3016780.

30. Hu, S., et al. (2020a). Weakly supervised deep learning for COVID-19 infection detection and classification from CT images. *IEEE Access, 8,* 118869–118883. https://doi.org/10.1109/ACCESS.2020.3005510.

31. Hu, Z., Ge, Q., Jin, L., & Xiong, M. (2020b). Artificial intelligence forecasting of Covid-19 in China. *arXiv preprint arXiv:2002.07112.*

32. Hussain, E., Hasan, M., Rahman, M. A., Lee, I., Tamanna, T., & Parvez, M. Z. (2020). CoroDet: A deep learning based classification for COVID-19 detection using chest X-ray images. *Chaos, Solitons & Fractals, 142,* 110495.

33. Islam, M., Karray, F., Alhajj, R., & Zeng, J. (2020). A review on deep learning techniques for the diagnosis of novel coronavirus (covid-19). *arXiv preprint arXiv:2008.04815.*

34. Jain, G., Mittal, D., Thakur, D., & Mittal, M. K. (2020a). A deep learning approach to detect Covid-19 coronavirus with X-ray images. *Biocybernetics and Biomedical Engineering, 40*(4), 1391–1405.

35. Jain, R., Gupta, M., Taneja, S., et al. (2020b). Deep learning based detection and analysis of COVID-19 on chest X-ray images. *ApplIntell.* https://doi.org/10.1007/s10489-020-01902-1.

36. Jamshidi, M., et al. (2020). Artificial intelligence and COVID-19: deep learning approaches for diagnosis and treatment. *IEEE Access, 8,* 109581–109595. https://doi.org/10.1109/ACCESS.2020.3001973.

37. Khalid, El Asnaoui, & Chawki, Y. (2020). Using X-ray images and deep learning for automated detection of coronavirus disease. *Journal of Biomolecular Structure and Dynamics.* https://doi.org/10.1080/07391102.2020.1767212.

38. Khalifa, N. E. M., Taha, M. H. N., Hassanien, A. E., & Elghamrawy, S. (2020). Detection of coronavirus (covid-19) associated pneumonia based on generative adversarial networks and a fine-tuned deep transfer learning model using chest x-ray dataset. *arXiv preprint arXiv:2004.01184.*

39. Li, L., et al. (2020). Artificial intelligence distinguishes COVID-19 from community acquired pneumonia on chest CT. *Radiology, 296*(2), 200905.

40. Luz, E., Silva, P. L., Silva, R., Silva, L., Moreira, G., & Menotti, D. (2020). Towards an effective and efficient deep learning model for covid-19 patterns detection in x-ray images. *arXiv preprint arXiv:2004.05717.*

41. Maghdid, H. S., Asaad, A. T., Ghafoor, K. Z., Sadiq, A. S., & Khan, M. K. (2020). Diagnosing COVID-19 pneumonia from X-ray and CT images using deep learning and transfer learning algorithms. *arXiv preprint arXiv:2004.00038.*

42. Majhi, R., Thangeda, R., Sugasi, R. P., & Kumar, N. (2020). Analysis and prediction of COVID19 trajectory: A machine learning approach. *Journal of Public Affairs,* e2537.

43. Meng, L., Dong, D., Li, L., Niu, M., Bai, Y., Wang, M., & Tian, J. (2020). A deep learning prognosis model help alert for COVID-19 patients at high-risk of death: A multi-center study. *IEEE Journal of Biomedical and Health Informatics, 24*(12), 3576–3584.

44. De Moura, J., Ramos, L., Vidal, P. L., Cruz, M., Abelairas, L., Castro, E., Novo, J., & Ortega, M. (2020). Deep convolutional approaches for the analysis of covid-19 using chest x-ray images from portable devices. *IEEE Access.* https://doi.org/10.1109/access.2020.3033762.

45. Narin, A., Kaya, C., & Pamuk, Z. (2020). Automatic detection of coronavirus disease (covid-19) using x-ray images and deep convolutional neural networks. *arXiv preprint arXiv:2003.10849.*

46. Nayak, S. R., Nayak, D. R., Sinha, U., Arora, V., & Pachori, R. B. (2020). Application of deep learning techniques for detection of COVID-19 cases using chest X-ray images: A comprehensive study. *Biomedical Signal Processing and Control, 64,* 102365.

47. Oh, Y., Park, S., & Ye, J. C. (2020). Deep learning COVID-19 features on CXR using limited training data sets. *IEEE Transactions on Medical Imaging, 39*(8), 2688–2700. https://doi.org/10.1109/TMI.2020.2993291.

48. Ouchicha, C., Ammor, O., & Meknassi, M. (2020). CVDNet: A novel deep learning architecture for detection of coronavirus (Covid-19) from chest x-ray images. *Chaos, Solitons & Fractals, 140*, 110245.
49. Ozturk, S., Ozkaya, U., & Barstugan, M. (2020). Classification of coronavirus images using shrunken features, *medRxiv*.
50. Pereira, R. M., Bertolini, D., Teixeira, L. O., Silla, C. N., Jr., & Costa, Y. M. (2020). COVID-19 identification in chest X-ray images on flat and hierarchical classification scenarios. *Computer Methods and Programs in Biomedicine, 194*, 105532.
51. Rahimzadeh, M., & Attar, A. (2020). A modified deep convolutional neural network for detecting COVID-19 and pneumonia from chest X-ray images based on the concatenation of Xceptionand ResNet50V2. *Informatics Med Unlocked, 19*, 100360.
52. Ramadhan, M., Faza, A., Lubis, L., Yunus, R., Salamah, T. et al. (2020). Fast and accuratedetection of covid-19-related pneumonia from chest x-ray images with novel deep learning model. *arXiv preprint arXiv:2005.04562*.
53. Razzak, I., Naz, S., Rehman, A., Khan, A., & Zaib, A. (2020). Improving coronavirus (Covid-19) diagnosis using deep transfer learning. *medRxiv*.
54. Sedik, A., Hammad, M., Abd El-Samie, F. E., Gupta, B. B., & Abd El-Latif, A. A. (2021). Efficient deep learning approach for augmented detection of Coronavirus disease. *Neural Computing and Applications*, 1–18.
55. Sethy, P. K., Behera, S. K., Ratha, P. K., & Biswas, P. (2020). Detection of coronavirus disease (COVID-19) based on deep features and support vector machine. *International Journal of Mathematical, Engineering and Management Sciences, 5*, 643.
56. Shahid, F., Zameer, A., & Muneeb, M. (2020). Predictions for COVID-19 with deep learning models of LSTM, GRU and Bi-LSTM. *Chaos, Solitons & Fractals, 140*, 110212.
57. Shan, F., Gao, Y., Wang, J., Shi, W., Shi, N., Han, M., ... Shi, Y. (2020). Lung infection quantification of covid-19 in CT images with deep learning. *arXiv preprint arXiv:2003.04655*.
58. Shastri, S., Singh, K., Kumar, S., Kour, P., & Mansotra, V. (2020). Time series forecasting of Covid-19 using deep learning models: India-USA comparative case study. *Chaos, Solitons & Fractals, 140*, 110227.
59. Soares, L. P., & Soares, C. P. (2020). Automatic detection of covid-19 cases on x-ray images using convolutional neural networks. *arXiv preprint arXiv:2007.05494*.
60. Sohrabi, C., Alsafi, Z., O'Neill, N., Khan, M., Kerwan, A., Al-Jabir, A., & Agha, R. (2020). World Health Organization declares global emergency: A review of the 2019 novel coronavirus (COVID-19). *International Journal of Surgery, 76*, 71–76. https://doi.org/10.1016/j.ijsu.2020.02.034.
61. Song, Y., Zheng, S., Li, L., Zhang, X., Zhang, X., Huang, Z., ... Yang, Y. (2020). Deep learning enables accurate diagnosis of novel coronavirus (COVID-19) with CT images. *MedRxiv*.
62. Tuli, S., Tuli, S., Tuli, R., & Gill, S. S. (2020). Predicting the growth and trend of COVID-19 pandemic using machine learning and cloud computing. *Internet of Things*, 100222. https://doi.org/10.1016/j.iot.2020.100222.
63. Wang, L., Lin, Z. Q., & Wong, A. (2020). COVID-Net: A tailored deep convolutional neural network Design for detection of COVID-19 cases from chest X-ray images. Available: https://arxiv.org/abs/2003.09871
64. WHO. (2020). https://www.who.int/emergencies/diseases/novel-coronavirus-2019/events-as-they-happen
65. Wu, X., et al. (2020). Deep learning-based multi-view fusion model for screening 2019 novel coronavirus pneumonia: A multicentre study. *European Journal of Radiology, 128*, 109041. https://doi.org/10.1016/j.ejrad.2020.109041.
66. Xu, X., Jiang, X., Ma, C., Du, P., Li, X., Lv, S., ... Li, L. (2020). A deep learning system to screen novel coronavirus disease 2019 pneumonia. *Engineering, 6*(10), 1122–1129.
67. Pathak, Y., Shukla, P. K., Tiwari, A., Stalin, S., Singh, S., & Shukla, P. K. (2020). Deep transfer learning based classification model for covid19 disease. *IRBM*. https://doi.org/10.1016/j.irbm.2020.05.003.

68. Yang, X., He, X., Zhao, J., Zhang, Y., Zhang, S., & Xie, P. (2020a). COVID-CT-dataset: A CT scan dataset about COVID-19. *arXiv e-prints 2003.13865.*

69. Yang, D., Yurtsever, E., Renganathan, V., Redmill, K. A., & Özgüner, Ü. (2020b). A vision-based social distancing and critical density detection system for covid-19. *arXiv preprint arXiv:2007.03578,* 24–25.

70. Yousaf, M., Zahir, S., Riaz, M., et al. (2020). Statistical analysis of forecasting COVID-19 for upcoming month in Pakistan. *Chaos, Solitons and Fractals, 138,* 109926. https://doi.org/10.1016/j.chaos.2020.109926.

71. Zeroual, A., Harrou, F., Dairi, A., & Sun, Y. (2020). Deep learning methods for forecasting COVID-19 time-series data: A comparative study. *Chaos, Solitons & Fractals, 140,* 110121.

72. Zhu, J. S., Ge, P., Jiang, C., Zhang, Y., Li, X., Zhao, Z., ... Duong, T. Q. (2020). Deep learning artificial intelligence analysis of clinical variables predicts mortality in COVID19 patients. *Journal of the American College of Emergency Physicians Open, 1,* 1364.

73. Arora, P., Kumar, H., & Panigrahi, B. K. (2020). Prediction and analysis of COVID-19 positive cases using deep learning models: A descriptive case study of India. *Chaos, Solitons & Fractals.* https://doi.org/10.1016/j.chaos.2020.110017.

Internet of Health Things (IoHT) for COVID-19

Charles Oluwaseun Adetunji (iD), **Olugbemi Tope Olaniyan,**
Olorunsola Adeyomoye, Ayobami Dare, Mayowa J. Adeniyi, Enoch Alex,
Maksim Rebezov, Ekaterina Petukhova, and Mohammad Ali Shariati

1 Introduction

Over the years, application of smart technology in healthcare has been on steady increase. In fact, with smart technology, it is now possible to have a virtual clinics

C. O. Adetunji (✉)
Applied Microbiology, Biotechnology and Nanotechnology Laboratory, Department of Microbiology, Edo University Iyamho, Auchi, Edo State, Nigeria
e-mail: adetunji.charles@edouniversity.edu.ng

O. T. Olaniyan
Laboratory for Reproductive Biology and Developmental Programming, Department of Physiology, Edo University Iyamho, Iyamho, Nigeria

O. Adeyomoye
Department of Physiology, University of Medical Sciences, Ondo City, Nigeria

A. Dare
Department of Physiology, School of Laboratory Medicine and Medical Sciences, College of Health Sciences, Westville Campus, University of KwaZulu-Natal, Durban, South Africa

M. J. Adeniyi
Department of Physiology, Edo State University Uzairue, Iyamho, Edo State, Nigeria

E. Alex
Department of Human Physiology, Ahmadu Bello University Zaria, Kaduna State, Nigeria

M. Rebezov
Prokhorov General Physics Institute, Russian Academy of Sciences, Moscow, Russia

K.G. Razumovsky Moscow State University of Technologies and Management (the First Cossack University), Moscow, Russia

E. Petukhova · M. A. Shariati
K.G. Razumovsky Moscow State University of Technologies and Management (the First Cossack University), Moscow, Russia

© The Author(s), under exclusive license to Springer Nature Switzerland AG 2022
S. K. Pani et al. (eds.), *Assessing COVID-19 and Other Pandemics and Epidemics using Computational Modelling and Data Analysis*,
https://doi.org/10.1007/978-3-030-79753-9_5

and laboratory almost everywhere [1, 2]. Today, our ecosystem harbors many smart devices such as asthma monitor and insulin pens and wearable devices such as smart watches, smart thermometers, privacy conundrum, biosensors (such as blood pressure, electrocardiogram, oxygen saturation, temperature, heart rate, and breathing rate), and many more, with which individuals can monitor their vital signs and biometrics, meet their own medical needs, and access treatment. These devices are known as Internet of Things. Internet of Things are technologically aided devices that exhibit the tendency of transmitting data without the involvement of man. Just last year, a wireless wearable sensor that is capable of tracking and monitoring coronavirus patients diagnosed or under suspicion was developed. In addition, the device demonstrates the tendency of identifying COVID-19-induced early deterioration and remotely monitor patients in isolation centers.

Development of smart contact tracing will assist immensely in identifying the population at risk of COVID-19 infection. In India, a smartphone application similar to mobile Close Contact developed in China called ArogyaSeta was created to bridge the gap between Indian people and healthcare delivery. In order to reduce the likelihood of infection, health personnel adopted Cloud Minded Artificial Intelligence-synched bracelet in Wuhan, China. With this bracelet, vital signs were monitored. Using Internet of Things, real-time data of COVID-19 infected patients can be taken [3, 4] from remote locations. In the course of COVID-19 treatment, Internet of Things was used to connect all medical gadgets via internet and with this real-time information were conveyed during treatment.

In addition, with Internet of Things gadgets such as biosensors and smart thermometers, COVID-19 patients can monitor their vital signs (heart rate, blood pressure, body temperature) and other biometrics. Although, it is not only COVID-19 that is characterized by fever, smart thermometer has been extensively deployed in the detection of COVID-19 cases [5]. Smart devices that monitor blood glucose are also helpful especially for diabetic patients, patients with chronic illnesses and the elderly who are being deprived of the opportunity to visit hospital and diagnostic center due to COVID-19 lockdown, restriction of movement, border closure, and social distancing [6]. A great surge was observed in the usage of pulse oximeters during COVID-19 to determine blood oxygen levels at homes [1]. Google, Samsung, and Apple have invented handwashing applications and reminders which are helpful in COVID-19 prevention and control. The aim of the application is to encourage and promote handwashing culture. Handwashing has long been identified as one of the ways of mitigating the spread of diseases. The chapter aims at providing a holistic information on how the concept of the Internet of Health Things (IoHT) could help in the management of COVID-19 diseases. The relevance of Internet of Things in the management of pandemic situations is illustrated in Fig. 1.

Fig. 1 Relevance of Internet of Things in the management of pandemic situations

2 Relevant Facts About COVID-19

Coronavirus 2019 (also known as severe acute respiratory syndrome coronavirus 2), a life-threatening and highly contagious disease, is a global pandemic with variable symptoms. It is caused by a novel virus that belongs to the same family as Middle East respiratory syndrome coronavirus and severe acute respiratory syndrome coronavirus (SARS-CoV). In December 2019, the first case was identified in China, specifically in a city called Wuhan and has grown in fame since then severe acute respiratory syndrome coronavirus 2 (SARS-CoV-2) [7]. In Nigeria, the first case was announced on February 27, 2020, and has grown tremendously since then [8]. The term "severe acute respiratory syndrome coronavirus 2" used in depicting the virus clearly indicate the lungs are the major hubs affected by the virus. For COVID-19, symptoms appear between 1 and 14 days following exposure to the virus. It is worth noting that 33% of people infected usually do not present with noticeable features. The virus spreads through contact, infected surfaces, aerosols, and respiratory droplets. Even though, symptoms appear variable, most of the widely reported symptoms include pneumonia, shortness of breath, cough, anosmia, ageusia, sore throat, and many more [7]. The pathophysiologic mechanism involves modulation of renin-angiotensin-aldosterone system [9]. Specifically, the virus invades the host cells through angiotensin converting enzyme II, an enzyme that is present in large quantity in pneumocytes [10]. Hence, therapeutic goals antagonizing angiotensin II activity might prove helpful [11].

As expected, innate immunity, specifically inflammatory response occurs during viral invasion. However, in COVID-19, there appears to be an association between COVID-19 severity and the levels of inflammatory cytokines such as monocyte chemoattractant protein-1, gamma-interferon, interleukin-IB [12], and many more.

This implies that barrage of diseases characterizing the condition is attributable to severe inflammatory responses.

3   COVID-19 and Mental Health

A Chinese study showed that 96.2% of COVID-19 inpatients experienced posttraumatic stress symptoms [13]. Even in healthcare providers, a study reported that the prevalence of anxiety and depression stands at 45% and 51%, respectively [14, 15]. A study conducted in Ibadan, Nigeria, on the health implication of COVID-19-induced lockdown on adult residents revealed that sleeplessness correlated positively with anxiety and depression [16]. In Spain, an elevated level of anxiety and stress was orchestrated by COVID-19 lockdown measures in young children [17]. About one quarter of older people with depression may be adversely affected by COVID-19 measures, most especially, social distancing measures.

4   COVID-19 Diagnosis and Management

Diagnosis of COVID-19 has been done using real-time polymerase chain reaction (TaqPath COVID-19 kit), reverse transcription loop-mediated isothermal amplification, and transcription-mediated amplification using nasopharyngeal swab [18]. Serology test which detects serum or plasma antibodies is also available [19]. This test detects plasma antibodies such as IgG, IgM, and IgA against the viral strain. Non-invasive techniques such as chest computerized tomography, computerized tomography, and X-rays are also beneficial as far as examination of the pulmonary structure is concerned.

One of the ways of COVID-19 prevention is social distancing (at least 6 feet). Other preventive strategies include regular hand washing, avoiding unnecessary movement, social gathering restrictions, and use of face mask. Face mask was found very effective as a preventive measure [20]. In addition, vaccines are now available to induce an artificial active immunity against the virus [21]. There are currently no drugs to cure the virus. However, COVID-19 patients benefit from candidate drugs such as vasodilators, antiviral drugs, corticosteroids, and many more.

5   Performance of Several Laboratory Diagnostic Evaluations

Diagnosis of COVID-19 involves confirmation of coronavirus using molecular technique. Also test that detects serum antibodies against the virus (serological tests) is also available. With respect to molecular test, the challenge is the time required for the result to be available. For serological tests, the issue is the sensitivity of the technique. Yet rapidity and sensitivity of diagnosis are crucial in the prevention of disease progression and community transmission. As part of measures to unravel the challenge, Biotechnology Companies are working hard to ensure rapid techniques are developed. An example of this is a new multiplex real-time polymerase chain reaction developed by Thermo Fisher Scientific called TaqPath COVID-19 kit. The kit contains both controls and assays that are required for detection of ribonucleic acid from coronavirus-2019 virus.

Many studies on COVID-19 diagnosis identified real-time polymerase chain reaction as a consensus technique for COVID-19 validation. Sensitivity of this technique is dependent on specimen type. For instance, analysis of sputum specimens yielded a sensitivity of 69%, while that of nasopharyngeal swabs yielded 63% [14, 15]. It is worth noting that timing of specimen collection is important. Samples collected earlier before disease onset or later may yield a false negative result. Evidence exists to prove that sensitivity of the test using sputum specimen is dependent on the timing of sample collection. COVID-19 viral load appears greatest at the first week after the onset of symptom and decline thereafter [14, 15].

6 Healthcare Systems

A healthcare system is a collection of people, institutions, and resources that provides healthcare services to meet a targeted population. Over the years, the healthcare system has experienced significant changes and development globally [22]. Both governmental and non-governmental organizations have played critical role in ensuring that there is significant improvement in the healthcare sector. Although several challenges still persist most especially in the developing countries however, with the advent of modern technology, the future of the world's healthcare system has greatly improved [23]. Studies have shown that people live healthier and better today than several decades ago with significant decrease in infant and maternal mortality rates [23, 24]. Furthermore, there has been an increase in the world's population and a decrease in health expenditure [23]. Despite all these improvements, the healthcare system is still faced with many challenges some of which include increase in population, urbanization, and rise in chronic diseases [25]. It has been projected that the world's population will experience a sharp increase over the next decades and if there are no plans to improve the healthcare system and delivery most especially in developing countries, the urban areas of these countries may likely be the major centers or hub for the outbreak of infectious diseases [26]. In a review paper presented by Almalki et al. [27], it was reported that the government of Saudi Arabia gave priority to their healthcare system at all levels and observed significant improvement in the healthcare of their citizens. However, a number of challenges pose great concern to their healthcare system which include shortage of healthcare professionals, shortage of financial resources, increase demand for free services, poor access to some healthcare facilities, and poor national crisis management policy. All these problems highlighted may not just be peculiar to

Saudi Arabia but, other countries of the world may also be faced with challenges in these areas.

7 IoT-Based Technologies

The Internet of Things (IoT) has to do with the use physical objects contained in a sensor, software, and other technologies to connect and exchange data. Over the past few decades, there were several challenges facing the healthcare system on the delivery of services to people most especially in developing countries [28]. However, with the advent of IoT, healthcare service delivery has greatly improved. A large amount of data is generated every moment with the advent of IoT, and these have been of great value to all stakeholder in the health sector. Most of the data obtained through IoT are stored in both public and private databases [29]. IoT-based devises can also share data between devices. Cyber security and privacy of data are the major challenges when sharing data as information may easily be hijacked [30]. However, studies have shown that with advanced technologies, such as the distributed ledger technologies (DLTs), data may easily be protected, as this provides blockchain, Ethereum, and IOTA tangle which help to prevent access to data in public and private domains [31]. Many of these technologies are integrated with IoT to improve data collection and sharing. IoT technologies are fast and reliable and, if intergrade successfully with other technologies, could boost the people's confidence and healthcare delivery [32]. With the global COVID-19 pandemic, the use of IoT has greatly help in the monitoring of SARS-CoV-2-infected persons [33]. The IoT device could also be used by patients to monitor and report the symptoms of COIVD-19 to appropriate agencies and also to seek for proper healthcare treatments [5]. Many individuals who are already infected with SARS-CoV-2 are asymptomatic and they walk about without even knowing their status. Many persons who summoned courage to undergo COVID-19 test to know their status discovered that they have had contact with several other people in their environment. Therefore, the contact tracing of these persons may be extremely difficult but, with IoT technologies, contact with infected persons can easily be traced and the individual isolated for treatment [5]. Furthermore, with the ongoing vaccination with COVID-19 vaccine, the data of persons who had received the vaccine can easily be stored on IoT devices, and this will help to know the number of people who had been vaccinated. It will also help to strategize on how much of the vaccines are needed and plan on how to extend vaccination to all COVID-19-prone areas [34].

8   Utilization of IoT-Based Technologies in Data Acquisition

The fundamental idea of IoT is to be able to synchronize information, idea, facts, and other useful resources over the internet. This idea has become highly influential in the twenty-first century and has been utilized to improve effectiveness in various sectors including health. The connectivity between thousands of computers and other electronic devices via proper application of internet facilities can transform the health sector by improving efficiency and effectivity of several process including data acquisition, which involves measuring and analyzing various data. Utilization of IoT allows effective synchronization of several devices to execute different tasks depending on the development and operation of these devices. Adequate synchronization between several devices has become the major drive for the development of artificial intelligence techniques, where these devices can be programmed to decide on their own without human expertise [35]. Over the past decades, most of the data are generated manually, such as physical interview, with much difficulty in processing such information. Recently, IoT usage enables large amount of data to be generated and processed in short time, making data acquisition to become a vital part of almost every sector. In fact, it is that more than 50 billion gadgets will be interconnected via the internet by the year 2020, and this will increase the net worth of IoT [36]. For adequate data acquisition, the IoT system, via an internet-enabled device, uses different hardware including the processor, sensor, and communicator, to gather information (in form of data) from different sources, transform, and send such data for storage or analysis. The IoT devices share data through their connection to the internet gateway or neighboring devices, such data are subsequently transferred to the cloud for analysis using sophisticated software or analyzed locally, with or without any human influence. In this regard, the connection, communication, and integration processes that enhance data acquisition are determined by the specific application used. Furthermore, IoT can utilize artificial intelligence techniques such as machine learning algorithms to improve data acquisition thereby making the process easy, faster, and efficient.

9   IoT-Based Technologies and Healthcare Systems

Before now, the major platform of interaction between patients and healthcare provider's visitation, with great difficulty in the progressive monitoring and prescriptions. Some decades ago, application of telecommunication and short message service brought some improvement to the medical field. Recently, sophisticated technologies that harmonized several information on multiple devices have led to the development of IoT-enabled devices [37]. These devices have improved remote monitoring of patient's health, keeping them safe and healthy as well as enabled the healthcare providers to give superlative care. Thus, application of IoT-based technologies in healthcare system can provide a platform for easy and efficient interaction between patients and physician, while remote health monitoring can prevent hospital admission and length of stay in the clinic [38], thus, reducing the use of hospital facilities and cost of healthcare services, with increased satisfaction and treatment outcomes. Application of IoT-based technologies is without prejudice revolutionizing the healthcare system bringing significantly benefits to the healthcare providers, patients, families, clinics, and health-related companies such as insurance [39, 40]. The commonly used IoT-based devices are categorized into the following: (1) Wearable external devices such as biosensor with wireless transmission (electrocardiogram, glucometer) that monitor biological data which are often used for telemedicine. (2) Implanted medical devices such as drug delivery devices, and neurostimulators which are used to replace damaged biological structures or enhance their function. (3) Stationary medical devices which include X-rays, CT scan, and other laboratory devices, connected to the internet to assist the physician with proper diagnosis of diseases. Application of IoT-based devices such as glucometer, blood pressure monitoring device, and other fitness items such as wearable hand band can enable patient attention to personalized healthcare [41], since these can be used at their comfort zones without travelling long distance to the clinic for checkup. Also, these devices can improve accurate measurement of health indices that easily fluctuates with several factors (e.g., blood pressure, glucose level, and heat rate), allowing remote diagnosis, proper monitoring, and treatment of patients [42]. Furthermore, adequate interaction between multiple devices via the internet can elicit suitable alert signals (especially among the elderly), which can be sent to relatives or healthcare provider for immediate and adequate treatment response. Using IoT-based technologies such as nebulizers, wheelchairs, and other equipment can help the physician to monitor patient adherence to recommended treatment modalities, thereby allowing proper monitoring of patient's health and prognosis to achieve the expected treatment outcomes. Also, IoT-based technologies can assist in reducing wastage and loss of medical equipment. For example, IoT-enabled technologies can enable proper filing and identification of equipment and patients as well as real-time analysis in the deployment medical facilities and staff to several locations. These technologies can help improve remote hospital operations such as checkups and visitation. This can reduce physical contact in the clinic and the transmission of infectious diseases (Laplante et al., [43].

10 Specific Authors That Have Worked on the Application of Internet of Things (IoT) in the Management of COVID-19 Diseases

It has been observed that Internet of Things (IoT) has empowered the healthcare system around the globe to effectively evaluate the challenges encountered by COVID-19 patients, through employing an interconnected network. This technology plays a crucial role in enhancing the patients' satisfaction and decrease the level of readmission rate in the hospital. In view of this, Singh et al. [38] wrote a comprehensive review on the application of Internet of Things for effective management of COVID-19. They discovered that the application of IoT minimize the cost involved in the management of numerous affected patients. They also discovered that 12 significant utilizations of IoT were recognized and discussed extensively. They also stated that the application of IoT has enhanced the productivity level of scientists, researchers, and academicians, in proffering several solutions that could help in mitigating the issue of COVID-19 pandemic. Also, the role IoT has been identified to play a crucial role in the recognition of symptoms and plays a role in improving affected surgeon, patient, physician, and hospital management system.

The incident of COVID-19 (coronavirus) pandemic has prompted a higher level of request for necessary healthcare equipment, most especially in the area of medicine with obligation for increase in advanced information technology applications. The application of Industry 4.0 has been recognized as the fourth industrial revolution, which has prospective to accomplish customized prerequisite during COVID-19 crisis. This begun with the utilization of current advances in digital information technologies and manufacturing of relevant technology. In view of the aforementioned, Swayamsiddha and Mohanty [44] wrote a comprehensive review on the technologies of Industry 4 and their utilization in the COVID-19 pandemic. They discovered that the application of technologies of Industry 4.0 has played diverse functions in the prevention, diagnosis, detection of COVID-19 pandemic symptom, and associated challenges. It was also stated that Industry 4.0 can enhance the role of personalized face gloves and masks, collect information, and collation of data for healthcare systems for adequate management of COVID-19 patients. The authors also highlighted ten significant technologies of Industry 4.0 that could help in effective management of COVID-19 patients and the virus responsible for this disease. Moreover, it was also stated that Industry 4.0 technologies could improve the level of education and dissemination of information relating to public health. They concluded that this technology could play a vital role in the development of innovative role that could be applied in the mitigation of several medical emergencies at global and local levels.

The relevance of IoT has been identified as one of the advanced digital technologies that has been deployed in the management of COVID-19 pandemic. In view of this, Javaid et al. [4] wrote a comprehensive review on the utilization of cognitive radio which depends on the IoT specific for the medical domain that is recognized as cognitive Internet of Medical Things (CIoMT) that could be applied in the management of diverse global challenges. The application of CIoMT has been applied in the management of COVID-19 pandemic for adequate monitoring via a massive network that necessitates efficient spectrum management. Therefore, Javaid et al. [4] also emphasize the impact of CR-based dynamic spectrum allocation methodology that can accommodate numerous applications and devices. Moreover, it was stated that the CIoMT platform permits real-time tracking, screening, surveillance, quick identification of diverse cases, remote health monitoring, clustering, and contact tracking which play a crucial function in decreasing the workload on the medical industry for mitigation and management of the infection. Hence, the application of CIoMT will play a crucial role in the prevention of this virus.

The recent epidemic outbreak of COVID-19 has led to higher level of request for medical accessories and medical equipment together with daily essentials for the safety of healthcare workers. In view of this, Kumar et al. [45] recognized several operational problems encountered by numerous re retailers in providing efficient services. They also emphasize the role of Industry 4.0 in decreasing the impact of impact of COVID-19. They recognized 12 significant trials for the retail sectors that could serve as operational challenges and provide the role of Industry 4.0 technologies.

Kelly et al. [28] deploy the application of IoT technology in healthcare and emphasize how IoT devices could be applied in order to achieve a sustainable healthcare delivery as well as provide a detailed information on how IoT technology could influence the rate of global healthcare in the next decade. It was stated that IoT could enhance the accessibility of preventative public health services and their eventual transmission of current secondary and tertiary healthcare in becoming more effective, coordinated, and continuous system. The authors also pointed out certain problems that are associated with IoT-based healthcare most especially to market adoption from healthcare professionals and patients, privacy and security, remuneration, acceptability, data storage, standardization, interoperability, control, and ownership. They also emphasize the function of IoT in current healthcare which leverage on cybersecurity-focused guidelines, careful strategic planning, and policy support among healthcare organizations. Hence, IoT-based healthcare portends a greater capacity to enhance effectiveness of health system and enhance health population.

11   Conclusion and Future Perspectives

At present, virtually every sector benefits from IoT-based technology including the healthcare system where IoT is being applied to streamline healthcare delivery processes and timely completion of medical task, thereby improving healthcare delivery system. Presently, most IoT devices come with some form of connectivity, ranging from simple biosensor devices such as fit band to complex machines such as X-ray and computed tomography (CT) scans with Wi-Fi or Bluetooth. IoT-enabled medical devices generate important data that assist healthcare providers during complex activities, thus reducing the stress during practices. Despite the substantial benefits of IoT-based technologies in healthcare system, the incorporation of this technology depends on proper and sufficient data-handling procedures to avoid compromising patient's privacy and confidentiality, adequate security of IoT acquired data against hackers, and preventing the misuse of equipment. Although, the application of IoT-based technology is already adopted in medical practices, there exist numerous questions that need to be addressed, relating to the incorporation of IoT technology in the healthcare system. These questions include standardization and validation of IoT device operations, reliability and reproducibility of result generated, patient compliance during remote application of IoT-based technology as well as sustainability and maintenance of the technology. Also, more effort is needed to increase the awareness, acceptability, and usage of this technology by all stakeholders (including patients, Caregivers, physician, health-related companies) in the health sector while using IoT to enhance the quality and delivery of health services.

References

1. Yang, B. L. D., Wang, X., Lin, T., Zhu, X., Zhong, N., Bai, C., Powell, C. A., Chen, R., Zhou, J., Song, Y., Xin, Z., Zhu, H., Han, B., Li, Q., Shi, G., Li, S., Wang, C., Qiu, Z., Yong Zhang, Y. X., Liu, J., Zhang, D., Wu, C., Li, J., Yu, J., Wang, J., Dong, C., Wang, Y., Wang, Q., Zhang, L., Zhang, M., Ma, X., Lin, Z., Yu, W., Xu, T., Yang, J., Wang, X., Wang, Y., Jiang, Y., Chen, H., Xiao, K., Zhang, X., Song, Z., Zhang, Z., Wu, X., Sun, J., Shen, Y., Ye, M., Chunlin, T., Jiang, J., Yu, H., & Tan, F. (2020). Chinese experts' consensus on the Internet of Things-aided diagnosis and treatment of coronavirus disease 2019. *Clinical eHealth, 3*, 7–15.
2. Haleem, A., Javaid, M., & Khan, I. H. (2019). Internet of things (IoT) applications in orthopaedics. *Journal of Clinical Orthopaedics and Trauma, In Press*. https://doi.org/10.1016/j.jcot.2019.07.003.
3. Allam, Z., & Jones, D. S. (2020). On the coronavirus (COVID-19) outbreak and the smart city network: Universal data sharing standards coupled with artificial intelligence (AI) to benefit urban health monitoring and management. *Healthcare, 8*(1), 46. Multidisciplinary Digital Publishing Institute.
4. Javaid, M., Haleem, A., Vaishya, R., Bahl, S., Suman, R., & Vaish, A. (2020). Industry 4.0 technologies and their applications in fighting COVID-19 pandemic. *Diabetes & Metabolic Syndrome, 14*(4), 419–422. https://doi.org/10.1016/j.dsx.2020.04.032.
5. Singh, R. P., Javaid, M., Haleem, A., & Suman, R. (2020). Internet of things (IoT) applications to fight against COVID-19 pandemic. *Diabetes & Metabolic Syndrome, 14*(4), 521–524.
6. Tope, O. O., Adetunji, C. O., Okotie, G. E., Adeyomoye, O., Anani, O. A., & Mali, P. C. (2021). Impact of COVID-19 on assisted reproductive technologies and its multifacet influence on global bioeconomy. *Journal of Reproductive Healthcare and Medicine, 2*(Suppl_1), 92–104. https://doi.org/10.25259/JRHM_44_2020.
7. Page, J., Hinshaw, D., & McKay, B. (2021). In Hunt for Covid-19 Origin, Patient Zero Points to Second Wuhan Market – The man with the first confirmed infection of the new coronavirus told the WHO team that his parents had shopped there. *The Wall Street Journal*. Retrieved 27 Feb 2021.
8. Ibeh, I. N., Enitan, S. S., Akele, R. Y., Isitua, C. C., & Omorodion, F. (2020). Global impacts and Nigeria responsiveness to the COVID-19 pandemic. *International Journal of Healthcare and Medical Sciences, 6*, 27–45.
9. Olaniyan Olugbemi, T., Ayobami, D., Okotie, G. E., Adetunji, C. O., Oluwaseun, I. B., Bamidele, O. J., & Olugbenga, E. O. (2020). Testis and blood-testis barrier in Covid-19 infestation: Role of angiotensin converting enzyme 2 in male infertility. *Journal of Basic and Clinical Physiology and Pharmacology. DEGRUYTER, 31*(6), 1-13. https://doi.org/10.1515/jbcpp-2020-0156.
10. Verdecchia, P., Cavallini, C., Spanevello, A., & Angeli, F. (2020). The pivotal link between ACE2 deficiency and SARS-CoV-2 infection. *European Journal of Internal Medicine., 76*, 14–20.

11. Gurwitz, D. (2020). Angiotensin receptor blockers as tentative SARS-CoV-2 therapeutics. *Drug Development Research, 81*(5), 537–540.
12. Quirch, M., Lee, J., & Rehman, S. (2020). Hazards of the cytokine storm and cytokine-targeted therapy in patients with COVID-19: Review. *Journal of Medical Internet Research., 22*(8), *e20193*.
13. Bo, H. X., Li, W., Yang, Y., Wang, Y., Zhang, Q., Cheung, T., Wu, X., & Xiang, Y. T. (2020). Posttraumatic stress symptoms and attitude toward crisis mental health services among clinically stable patients with COVID-19 in China. *Psychological Medicine, 27*, 1-2. https://doi.org/10.1017/S0033291720000999. Epub ahead of print. PMID: 32216863; PMCID: PMC7200846.
14. Liu, R., Han, H., Liu, F., Lv, Z., Wu, K., Liu, Y., Feng, Y., & Zhu, C. (2020a). Positive rate of RT-PCR detection of SARS-CoV-2 infection in 4880 cases from one hospital in Wuhan, China, from Jan to Feb 2020. *Clinica Chimica Acta, J505*, 172–175.
15. Liu, S., Yang, L., Zhang, C., Xiang, Y. T., Liu, Z., Hu, S., & Zhang, B. (2020b). Online mental health services in China during the COVID-19 outbreak. *Lancet Psychiatry, 7*(4), e17–e18. https://doi.org/10.1016/S2215-0366(20)30077-8. Epub 2020 Feb 19. PMID: 32085841; PMCID: PMC7129099.
16. Abimbola, A. (2020). Mental health implications of lockdown during coronavirus pandemic among adult resident in Ibadan, Nigeria. *AJSW, 10*(3), 50–58. Special Issue on COVID-19.
17. Semo, B., & Frissa, S. M. (2020). The mental health impact of the COVID-19 pandemic: Implications for Sub-Saharan Africa. *Psychology Research and Behavior Management, 13*, 713–720.
18. Ai, T., Yang, Z., Hou, H., Zhan, C., Chen, C., Lv, W., Tao, Q., Sun, Z., & Xia, L. (2020). Correlation of chest CT and RT-PCR testing for Coronavirus disease 2019 (COVID-19) in China: A report of 1014 cases. *Radiology, 296*(2), E32–E40. https://doi.org/10.1148/radiol.2020200642. Epub 2020 Feb 26. PMID: 32101510; PMCID: PMC7233399.
19. Salehi, S., Abedi, A., Balakrishnan, S., & Gholamrezanezhad, A. (2020). Coronavirus disease 2019 (COVID-19): A systematic review of imaging findings in 919 patients. *AJR. American Journal of Roentgenology., 215*(1), 87–93.
20. Jeremy, H., Huang, A., Li, Z., Tufekci, Z., Zdimal, V., van der Westhuizen, H.-M., von Delft, A., Price, A., Fridman, L., Tang, L.-H., Tang, V., Watson, G. L., Bax, C. E., Shaikh, R., Questier, F., Hernandez, D., Chu, L. F., Ramirez, C. M., & Rimoin, A. W. (2021). *Proceedings of the National Academy of Sciences, 118*(4), e2014564118. https://doi.org/10.1073/pnas.2014564118.
21. Anderson, R. M., Heesterbeek, H., Klinkenberg, D., & Hollingsworth, T. D. (2020). How will country-based mitigation measures influence the course of the COVID-19 epidemic? *Lancet, 395*(10228), 931–934.
22. Durrani, H. (2016). Healthcare and healthcare systems: Inspiring progress and future prospects. *mHealth, 2*, 3.
23. Lichtenstein, R. L. (1993). The United States' health care system: Problems and solutions. *Survey of Ophthalmology, 38*(3), 310–317.
24. Hoffer, E. P. (2019). America's Health Care System is broken: What went wrong and how we can fix it. Part 3: Hospitals and doctors. *The American Journal of Medicine, 132*(8), 907–911.
25. Obama, B. (2016). United States health care reform: Progress to date and next steps. *JAMA, 316*(5), 525–532.
26. Trastek, V. F., Hamilton, N. W., & Niles, E. E. (2014). Leadership models in health care - a case for servant leadership. *Mayo Clinic Proceedings, 89*(3), 374–381.
27. Almalki, M., Fitzgerald, G., & Clark, M. (2011). Health care system in Saudi Arabia: An overview. *Eastern Mediterranean health journal = La revue de sante de la Mediterranee orientale = al-Majallah al-sihhiyah li-sharq al-mutawassit, 17*(10), 784–793.
28. Kelly, J. T., Campbell, K. L., Gong, E., & Scuffham, P. (2020). The Internet of Things: Impact and implications for health care delivery. *Journal of Medical Internet Research, 22*(11), e20135.

29. Issel, L. M. (2019). Managing the Internet of Things in health care organizations. *Health Care Management Review, 44*(3), 195.
30. Rajan, J. P., & Rajan, S. E. (2018). An Internet of Things based physiological signal monitoring and receiving system for virtual enhanced health care network. *Technology and Health Care: official journal of the European Society for Engineering and Medicine, 26*(2), 379–385.
31. Zheng, X., Sun, S., Mukkamala, R. R., Vatrapu, R., & Ordieres-Meré, J. (2019). Accelerating health data sharing: A solution based on the Internet of Things and distributed ledger technologies. *Journal of Medical Internet Research, 21*(6), e13583.
32. Jamil, F., Ahmad, S., Iqbal, N., & Kim, D. H. (2020). Towards a remote monitoring of patient vital signs based on IoT-based Blockchain integrity management platforms in smart hospitals. *Sensors (Basel, Switzerland), 20*(8), 2195.
33. Sarbadhikari, S., & Sarbadhikari, S. N. (2020). The global experience of digital health interventions in COVID-19 management. *Indian Journal of Public Health, 64*(Supplement), S117–S124.
34. Ting, D., Carin, L., Dzau, V., & Wong, T. Y. (2020). Digital technology and COVID-19. *Nature Medicine, 26*(4), 459–461.
35. Khanna, A., & Kaur, S. (2020). Internet of Things (IoT), applications and challenges: A comprehensive review. *Wireless Personal Communications, 114*, 1687.
36. Mainetti, L., Patrono, L., & Vilei, A. (2011). Evolution of wireless sensor networks towards the internet of things: A survey. In *2011 19th international conference on software, telecommunications and computer networks (SoftCOM)* (pp. 1–6). IEEE.
37. Dang, L. M., Piran, M. J., Han, D., Min, K., & Moon, H. (2019). A survey on internet of things and cloud computing for healthcare. *Electronics, 8*(7), 768.
38. Pacheco, R., Dias, Santinha, Rodrigues, Queirós, & Rodrigues. (2019). Smart cities and healthcare: A systematic review. *Technologies, 7*(3), 58.
39. Rohokale, V. M., Prasad, N. R., & Prasad, R. (2011, Feb). A cooperative Internet of Things (IoT) for rural healthcare monitoring and control. In *Proceedings of the2nd international conference on wireless communication, vehicular technology, information theory and aerospace & electronic systems technology (wireless VITAE)* (pp. 1–6). IEEE.
40. Tarouco, L. M. R., Bertholdo, L. Z. G., Arbiza, L. M. R., Carbone, F., Marotta, M., & de Santanna, J. J. C. (2012). Internet of Things in healthcare: Interoperatibility and security issues. In *Proceedings of theIEEE international conference on communications (ICC), 2012* (pp. 6121–6125). IEEE.
41. Schofield, P., Shaw, T., & Pascoe, M. (2019). Toward comprehensive patient-centric care by integrating digital health technology with direct clinical contact in Australia. *Journal of Medical Internet Research, 21*(6), e12382.
42. Veazie, S., Winchell, K., Gilbert, J., Paynter, R., Ivlev, I., Eden, K. B., Nussbaum, K., Weiskopf, N., Guise, J. M., & Helfand, M. (2018). Rapid evidence review of mobile applications for self-management of diabetes. *Journal of General Internal Medicine, 33*(7), 1167–1176. https://doi.org/10.1007/s11606-018-4410-1. Epub 2018 May 8. PMID: 29740786; PMCID: PMC6025680.
43. Laplante, P. A., Kassab, M., Laplante, N. L., & Voas, J. M. (2018). Building caring healthcare systems in the Internet of Things. *IEEE Systems Journal, 12*(3). https://doi.org/10.1109/JSYST.2017.2662602.
44. Swayamsiddha, S., & Mohanty, C. (2020). Application of cognitive Internet of Medical Things for COVID-19 pandemic. *Diabetes & metabolic syndrome, 14*(5), 911–915. https://doi.org/10.1016/j.dsx.2020.06.014.
45. Kumar, M. S., Raut, D., Narwane, D., & Narkhede, D. (2020). Applications of industry 4.0 to overcome the COVID-19 operational challenges. *Diabetes & Metabolic Syndrome, 14*(5), 1283–1289. https://doi.org/10.1016/j.dsx.2020.07.010.

Diagnosis for COVID-19

Ashish Tripathi, Anand Bhushan Pandey, Arun Kumar Singh, Arush Jain, Vaibhav Tyagi, and Prem Chand Vashist

1 Introduction

According to the study, a rare and unexplained case of pneumonia was reported in Wuhan, China, at the end of December 2019. Wuhan is a Chinese city in the Hubei Province. At first, the symptoms of this illness seemed to be somewhat similar to those of pneumonia. However, experts from the "Chinese Center for Disease Control and Prevention (China CDC)" said that this is a new kind of virus-infected disease. The report was focused on the examination of a number of respiratory samples (throat swab samples). It is caused by a novel coronavirus that resembles the native SARS-CoV and MERS-CoV families [3]. The illness was later dubbed COVID-19 by the World Health Organization (WHO) (coronavirus disease). According to a WHO survey, the pandemic affected 205 countries until April 2, 2020, with 900,306 reported cases and 45,692 deaths [4]. COVID-19 has turned into a global threat to mankind. According to the study, the coronavirus spreads among humans through a variety of routes, including direct contact or inhalation of infected droplets and contact with polluted hands/places/objects/surfaces [3]. However, in some circumstances, the presence of gastrointestinal symptoms or a live infectious virus in the feces may be another way for infections to spread in humans through fecal–oral transmission [5]. The droplets are divided into two groups depending on their size. Respiratory droplets are described as droplets that are larger than 5–10 m in diameter. When the scale is less than 5 m, it is referred to as a droplet nucleus [6]. According to current studies, respiratory droplets are largely responsible for

A. Tripathi (✉) · A. B. Pandey · A. K. Singh · A. Jain · V. Tyagi · P. C. Vashist
Department of Information Technology, G. L. Bajaj Institute of Technology and Management, Greater Noida, UP, India

© The Author(s), under exclusive license to Springer Nature Switzerland AG 2022
S. K. Pani et al. (eds.), *Assessing COVID-19 and Other Pandemics and Epidemics using Computational Modelling and Data Analysis*,
https://doi.org/10.1007/978-3-030-79753-9_6

coronavirus transmission in humans [7–9]. SARS-CoV-2, SARS-CoV, and MERS-CoV all belong to the Orthocoronavirinae, according to phylogenetic analysis. They are also members of the Coronaviridae family [10]. SARS-CoV and MERS-CoV, on the other hand, are not the same as SARS-CoV-2 [11, 12].

The four genera of Orthocoronavirinae subfamily indulged in affecting living objects, for example, mammals are infected through αCoV and βCoV while δCoV and γ-CoV are the main causes of infection in birds. Recently, two outbreaks were found by CoV in humans. This was the case of viral pneumonia caused by SARS and MERS. The first case of SARS-CoV in China was reported in the year 2002, a new kind of CoV, which spread worldwide very quickly. The mortality rate of this outbreak was 11% [13, 14]. This outbreak fetched the attention of researchers on evolution, replication, transmission, severity, biological structure, and development manner of the disease (i.e., microbial infection, inflammation, malignancy, and tissue breakdown) of CoV. After a few years, another virus-related severe respiratory disease outbreak was reported in the Middle East, in 2012 [15]. The starting point of this outbreak was Saudi Arabia and after that spread to other countries, with 37% fatality rate [16]. As this virus was emerged in the Middle East, it was named MERS-CoV. In both the epidemic cases of virus infection, one thing was very clear that they likely came from bats and reached to humans through any intermediate carrier [17]. In a research, it was found that the camel and the civet, respectively, were used as carriers for MERS-CoV and SARS-CoV [18, 19]. Like SARS-CoV and MERS-CoV, the newly introduced SARS-CoV-2 is also related to βCoV lineage.

This paper presents the systematic report on origin of disease, epidemiology, genome structure and life cycle, clinical features, diagnosis, isolation, treatment, prevention, societal and economical impact, and global response.

1.1 Classification of Coronavirus (CoVs)

Coronaviridae has a subfamily called Orthocoronavirinae. Coronaviruses (CoVs) are classified as Orthocoronavirinae. Coronaviruses (CoVs) are divided into four classes. Coronaviridae, Arteriviridae, Mesoniviridae, and Roniviridae are the four families. They are members of the Nidovirales order. As seen in Fig. 1, the Orthocoronavirinae subfamily contains four genera: alphacoronavirus (CoV), beta-coronavirus (CoV), deltacoronavirus (CoV), and gammacoronavirus (-CoV). Coronaviridae viruses are single-stranded, enveloped, and very large ribonucleic acid (RNA) viruses with a positive single-stranded, enveloped genome. In contrast to other recognized RNA viruses [10], the genome size of Coronaviridae family viruses is very high which ranges from 25,000 to 32,000 base pairs. The diameter of a virion will range from 118 to 136 nm. Coronavirinae and Torovirinae are the two subfamilies of the Coronaviridae family [19].

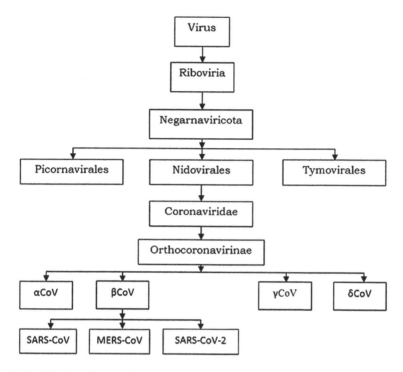

Fig. 1 Classification of coronavirus

1.2 Genomic Organization and Structure of Coronavirus (SARS-CoV-2)

The SARS-CoV-2 contains large, non-segmented, positive single-stranded RNA genome of 30 kb in size. The RNA genome contains a $3'$ polyadenylated (A) tail with the $5'$ capped structure to translate (encode) replicase polyproteins (large and nonstructural proteins), and for this translation, RNA genome acts as an mRNA. In the whole translation process, discontinuous transcription process is used to synthesize the mRNA [20, 21]. To continue the transcription and replication of RNA, it requires multiple stem-loop structures. These structures are contained by untranslated region (UTR) of the $5'$ end of genome. Also the transcriptional regulatory sequences (TRSs) are required for the expression of every genes of the structural gene. Additionally, viral RNA imitation and synthesis are executed by RNA structures of $3'$ end UTR.

The RNA genome contains open reading frame (ORF), 1a/1b, which further generates five accessory proteins (i.e., ORF9, ORF8, ORF7, ORF6, and ORF3a), four structural proteins (nucleocapsid protein, envelope protein, membrane protein, and spike glycoprotein) (Fig. 2), and 15/16 nonstructural proteins (majority of them

Fig. 2 SARS-CoV-2
structure with four structural
proteins. Here the full name
of the mnemonics is as
follows: N (nucleocapsid
protein), E (envelope protein),
M (membrane protein), and S
(spike glycoprotein)

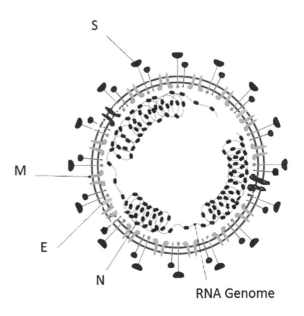

are involved in the synthesis of viral RNA) [22, 23]. These proteins support in spreading infection of SARS-CoV-2.

In a recent research, it was found that the coronavirus is spherically shaped and virions are of approximately 125 nm in diameter [24, 25]. Another report says that the diameter of virions varies between 118 and 140 nm [19]. The genome size of this virus lies between 26.4 and 31.7 kb, which is the largest size among all known genomes of RNA viruses [26]. The distinct feature of the coronavirus is its club-shaped spike projections, which originated or emitted from the virion surface. A virion envelop contains nucleocapsids, which come under the category of structural proteins. Presence of the nucleocapsids is very common in all known negatively sensed RNA viruses, but this is very uncommon in positively sensed RNA viruses. Among four structural proteins of coronavirus, spike glycoprotein (S) plays a significant role to host cells in the attachment to itself, and further host proteases (protease like furin) are used to cleave S into two distinct polypeptides, that is, S1 and S2 [27, 28]. For gating access to endoplasmic reticulum (ER), the S protein uses signal sequences of N-terminal. Homotrimers of the S (virus-encoded) protein makes a special spike shape on virus surface [29, 30]. The binding of host receptor and S1 can unstable the prefusion trimer. As a result, highly stable postfusion conformation is formed due to S1 shedding and S2 transition [31]. S1's receptor binding domain goes through conformational movements to engage a host receptor to cover (down conformation) or uncover (up conformation) the determinants of receptor binding transiently [32]. In this context, receptor's inaccessible state can be represented by down conformation, while the up conformation represents receptor's accessible state.

The coronavirus virion contains a very small quantity of envelope protein (E) and protein E follows a common architecture but it is very much divergent in nature [33]. This protein has ion channel activity as well as it contains C-terminal endo-domain and N-terminal ecto-domain. E protein plays the significant role in the facilitation of virus assembly and release. The role of E protein is necessary for pathogenesis [34].

Virion of coronavirus has rich storage of membrane protein (M). M protein is small in size and contains three transmembrane domains to give the shape of the virion [33]. The M protein has the ability to extend between 6 and 8 nm due to the presence of C-terminal endo-domain (small in size) and N-terminal glycosylated ecto-domain (very large in size). In a recent study, it has been found that two distinct conformations may be adopted by M protein to allow it to bind the nucleocapsid and also to boost the membrane curvature [35].

The nucleocapsid protein (N) of coronavirus is a kind of multifunctional protein that plays a vital role in enhancing the effectiveness of assembly and transcription of virus. During the assembly of virion, N protein establishes interaction with the viral M protein. The N protein has two distinct domains, that is, N-terminal domain (NTD) and C-terminal domain (CTD). Both domains are separated by RNA-binding domain. Both domains are separately attached with viral RNA. The optimality of RNA binding depends on the same effort from both domains [36].

Transcriptional regulatory sequence (TRS) and genome packaging signal are the two RNA substrates related to N protein [37, 38]. To bind the RNA-binding domain of C-terminal, genomic packaging signal is used [39].

Hence, detailed understanding of the proteins and genomic structure of coronavirus will help to design and develop vaccines and monoclonal antibody drugs.

2 Manifestations and Epidemiology of SARS-CoV-2

2.1 Origin of SARS-CoV-2

The source of origin plays a significant role in understanding any disease. In the case of COVID-19, primarily the researchers found that a large number of people were infected from the market of live animals (Huanan seafood market) in Wuhan city of China. Initially, the infected people were either shop employees, owner of the shop, or customers to this market. After this incident, the market was closed on January 1, 2020. On the basis of the efforts taken to understand the cause of the spreading of coronavirus, it was found that this virus was transmitted through intermediate carriers in humans. In the primary report of researchers, environmental samples taken from the seafood market depicted the positive evidence of viral RNA, and on that basis, it was recognized that Asian palm civet might be the intermediate host of coronavirus [40]. After several investigations, still there has been doubt about the potential host of the SARS-CoV-2 [41]. Recently, on the basis of several studies, it

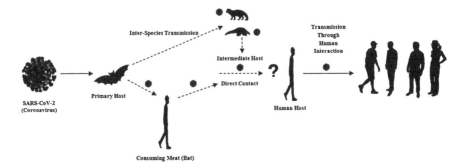

Fig. 3 Illustration of the coronavirus's origin and dissemination into humans through various mediums. The most common way for an infection to spread from an animal to a human is by eating the animal's meat (primary host) or near contact with an intermediate host. The infection is then transmitted from one sick person to another healthy person. The virus is then spread to the general population through human-to-human contact [44]

has been identified that the bats might be the potential host of this virus [11, 12, 42]. Based on the clinical research, it was found that the genetic sequence of coronavirus has more than 95% similarity with the bat coronavirus and more than 70% similarity with SARS-CoV [42, 43]. Figure 3 shows the origin of coronavirus and its spread to humans through potential mediums.

A report published in "The Lancet" says that the origin of the coronavirus is the wildlife. Some studies say that the virus had been brought by an infected human in the seafood market and then it spread out in the market environment [45]. Some reports say that there was a conspiracy behind the origin of this deadly virus. In this conspiracy, the role of the Wuhan Institute of Virology has been found suspicious. There is a rumor that this institute is involved in developing human-generated bioweapon. In another investigation, the virologists stated that this virus is new and has originated from the environment. However, after many investigations, still there is a question about what is the primary source of virus amplification in humans. In this context, more investigations are required to understand the real source of spreading of the virus in humans.

2.2 Symptoms of SARS-CoV-2

Symptoms of COVID-19 infection can take anywhere from 1 to 14 days to appear. Dry cough, respiratory signs (like the flu), acute fever, tiredness, and headache are all typical COVID-19 symptoms [3, 46–48]. The disease can manifest itself in a variety of ways, from negative to positive. Good signs indicate the presence of a disease outbreak in this case. A new research looked at 41 patients who were diagnosed with COVID-19 at the outset. Cough, fever, and nausea were the most frequent symptoms in the majority of the patients. Although there was trouble in breathing, headache, hemoptysis, dyspnea, diarrhea, and lymphopenia were all serious and even fatal symptoms in some instances [3]. During a chest CT scan, it was discovered that both the patients had pneumonia. Acute respiratory distress syndrome was seen in almost 29% of the cases, acute heart attack in 12% of the cases, and other infections in 10% of the cases [3].

In another clinical report, patients were found to have unusual symptoms such as acute cardiac injury and respiratory distress syndrome [3]. Gastrointestinal symptoms (i.e., vomiting and diarrhea, and nausea) were observed in the patients during the illness [3, 49, 50]. The first case of the USA was confirmed for respiratory illness followed by vomiting and diarrhea, nausea, and abdominal pain [49].

2.3   Incubation Period of SARS-CoV-2

The time between the symptom onset and getting exposure to the virus is known as the incubation period for SARS-CoV-2. Average time of the incubation period is 5–6 days; however, this period can extend up to 14 days and known as the presymptomatic period. Contagiousness can be spread out by a few of the infected persons. In the earlier studies, the estimated mean incubation period of COVID-19 was 3–7 days within 2–14 days, which shows that this virus had a long transmission period [51, 52]. The transmission period of CoVs was found a little bit different from its counterpart viruses like mean incubation period for SARS-CoVs, it was 5 days within 2–14 days [53], and for MERS-CoVs, it was 5–7 days within 2–14 days [54]. In a recent clinical report on 138 cases, the median incubation period from the initial symptoms to acute respiratory disorder and finally hospitalization was 5 days within the range of 1–10 days, 7 days within the range of 4–8 days, and 8 days within the range of 6–12 days, respectively [55]. In another confirmed case of 425 patients the mean incubation period was 5.2 days (range 4.0–7.0 days) for some patients and for others it was 7.5 days (range 5.0–19 days). The confidence interval (CI) was observed to be 95% for all 425 patients [9]. While the mean incubation period estimated over 1099 patients was 3.0 days, range within 0–24 [56]. In another report, the mean incubation period of the infection was 4.6 days within 3.8–5.8 days with 95% CI and the beginning of 95% disease occurred within 10 days [57, 58]. In comparison to SARS-CoVs and MERS-CoVs, the fatality rate of COVID-19 is less due to the lower incubation period which was estimated to be 3.0 days [55]. Days between 2 and 8 has been estimated as a mean time from primary symptom to admission in hospital [59]. The time duration from the initial symptoms of COVID-19 to death ranged between 6 and 41 days, and the median of the incubation period was 14 days [47]. Age and immune system of the patient decide this period. In critical cases of COVID-19, this can take 3–6 weeks to recover or die out from the disease. Generally, the period is smaller for the patient, whose age is less than 70 years as compared to patients with age greater than 70 years [47]. In symptomatic transmission, the repeated biological samples based on virologic and clinical activities observed that the expulsion and release of CoVs are highest in the upper respiratory tract in the initial phase of the infection of confirmed cases [60–63], which happened in the first 3 days from the starting of the symptoms [62, 63].

2.4 Transmission of SARS-CoV-2

As per the current studies on the pathophysiological properties of SARS-CoV-2, it had been found that the spreading mechanism of this virus is very uncertain. This study was based on the human-to-human transmission of coronavirus through respiratory droplets [64]. The size of the respiratory droplets lies between 5 and 15 μm, and they remain to sustain in the air for several minutes before dropping on the floor. However, the droplet particles whose size is less than 5 μm remain to sustain in the air up to hours due to their small size and less weight. During this time, if any person comes in the close contact with the infected droplets, then he/she becomes infected.

In the case of symptomatic symptoms, the contagiousness of the viruses related to respiratory infection becomes very high. In symptomatic transmission, a person transmits the infection to another person, while he/she is experiencing the symptoms of coronavirus. The virologic and epidemiologic studies say that the symptomatic transmission occurred through direct contact with the infected patients, or close contact with surfaces and objects contaminated by the virus, or the close contact with respiratory droplets [7–9]. In asymptomatic transmission, the virus transmission from the infected person to another healthy person occurs in such a way that it does not develop any symptom of COVID-19. In some laboratory reports, some asymptomatic cases have been confirmed. However, there has not been any confirmed case of asymptomatic transmission found [65].

Generally, the symptomatic patient is more infectious due to the presence of respiratory viruses. However, the evidence of virus transmission is increasing during human-to-human interaction in the incubation period of asymptomatic transmission of COVID-19, and the probability estimation of this period has been between 2 and 10 days [64].

A lot of analysis has been done to understand the source of transmission of SARS-CoV-2 in humans. In spite of respiratory droplets, close contact is also a big source of transmission of SARS-CoV-2. Typically, three organs of human body like mouth, nose, and eyes are responsible for virus outbreak and transmission. Another possibility of infection by aerosol transmission to the people through airborne particles of sneezing, coughing, talking, and breathing, those remain active in the air for a long time. The aerosol has two distinct sizes. The first size varies between 0.25 and 1 μm and the second size is bigger than 2.5 μm. In recent studies on SARS-CoV-2, aerosol remains active in the air for up to 3 hours, which is very much similar to SARS-CoV and MERS-CoV [66].

3   Diagnostic for SARS-CoV-2

The seventh coronavirus, SARS-CoV-2, can infect humans (HCoV). Its infections can be asymptomatic or severe which in case of early diagnosis can aid clinical management and hence control the outbreaks. The diagnostic detects the virus itself or detects the response of the immune system to the infection, that is, antibodies, biomarkers, etc. In Table 1, symptoms of different viral disease are compared. Table 2 mentions the systematic disorders and respiratory disorders caused by COVID-19 infections.

3.1   SARS-CoV-2 RNA Detection

It is based on detecting the unique viral sequences using nucleic acid amplification tests (NAATs), like real-time reverse-transcription polymerase chain reaction abbreviated as rRT-PCR.

Once someone is infected by the virus, it takes some time known as incubation period to develop symptoms (about a week), with a range of 1 and 14 days after exposure and can be detected 1–3 days before the symptom onset in the upper respiratory tract (URT) [9, 51]. In URT, the concentration of the virus is highest at the time of symptom onset and it declines thereafter. The presence of viral RNA

Table 1 Comparison of symptoms in different diseases

S.N	Diseases	Symptoms
1	COVID-19 (Coronavirus disese-19)	Respiratory system failure, severe breathlessness, fever, muscle ache, headache
2	Mild COVID-19	Cough, sore throat, fever
3	SARS (Severe acute respiratory syndrome)	Dyspnea, cough, fever, headache, diarrhea, malaise, chill
4	MERS (Middle East respiratory syndrome)	Dyspnea, sore throat, dry cough, fever, rigor, chill
5	Influenza	Runny nose, stuffy nose, dry cough, sore throat, high fever, malaise, muscle ache
6	Cold	Sneeze, runny nose, stuffy nose

Table 2 Systemic and respiratory disorders in COVID-19

Systemic disorders	Respiratory disorders
Fever, cough, fatigue, sputum production, headache	Rhinorrhea, sneezing, sore throat
Hemoptysis	Pneumonia
Acute cardiac injury	Ground-glass opacities
Hypoxemia, dyspnea, lymphopenia, diarrhea	RNAaemia, acute respiratory distress syndrome

is increased during the second week of the illness in the lower respiratory tract (LRT) [67]. In some patients, it is detectable for many days, while in others for several weeks or months but a prolonged presence does not signify a prolonged infectiousness.

Respiratory secretions vary in composition, and sampling adequacy may also vary, resulting in false-negative PCR results. Viral RNA is detected in LRT secretions for the patient who is suspected of SARS-CoV-2 infection and the URT swab is negative. Rectal swabs or feces were positive for SARS-CoV-2 RNA in some of the patients, and some studies suggest that the positivity is prolonged in comparison to that of respiratory specimens. In some cases, SARS-CoV-2 RNA is detected in blood samples, and some studies show that it depends on the severity of the disease, but more studies on this are needed. In oral fluid specimens, that is, induced saliva, the detection rates vary widely when compared with URT specimens of that very same patient, and as of now very limited data is available on detection adequacy of the virus in mouth washes [68]. SARS-CoV-2 was detected in ocular fluids, urine, and semen samples of some of the patients, while positive RNA detection for cerebrospinal fluid and brain tissue has also been reported.

To conclude, SARS-CoV-2 can be detected in respiratory material and at the same time in other body fluids and compartments, but the respiratory samples are taken for diagnostics as the virus is most frequently detected in respiratory material.

Rapid collections of the specimens as well as accurate diagnosis are required to support clinical management of the patients and infection controlling measures to be taken. The trained and competent operators should collect and perform laboratory diagnosis because of the complexity of sampling, analysis done in laboratory, and interpretation of the final results. Those infected with SARS-CoV-2 may or may not develop symptoms, and the most robust evidence of infection may be detected from the fragments of the virus, nucleic acids, and proteins through virological testing. Reports show the cases of infection of COVID-19 with other pathogens either, so if other pathogens test positive, it does not rule out SARS-CoV-2 infection and vice versa.

3.2   Adequate Specimen Collection

While specimen collection, testing, storage, and research, adequate standard operating procedures (SOP) should be there and staff must be trained for all this. Testing combined nasopharyngeal and oropharyngeal swabs has been shown to enhance sensitivity in detection of viruses and the reliability of the result has improved. Swabs from two individuals can be combined, and nasopharyngeal and oropharyngeal swabs may be taken, but some studies show that nasopharyngeal swabs provide results more reliable as compared to oropharyngeal swabs. In mild or asymptomatic cases, the upper respiratory specimens are suitable for the test only in the early-stage of infections. Lower respiratory specimens are collected from patients having a negative URT sampling or if sample collection is delayed.

Because of high risk of aerosolization, caution should be exercised and strictly abiding by IPC procedures is required during collection of the samples. Other oral or respiratory fluids sampling procedures must first clear validation test in the laboratory before implementation for the targeted groups of the patients.

3.3 Simplified Specimen Collection

Some studies on combined oropharyngeal and nares/nasal swab, midturbinate or lower nasal or nares swabs or tongue swab by a trained staff or by self-sampling show that these approaches perform well, but their sample sizes are limited [69, 70]. Further assessment as well as validation is required before implementation of these alternatives and to determine for what purpose these collection methods can serve as alternatives. In case of elderly people and young child, oral fluids can be a suitable specimen in schools or nursing homes for mass screening as collecting nasopharyngeal and oropharyngeal swabs are problematic as compared with URT specimens [71, 72].

Oral fluid collection methods have a wide range of the performances compared with naso- and/or oropharyngeal sample collection; they also show variations from posterior oropharyngeal fluids/saliva that may be collected by spitting or drooling, with pipet or special sponges, and gargling with saline solutions is another alternative.

3.4 Serum Specimens

A person in whom COVID-19 infection is strongly suspected but obtained NAAT results are negative, a paired serum specimen can be collected. First specimen may be collected in the acute phase and the other one in the convalescent phase which can be used to test for seroconversion or an increase in antibody titers.

3.5 Fecal Specimens

While testing feces make sure that for this type of sample, the extraction method and NAAT have been validated. If URT and LRT both are tested negative but clinical suspicion on COVID-19 infection is there, NAAT can be considered for fecal specimens [73].

3.6   Postmortem Specimens

The postmortem swab like needle biopsy or specimens from tissues of the dead body from the autopsy may be taken which also includes lung tissues for pathological and microbiological tests [74].

After collection, the specimens must reach the laboratory quickly and a correct specimen handling in the laboratory and during transportation is very essential.

4   Testing of COVID-19 (SARS-CoV-2)

4.1   Nucleic Acid Amplification Test (NAAT)

NAAT such as rRT-PCR should be used to test suspected COVID-19 infections and SARS-CoV-2 genome should be targeted by such tests. Optimal diagnostics has a NAAT assay and minimum of two independent targets on SARS-CoV-2 genome, but a simple algorithm may be adopted in widespread transmission areas. While using one-target assay, a strategy should be in place so as to monitor the mutations which may affect the overall performance. Some of the NAAT systems are having the capacity for fully automated tests by which sample processing, the capacity for RNA extraction, amplification, and reporting is integrated. These systems provide access of tests in areas having limited laboratory capacity with rapid turnaround time. The amplification/detection methods, like isothermal nucleic acid amplification technologies, clustered regularly interspaced short palindromic repeats (CRISPR), and molecular microarray assays are going to be commercialized [75–77].

Some negative test results do not actually rule out infection because there are factors that can lead to a negative result even if the person is infected [78, 79]. Such factors are:

- The specimen of poor quality
- The specimen was collected from a body compartment which was not containing SARS-CoV-2 at that time
- Handling or shipping of the specimen was not appropriately done, etc.

Figure 4 is a pictorial representation of diagnostic flow of SARS-CoV-2 infection detection.

4.2   NAAT by Pooling Specimens

Diagnostic capacity can be increased by pooling of samples from different patients to detect SARS-CoV-2 if the testing rate is not meeting the demands [80–82]. Specimen pooling may be considered in population groups having very low

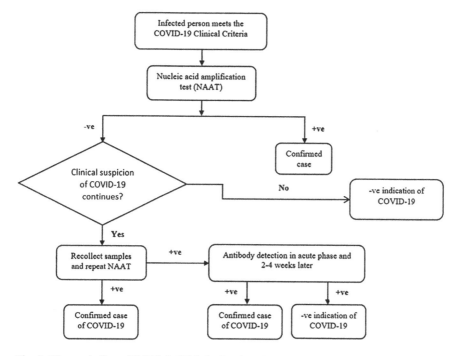

Fig. 4 Diagnostic flow of SARS-CoV-2 infection detection

prevalence of COVID-19 infection. Adequate automation is the key to reliable pooling (software-based algorithms, robotic systems, and laboratory middle-ware which works with sample pooling).

Strategies for pooling the specimens [83, 84] are:

- The individual specimens are treated to be negative if the result of the pool is negative.
- Strategies may be different in case of positive pool tests but every specimen must be tested (pool deconvolution) generally to find out positive specimen(s).
- In matrix pooling, pool is made per row and per column and finally tested using PCR.

4.3   Testing Antibodies

In a response to the infection, the human body produces antibodies which are detected by serological assays. SARS-CoV-2 is a novel pathogen and the understanding of the antibody response is in emerging phase, hence the antibody detection tests should not be used to detect acute infections [85, 86]. Lateral flow antibody detection assays are not suitable for diagnosis because they cannot detect the increase in antibody titers, while (semi)quantitative or quantitative assays can. Antibodies can be detected in the first week of illness in some of the patients while in others with subclinical/mild infection; it can take weeks to develop. The duration of the antibodies generated is still under study, hence serology is not an alternate for virological assays and the presence of antibodies does not ensure that they offer protective immunity. The target viral protein and timing of the testing affect performance of serologic assays in various testing groups (mild disease vs moderate-to-severe disease or young patient vs old patient) [87, 88].

4.4 Antigen Detection (Rapid Diagnostic Tests—RDTs)

The presence of SARS-CoV-2 is tested in viral proteins (antigens) collected from respiratory specimens for rapid (within 30 minutes) diagnostic tests. False-positive (showing a person is infected when actually he is not) results may be obtained if the antibodies also recognize antigens of other human coronaviruses which are non-SARS-CoV-2 viruses. Pairing of NAAT and antigen test validations are being encouraged in clinical studies. High viral loads improve antigen test performance, but antigen RDTs can be implemented in diagnostic algorithms only if the test performance is acceptable.

5   Treatments for SARS-CoV-2

Most of the COVID-19 patients are able to recover at home but at the same time scientists are leaving no stone unturned to develop an effective treatment. Therapies that are being investigated include the drugs that are used to treat autoimmune diseases, antibodies of recovered patients from COVID-19, and additional antiviral drugs.

5.1 Monoclonal Antibodies

FDA (Food and Drug Administration) granted emergency use authorization (EUA) for treatment named as Bamlanivimab which is a monoclonal antibody [89]. It is approved to treat non-hospitalized, recently tested positive for COVID-19 children and adults having mild-to-moderate symptoms but are at risk of severe COVID-19 disease. Within 10 days of symptom development, the patient should be given a single dose of the treatment.

5.2   Chloroquine and Hydroxychloroquine

Chloroquine has been reported as a broad-spectrum antiviral drug which is a commonly used autoimmune disease drug and anti-malarial drug. Chloroquine controls virus infection by interfering in the glycosylation of cellular receptors related to SARS-CoV and increases endosomal pH needed for virus fusion. Chloroquine was recommended for the prevention as well as treatment of COVID-19 pneumonia [90]. Hydroxychloroquine, an analog of chloroquine, has lesser concerns related to drug–drug interactions and has been found to be more effective than chloroquine for Vero cells infected by SARS-CoV-2. Both chloroquine and hydroxychloroquine have immunomodulatory effects and are capable of suppressing the immune response.

5.3   Plasma Transfusion

For SARS treatment, convalescent plasma was administered very early after symptom onset, and the pooled odd of mortality treatment was reduced compared with no therapy. For COVID-19 treatment, the National Health Commission of China appealed convalescent patients for donating blood [42]. Plasma should be collected within 2 weeks after recovery from COVID-19 which will ensure high neutralization; this complexity of obtaining plasma limits its clinical application.

5.4   Corticosteroids

Though corticosteroids were not recommended for viral pneumonia or acute respiratory distress syndrome (ARDS) but in case of severe CARDS such drugs are generally used.

Recently, a large-size RCT (the RECOVERY trial) shows that dexamethasone decreased the death rates by one-third in critical COVID-19 patients [91].

5.5   Vaccines

Vaccines deliver immunogen (specific type of antigen) to train immune system to recognize the pathogens like bacteria or viruses which reduces the risk of infections. Injections in muscles or under skin and oral route are different methods of administering vaccines. More than one dose of vaccine is given to build complete immunity, as a "booster" in case of immunity wears off and against the disease that is a seasonal disease like the yearly flu. Table 3 shows the technologies used for viral vaccines with their advantages and disadvantages.

Table 3 Technologies used for viral vaccines

S.N	Immunogen and working	Advantages and disadvantages	Vaccine examples
1	Protein subunit: A protein which is derived from a pathogen	Lesser side effects than whole virus. It is a complex process and poorly immunogenic too.	Influenza
2	Whole inactivated virus: Inactivated dead virus	Strong antibody response is induced. A large quantity of virus is needed.	Hepatitis A, Rabies and Influenza
3	Attenuated live virus: Live virus, doesn't cause disease [92]	Same response as natural infection is induced. Not suitable for immunocompromised person and pregnant women.	Rubella, Mumps, Measles, Smallpox, and Yellow fever
4	Nucleic acid: DNA or RNA coding for a viral protein	Rapid development and very strong cellular immunity. Low antibody responses.	COVID-19 vaccines
5	Replicating or non-replicating viral vector: Viral pathogen expressed on a safe virus, doesn't cause disease [93].	Rapid development. Immunogenicity may be reduced due to prior exposure to vector virus (e.g. Adenovirus).	Ebola
6	Recombinant: An antigen is expressed using host cell [94]	There is not any need to produce the whole virus. High cost and poorly immunogenic.	Hepatitis B
7	Peptides: Synthetic produced fragment of an antigen	Rapid development. Poorly immunogenic in spite of a very high cost.	COVID-19 vaccines

6 Vaccine Development

Infectious diseases are prevented by vaccines. Some of the diseases that are vaccine preventable are polio, measles, Hepatitis B, influenza, and many more. The ability of the pathogen to spread can be limited by "herd" or "indirect" or "population" immunity developed when most of the people are vaccinated against a disease. When many people have immunity, the people who cannot be vaccinated or have low immunity like young babies are protected by infectious disease. Peptide and nucleic acid like new technologies for human vaccines are being used along with the conventional and well-known technologies to develop the vaccines [91–93]. Table 4 shows the mechanism of actions for different types of vaccines, while Table 5 represents the steps taken in effective and safe vaccine development.

As of 2 October 2020, out of 42 candidate vaccines of COVID-19 which are in clinical evaluation, 10 are in Phase-III trials. In preclinical evaluation, 151 candidate vaccines are there [93]. Most of the candidate vaccines are for intramuscular injection and designed for a two-dose schedule. The WHO has launched a coordinated international Phase-III trial of candidate vaccines for speedy evaluation and to ensure that the vaccines are tested in different populations. Table 6

Table 4 Mechanism of actions for different types of vaccines

S.N	Type of vaccine	Mechanism of actions
1	Protein-based vaccines	A protein is extracted from alive or inactivated virus, purified, and then injected as a vaccine. In case of corona virus, it is spike protein which is most commonly used.
2	Viral vector vaccines	The gene for a pathogen needs to be inserted into a virus which can infect a person without causing any disease. The safe virus is a vector or a platform to deliver a protein which triggers an immune response. Then the safe virus is injected as a vaccine.
3	Virus vaccines	The virus is selected, weakened, or even completely in activated so that it becomes in capable to cause a disease.
4	Nucleic acid vaccines	The nucleic acid coding for the antigen is injected. It is a completely new technology and have never been used for human vaccine.

Table 5 Steps taken in vaccine (effective and safe) development

S.N	Action/step taken	Description
1	Preclinical studies	In animal studies, the vaccine is tested for efficacy and safety.
2	Phase-I clinical trial	Healthy adult volunteers in small groups receive the vaccine for testing the safety.
3	Phase-II clinical trial	The vaccine is given to people having similar characteristics to the people whom this new vaccine is actually intended for.
4	Phase-III clinical trial	The efficacy and safety is tested by giving the vaccine to thousands of people.
5	Phase-IV post marketing surveillance	Studies to monitor the adverse events and long-term effects in the population only after the vaccine is approved and licensed.
6	Human challenge studies	A vaccine is given followed by the pathogen against which the vaccine is designed to protect. These trials are not very common in people because of the ethical challenges posed.

is a list of the COVID-19 vaccines, which are in Phase-III clinical trial, along with the locations of their trials.

As of 4 October 2020, the USA, India, Brazil, Russian Federation, and Colombia are five countries which have highest number of COVID-19 cases. The countries having highest number of deaths are the USA, Brazil, India, Mexico, and the United Kingdom [93].

Table 6 COVID-19 vaccines in Phase-III trial

S.N	Vaccines and platform	Location
1	University of Oxford/AstraZeneca (Viral vector)	USA
2	CanSino Biological Inc./Beijing Institute of Biotechnology (Viral vector)	Pakistan
3	Gamaleya Research Institute (Viral vector)	Russia
4	Janssen Pharmaceutical Companies (Viral vector)	USA, Peru, Mexico, Philippines, Colombia, South Africa, Brazil
5	Sinovac (Inactivated virus)	Brazil
6	Wuhan Institute of Biological Products/Sinopharm (Inactivated virus)	UAE
7	Beijing Institute of Biological Products/Sinopharm (Inactivated virus)	China
8	BioNTech/Fosun Pharma/Pfizer (RNA)	Brazil, USA, Argentina
9	Moderna/NIAID (RNA)	USA
10	Novavax (Protein subunit)	UK

7 Conclusions

In spite of day and night efforts done by researchers, doctors, and other medical staff, everything about this disease is uncertain, be it the symptoms, the outcomes of the tests, or the success rate of the under trial vaccines. The shortage of resources and facilities as compared to the patients are the main reasons behind the surrender of medical infrastructure of a country. So a collaborative international effort is required to face the COVID-19 pandemic and establish pathways to manage this crisis. To face a pandemic like COVID-19, none of us was prepared so the patients, their relatives, and finally the community must be provided authentic and understandable information to face this devil. This paper presents state-of-the-art research work going on SARS-CoV-2 virus from its origin, its mutations, and diagnosis to clinical trials on vaccine developments.

References

1. Lu, G., Wang, Q., & Gao, G. F. (2015). Bat-to-human: Spike features determining "host jump" of coronaviruses SARS-CoV, MERS-CoV, and beyond. *Trends in Microbiology, 23*, 468–478.
2. Nieto-Torres, J. L., Dediego, M. L., Verdia-Baguena, C., et al. (2014). Severe acute respiratory syndrome coronavirus envelope protein ion channel activity promotes virus fitness and pathogenesis. *PLoS Pathogens, 10*, e1004077. https://doi.org/10.1371/journal.ppat.1004077.
3. Huang, C., Wang, Y., Li, X., Ren, L., Zhao, J., Hu, Y., et al. (2020). Clinical features of patients infected with 2019 novel coronavirus in Wuhan, China. *Lancet, 395*(10223), 497–506. https://doi.org/10.1016/S0140-6736(20)30183-5.
4. https://experience.arcgis.com/experience/685d0ace521648f8a5beeeee1b9125cd

5. Gu, J., Han, B., & Wang, J. (2020). COVID-19: Gastrointestinal manifestations and potential fecal-oral transmission. *Gastroenterology, 158*(6), 1518–1519.
6. World Health Organization. (2014). *Infection prevention and control of epidemic- and pandemic-prone acute respiratory infections in health care.* Geneva: World Health Organization. Available from: https://apps.who.int/iris/bitstream/handle/10665/112656/9789241507134_eng.pdf?sequence=1.
7. Liu, J., Liao, X., Qian, S., et al. (2020). Community transmission of severe acute respiratory syndrome coronavirus 2, Shenzhen, China, 2020. *Emerging Infectious Diseases, 26*(6), 1320–1323. https://doi.org/10.3201/eid2606.200239.
8. Chan, J., Yuan, S., Kok, K., et al. (2020). A familial cluster of pneumonia associated with the 2019 novel coronavirus indicating person-to-person transmission: A study of a family cluster. *Lancet, 395*(10223), 514–523. https://doi.org/10.1016/S0140-6736(20)30154-9.
9. Li, Q., Guan, X., Wu, P., Wang, X., Zhou, L., Tong, Y., Ren, R., Leung, K. S., Lau, E. H., Wong, J. Y., & Xing, X. (2020). Early transmission dynamics in Wuhan, China, of novel coronavirus-infected pneumonia. *The New England Journal of Medicine, 382*, 1199–1207. https://doi.org/10.1056/NEJMoa2001316.
10. Li, H., Liu, S. M., Yu, X. H., Tang, S. L., & Tang, C. K. (2020). Coronavirus disease 2019 (COVID-19): Current status and future perspective. *International Journal of Antimicrobial Agents, 55*(5), 105951.
11. Zhu, N., Zhang, D., Wang, W., Li, X., Yang, B., Song, J., et al. (2020). A novel coronavirus from patients with pneumonia in China, 2019. *The New England Journal of Medicine, 382*, 727–733.
12. Lu, R., Zhao, X., Li, J., Niu, P., Yang, B., Wu, H., et al. (2020). Genomic characterisation and epidemiology of 2019 novel coronavirus: Implications for virus origins and receptor binding. *Lancet, 395*(10224), 565–574.
13. Song, Z., Xu, Y., Bao, L., et al. (2019). From SARS to MERS, thrusting coronaviruses into the spotlight. *Viruses, 11*(1), 59.
14. Graham, R. L., Donaldson, E. F., & Baric, R. S. (2013). A decade after SARS: Strategies for controlling emerging coronaviruses. *Nature Reviews Microbiology, 11*, 836–848.
15. Zumla, A., Hui, D. S., & Perlman, S. (2015). Middle East respiratory syndrome. *Lancet, 386*, 995–1007.
16. Hui, D. S., Azhar, E. I., Kim, Y. J., et al. (2018). Middle East respiratory syndrome coronavirus: Risk factors and determinants of primary, household, and nosocomial transmission. *The Lancet Infectious Diseases, 18*, e217–e227.
17. Reusken, C. B., Haagmans, B. L., Muller, M. A., et al. (2013). Middle East respiratory syndrome coronavirus neutralising serum antibodies in dromedary camels: A comparative serological study. *The Lancet Infectious Diseases, 13*, 859–866.
18. de Wit, E., van Doremalen, N., Falzarano, D., et al. (2016). SARS and MERS: Recent insights into emerging coronaviruses. *Nature Reviews Microbiology, 14*, 523–534.
19. de Groot, R. J., Baker, S. C., Baric, R., Enjuanes, L., Gorbalenya, A. E., Holmes, K. V., Perlman, S., Poon, L., Rottier, P. J. M., Talbot, P. J., & Woo, P. C. Y. (2012). Family Coronaviridae. In *Virus taxonomy* (pp. 806–828). Amsterdam: Elsevier Academic Press.
20. Yang, D., & Leibowitz, J. L. (2015). The structure and functions of coronavirus genomic 3′ and 5′ ends. *Virus Research, 206*, 120–133.
21. Nakagawa, K., Lokugamage, K. G., & Makino, S. (2016). Viral and cellular mRNA translation in coronavirus-infected cells. In *Advances in virus research* (Vol. 96, pp. 165–192). Amsterdam: Academic Press.
22. Ramaiah, A., & Arumugaswami, V. (2020). Insights into cross-species evolution of novel human coronavirus 2019-nCoV and defining immune determinants for vaccine development. *bioRxiv*, 1–15. https://doi.org/10.1101/2020.01.29.925867.
23. Wu, A., Peng, Y., Huang, B., et al. (2020). Genome composition and divergence of the novel coronavirus (2019-nCoV) originating in China. *Cell Host and Microbe, 27*(3), 325–328.

24. Barcena, M., Oostergetel, G. T., Bartelink, W., et al. (2009). Cryo-electron tomography of mouse hepatitis virus: Insights into the structure of the coronavirion. *Proceedings of the National Academy of Sciences of the United States of America, 106*, 582–587.

25. Neuman, B. W., Adair, B. D., Yoshioka, C., et al. (2006). Supramolecular architecture of severe acute respiratory syndrome coronavirus revealed by electron cryomicroscopy. *Journal of Virology, 80*, 7918–7928.

26. Mousavizadeh, L., & Ghasemi, S. (2021). Genotype and phenotype of COVID-19: Their roles in pathogenesis. *Journal of Microbiology, Immunology and Infection, 54*(2), 159–163.

27. Abraham, S., Kienzle, T. E., Lapps, W., et al. (1990). Deduced sequence of the bovine coronavirus spike protein and identification of the internal proteolytic cleavage site. *Virology, 176*, 296–301.

28. Luytjes, W., Sturman, L. S., Bredenbeek, P. J., et al. (1987). Primary structure of the glycoprotein E2 of coronavirus MHV-A59 and identification of the trypsin cleavage site. *Virology, 161*, 479–487.

29. Beniac, D. R., Andonov, A., Grudeski, E., et al. (2006). Architecture of the SARS coronavirus prefusion spike. *Nature Structural & Molecular Biology, 13*, 751–752. https://doi.org/10.1038/nsmb1123.

30. Delmas, B., & Laude, H. (1990). Assembly of coronavirus spike protein into trimers and its role in epitope expression. *Journal of Virology, 64*, 5367–5375.

31. Walls, A. C., Xiong, X., Park, Y. J., et al. (2019). Unexpected receptor functional mimicry elucidates activation of coronavirus fusion. *Cell, 176*, 1026–1039.e1015.

32. Wrapp, D., Wang, N., Corbett, K. S., et al. (2020). Cryo-EM structure of the 2019-nCoV spike in the refusion conformation. *Science, 367*(6483), 1260–1263.

33. Armstrong, J., Niemann, H., Smeekens, S., et al. (1984). Sequence and topology of a model intracellular membrane protein, E1 glycoprotein, from a coronavirus. *Nature, 308*, 751–752.

34. Neuman, B. W., Kiss, G., Kunding, A. H., et al. (2011). A structural analysis of M protein in coronavirus assembly and morphology. *Journal of Structural Biology, 174*, 11–22. https://doi.org/10.1016/j.jsb.2010.11.021.

35. Godet, M., L'Haridon, R., Vautherot, J. F., et al. (1992). TGEV corona virus ORF4 encodes a membrane protein that is incorporated into virions. *Virology, 188*, 666–675.

36. Chang, C. K., Sue, S. C., Yu, T. H., et al. (2006). Modular organization of SARS coronavirus nucleocapsid protein. *Journal of Biomedical Science, 13*, 59–72. https://doi.org/10.1007/s11373-005-9035-9.

37. Stohlman, S. A., Baric, R. S., Nelson, G. N., et al. (1988). Specific interaction between coronavirus leader RNA and nucleocapsid protein. *Journal of Virology, 62*, 4288–4295.

38. Molenkamp, R., & Spaan, W. J. (1997). Identification of a specific interaction between the coronavirus mouse hepatitis virus A59 nucleocapsid protein and packaging signal. *Virology, 239*, 78–86.

39. Kuo, L., & Masters, P. S. (2013). Functional analysis of the murine coronavirus genomic RNA packaging signal. *Journal of Virology, 87*, 5182–5192. https://doi.org/10.1128/JVI.00100-13.

40. Kan, B., Wang, M., Jing, H., Xu, H., Jiang, X., Yan, M., et al. (2005). Molecular evolution analysis and geographic investigation of severe acute respiratory syndrome coronavirus-like virus in palm civets at an animal market and on farms. *Journal of Virology, 79*(18), 11892–11900.

41. Gralinski, L. E., & Menachery, V. D. (2020). Return of the coronavirus: 2019-nCoV. *Viruses, 12*(2), 135.

42. Zhou, P., Yang, X. L., Wang, X. G., Hu, B., Zhang, L., Zhang, W., Si, H. R., Zhu, Y., Li, B., Huang, C. L., & Chen, H. D. (2020). A pneumonia outbreak associated with a new coronavirus of probable bat origin. *Nature, 579*(7798), 270–273. https://doi.org/10.1038/s41586-020-2012-7.

43. Hui, D. S., Azhar, E. I., Madani, T. A., et al. (2020). The continuing 2019-nCoV epidemic threat of novel coronaviruses to global health – The latest 2019 novel coronavirus outbreak in Wuhan, China. *International Journal of Infectious Diseases, 91*, 264–266.

44. Tripathi, A., Jain, A., Mishra, K. K., Pandey, A. B., & Vashist, P. C. (2020). MCNN: A deep learning based rapid diagnosis method for COVID-19 from the X-ray images. *Revue d'Intelligence Artificielle, 34*(6), 673–682.
45. https://www.who.int/health-topics/coronavirus/who-recommendations-to-reduce-risk-of-transmission-of-emerging-pathogens-from-animals-to-humans-in-live-animal-markets
46. Ren, L. L., Wang, Y. M., Wu, Z. Q., Xiang, Z. C., Guo, L., Xu, T., et al. (2020). Identification of a novel coronavirus causing severe pneumonia in human: A descriptive study. *Chinese Medical Journal, 133*(9), 1015–1024. https://doi.org/10.1097/CM9.0000000000000722.
47. Wang, W., Tang, J., & Wei, F. (2020). Updated understanding of the outbreak of 2019 novel coronavirus (2019-nCoV) in Wuhan, China. *Journal of Medical Virology, 92*(4), 441–447. https://doi.org/10.1002/jmv.25689.
48. Carlos, W. G., Dela Cruz, C. S., Cao, B., Pasnick, S., & Jamil, S. (2020). Novel Wuhan (2019-nCoV) coronavirus. *American Journal of Respiratory and Critical Care Medicine, 201*(4), 7–8. https://doi.org/10.1164/rccm.2014P7.
49. Holshue, M. L., DeBolt, C., Lindquist, S., Lofy, K. H., Wiesman, J., Bruce, H., et al. (2020). First case of 2019 novel coronavirus in the United States. *The New England Journal of Medicine, 382*, 929–936. https://doi.org/10.1056/NEJMoa2001191.
50. Liu, K., Fang, Y. Y., Deng, Y., Liu, W., Wang, M. F., Ma, J. P., et al. (2020). Clinical characteristics of novel coronavirus cases in tertiary hospitals in Hubei Province. *Chinese Medical Journal, 133*(9), 1025–1031. https://doi.org/10.1097/CM9.0000000000000744.
51. Backer, J. A., Klinkenberg, D., & Wallinga, J. (2020). Incubation period of 2019 novel coronavirus (2019-nCoV) infections among travellers from Wuhan, China, 20–28 January 2020. *Euro Surveillance, 25*(5), 2000062.
52. Lauer, S. A., Grantz, K. H., Bi, Q., et al. (2020). The incubation period of 2019-nCoV from publicly reported confirmed cases: Estimation and application. *medRxiv*. https://www.medrxiv.org/content/10.1101/2020.02.02.20020016v1 2020
53. Varia, M., Wilson, S., Sarwal, S., et al. (2003). Investigation of a nosocomial outbreak of severe acute respiratory syndrome (SARS) in Toronto, Canada. *CMAJ, 169*, 285–292.
54. Assiri, A., Al-Tawfiq, J. A., Al-Rabeeah, A. A., et al. (2013). Epidemiological, demographic, and clinical characteristics of 47 cases of Middle East respiratory syndrome coronavirus disease from Saudi Arabia: A descriptive study. *The Lancet Infectious Diseases, 13*, 752–761.
55. Wang, D., Hu, B., Hu, C., et al. (2020). Clinical characteristics of 138 hospitalized patients with 2019 novel coronavirus-infected pneumonia in Wuhan, China. *JAMA, 323*(11), 1061–1069.
56. Guan, W. J., Ni, Z. Y., Hu, Y., et al. (2020). Clinical characteristics of coronavirus disease 2019 in China. *The New England Journal of Medicine, 382*, 1708–1720.
57. Chiu, W. K., Cheung, P. C., Ng, K. L., et al. (2003). Severe acute respiratory syndrome in children: Experience in a regional hospital in Hong Kong. *Pediatric Critical Care Medicine, 4*, 279–283.
58. Donnelly, C. A., Ghani, A. C., Leung, G. M., et al. (2003). Epidemiological determinants of spread of causal agent of severe acute respiratory syndrome in Hong Kong. *Lancet, 361*, 1761–1766.
59. Leung, G. M., Hedley, A. J., Ho, L.-M., & Chau, P. (2004). The epidemiology of severe acute respiratory syndrome in the 2003 Hong Kong epidemic: An analysis of all 1755 patients. *Annals of Internal Medicine, 141*, 662–673.
60. Wang, W., Xu, Y., Ruqin, G., et al. (2020). Detection of SARS-CoV-2 in different types of clinical specimens. *JAMA, 323*(18), 1843–1844. https://doi.org/10.1001/jama.2020.3786.9.
61. Lauer, S. A., Grantz, K. H., Bi, Q., et al. (2020). The incubation period of coronavirus disease 2019 (COVID-19) from publicly reported confirmed cases: Estimation and application. *Annals of Internal Medicine, 172*(9), 577–582. https://doi.org/10.7326/M20-0504.10.
62. Liu, Y., Yan, L. M., Wan, L., et al. (2020). Viral dynamics in mild and severe cases of CVOID-19. *The Lancet Infectious Diseases, 20*(6), 656–657. https://doi.org/10.1016/S1473-3099(20)30232-2.

63. Wolfel, R., Corman, V., Guggemos, W., et al. (2020). Virological assessment of hospitalized cases of coronavirus disease. *Nature, 581,* 465–469. https://doi.org/10.1101/2020.03.05.20030502.

64. Centers for Disease Control and Prevention. (2020). *2019 novel coronavirus.* https://www.cdc.gov/coronavirus/2019-ncov/about/transmission.html

65. https://www.who.int/docs/default-source/coronaviruse/situation-reports/20200402-sitrep-73-covid-19.pdf?sfvrsn=5ae25bc7_2

66. https://www.cidrap.umn.edu/news-perspective/2020/03/commentary-covid-19-transmission-messages-should-hinge-science

67. Weiss, A., Jellingsø, M., & Sommer, M. O. A. (2020). Spatial and temporal dynamics of SARS-CoV-2 in COVID-19 patients: A systematic review and meta-analysis. *eBioMedicine, 58,* 102916.

68. Guo, W. L., Jiang, Q., Ye, F., Li, S. Q., Hong, C., Chen, L. Y., & Li, S. Y. (2020). Effect of throat washings on detection of 2019 novel coronavirus. *Clinical Infectious Diseases, 71*(8), ciaa416. https://doi.org/10.1093/cid/ciaa416.

69. Vlek, A. L. M., Wesselius, T. S., Achterberg, R., & Thijsen, S. F. T. (2021). Combined throat/nasal swab sampling for SARS-CoV-2 is equivalent to nasopharyngeal sampling. *European Journal of Clinical Microbiology & Infectious Diseases, 40*(1), 193–195.

70. LeBlanc, J. J., Heinstein, C., MacDonald, J., Pettipas, J., Hatchette, T. F., & Patriquin, G. (2020). A combined oropharyngeal/nares swab is a suitable alternative to nasopharyngeal swabs for the detection of SARS-CoV-2. *Journal of Clinical Virology, 128,* 104442.

71. Tu, Y. P., Jennings, R., Hart, B., Cangelosi, G. A., Wood, R. C., Wehber, K., Verma, P., Vojta, D., & Berke, E. M. (2020). Swabs collected by patients or health care workers for SARS-CoV-2 testing. *The New England Journal of Medicine, 383*(5), 494–496.

72. Altamirano, J., Govindarajan, P., Blomkalns, A. L., Kushner, L. E., Stevens, B. A., Pinsky, B. A., & Maldonado, Y. (2020). Assessment of sensitivity and specificity of patient-collected lower nasal specimens for sudden acute respiratory syndrome coronavirus 2 testing. *JAMA Network Open, 3*(6), e2012005.

73. Ng, S. C., Chan, F. K., & Chan, P. K. (2020). Screening FMT donors during the COVID-19 pandemic: A protocol for stool SARS-CoV-2 viral quantification. *The Lancet. Gastroenterology & Hepatology, 5*(7), 642–643.

74. Tang, J. W., To, K. F., Lo, A. W., Sung, J. J., Ng, H. K., & Chan, P. K. (2007). Quantitative temporal-spatial distribution of severe acute respiratory syndrome-associated coronavirus (SARS-CoV) in post-mortem tissues. *Journal of Medical Virology, 79*(9), 1245–1253.

75. Carter, L. J., Garner, L. V., Smoot, J. W., Li, Y., Zhou, Q., Saveson, C. J., Sasso, J. M., Gregg, A. C., Soares, D. J., Beskid, T. R., & Jervey, S. R. (2020). Assay techniques and test development for COVID-19 diagnosis. *ACS Central Science, 6*(5), 591–605.

76. World Health Organization. (2020). *Diagnostic testing for SARS-CoV-2: Interim guidance, 11 September 2020* (No. WHO/2019-nCoV/laboratory/2020.6). Geneva: World Health Organization.

77. Esbin, M. N., Whitney, O. N., Chong, S., Maurer, A., Darzacq, X., & Tjian, R. (2020). Overcoming the bottleneck to widespread testing: A rapid review of nucleic acid testing approaches for COVID-19 detection. *RNA, 26*(7), 771–783.

78. Zou, L., et al. (2020). SARS-CoV-2 viral load in upper respiratory specimens of infected patients. *The New England Journal of Medicine, 382*(12), 1177–1179.

79. Young, B. E., Ong, S. W. X., Kalimuddin, S., Low, J. G., Tan, S. Y., Loh, J., Ng, O. T., Marimuthu, K., Ang, L. W., Mak, T. M., & Lau, S. K. (2020). Epidemiologic features and clinical course of patients infected with SARS-CoV-2 in Singapore. *JAMA, 323*(15), 1488–1494.

80. Yelin, I., Aharony, N., Shaer-Tamar, E., Argoetti, A., Messer, E., Berenbaum, D., Shafran, E., Kuzli, A., Gandali, N., Hashimshony, T., & Mandel-Gutfreund, Y. (2020). Evaluation of COVID-19 RT-qPCR test in multi-sample pools. *medRxiv.*

81. Sawarkar, S. S., Victor, A., Viotti, M., Haran, S. P., Verma, S., Griffin, D., & Sams, J. (2020). Sample pooling, a population screening strategy for SARS-CoV2 to prevent future outbreak and mitigate the second-wave of infection of the virus. *medRxiv.*

82. Abdalhamid, B., Bilder, C. R., McCutchen, E. L., Hinrichs, S. H., Koepsell, S. A., & Iwen, P. C. (2020). Assessment of specimen pooling to conserve SARS CoV-2 testing resources. *American Journal of Clinical Pathology, 153*(6), 715–718.

83. Aragón-Caqueo, D., Fernández-Salinas, J., & Laroze, D. (2020). Optimization of group size in pool testing strategy for SARS-CoV-2: A simple mathematical model. *Journal of Medical Virology, 92*(10), 25929. https://doi.org/10.1002/jmv.25929.

84. Ben-Ami, R., Klochendler, A., Seidel, M., Sido, T., Gurel-Gurevich, O., Yassour, M., Meshorer, E., Benedek, G., Fogel, I., Oiknine-Djian, E., & Gertler, A. (2020). Large-scale implementation of pooled RNA extraction and RT-PCR for SARS-CoV-2 detection. *Clinical Microbiology and Infection, 26*(9), 1248–1253.

85. World Health Organization. (2020, April 8). *Advice on the use of point-of-care immunodiagnostic tests for COVID-19.* Available from: https://apps.who.int/iris/handle/10665/331713

86. World Health Organization. (2020, July 27). *The unity studies: Early investigations protocols.* Available from: https://www.who.int/emergencies/diseases/novel-coronavirus-2019/technical-guidance/early-investigations

87. Okba, N. M., Müller, M. A., Li, W., Wang, C., Geurtsvan Kessel, C. H., Corman, V. M., Lamers, M. M., Sikkema, R. S., de Bruin, E., Chandler, F. D., & Yazdanpanah, Y. (2020). Severe acute respiratory syndrome coronavirus 2–specific antibody responses in coronavirus disease patients. *Emerging Infectious Diseases, 26*(7), 1478–1488.

88. Lou, B., Li, T. D., Zheng, S. F., Su, Y. Y., Li, Z. Y., Liu, W., Yu, F., Ge, S. X., Zou, Q. D., Yuan, Q., & Lin, S. (2020). Serology characteristics of SARS-CoV-2 infection since exposure and post symptom onset. *European Respiratory Journal, 56*(2), 2000763.

89. https://www.medicinenet.com/monoclonal_antibodies/article.htm#what_are_human_monoclonal_antibodies

90. Vincent, M. J., Bergeron, E., Benjannet, S., Erickson, B. R., Rollin, P. E., Ksiazek, T. G., Seidah, N. G., & Nichol, S. T. (2005). Chloroquine is a potent inhibitor of SARS coronavirus infection and spread. *Virology Journal, 2*(1), 1–10.

91. https://www.ncbi.nlm.nih.gov/books/NBK554776/

92. https://www.who.int/publications/i/item/an-international-randomised-trial-of-candidate-vaccines-against-covid-19

93. https://www.who.int/docs/default-source/coronaviruse/risk-comms-updates/update37-vaccine-development.pdf?sfvrsn=2581e994_6

94. https://www.nature.com/articles/d41586-020-01221-y

IoT in Combating COVID-19 Pandemics: Lessons for Developing Countries

Oyekola Peter, Suchismita Swain, Kamalakanta Muduli, and Adimuthu Ramasamy

1 Introduction

The outbreak of COVID-19 from Wuhan, China, has led to several countries imposing lockdown in the hopes of containing the spread of the virus. In most countries, schools, businesses and several industries have been severely affected, and this have created an atmosphere of economic instability with most countries reviewing their initial budget plans for the year to consider the impact of the virus spread [18, 54]. After the declaration of the situation as being a pandemic, the race to contain as well as develop technologies and vaccines to combat COVID-19 became fiercer. Several techniques have been proposed in dealing with the impact of the virus, some vaccines have been tried with both positive and negative outcomes, tracing applications have been developed as well as other developments which will be detailed in proceeding sections [11, 21]. However, till date, there has been no confirmed vaccine that can alleviate the current condition. For countries with very little technological might, more focus is being shifted towards damage control, contact tracing and prevention of further spread through leveraging on technologies such as smart systems and Internet of Things (IoT) [48]. COVID-19, which is a virus

O. Peter · K. Muduli (✉)
Department of Mechanical Engineering, Papua New Guinea University of Technology, Lae, Papua New Guinea

S. Swain
Department of Mechanical Engineering, Temple City Institute of Technology and Engineering, Khordha, Odisha, India

A. Ramasamy
Department of Business Studies, Papua New Guinea University of Technology, Lae, Papua New Guinea
e-mail: ramasamy.adimuthu@pnguot.ac.pg

Fig. 1 Total confirmed virus cases as on June 2020 [65]

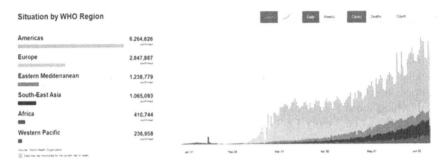

Fig. 2 Total infection by region (June 2020) [65]

that attacks the respiratory system [1], has spread round the globe with the United States being the worse hit region as of the time of this report. According to the World Health Organization (WHO), the total global confirmed cases in April 2021 reached approximately 135 million with 108 million recovered and about 2,927,442 deaths [65], which was on record as seen in Fig. 1. Of the total cases, the United States alone accounts for 32 million cases which translate to a whopping 24% of the total world cases. This unprecedented increase in cases therefore requires that immediate action needs to be taken as a countermeasure of the side effect of the outbreak (Fig. 2).

Mobility has become universally prohibited in a view to limit the spread of COVID-19. Physical contact too is considered as a heinous crime [45, 49]. As a result, people were forced to meet and chat with each other through virtual means like skype, zoom, webax and other convenient methods to interact and comfort each other instead of having physical meetings and get togethers. Healthcare professionals too have sought the assistance of technology in monitoring and controlling the infectious spread [21, 55, 56]. Their fight on COVID-19 has been propelled hugely thanks to IoT which has eliminated the risk of infection as humans get exposed to infected patients. IoT easily tracks the spread of the virus and pinpoints high-risk patients crucial in containing infections instantaneously. One advantage of IoT is that it forecasts the risks of death by sufficiently examining data previously provided by the patient. The main applications of IoT in the fight against COVID-19 are characterized as (1) early detection and diagnosis where inconsistent symptoms and other critical signs are communicated to health professionals to take action, (2) monitoring the treatment where IoT builds data platforms for automated predictions and tracking on a daily basis provides vital information on infected patients, (3) contact tracing of individuals where patients released into the communities are traced to identify who has contact with the patient, (4) projection of cases and mortality where IoT can monitor and predict the virus based on the information at hand and other social and mainstream media outlets, (5) reducing work hours for healthcare providers where information is gathered through digital modes by utilization of IoT devices resulting in less hours worked and limited exposure and contact with patients maintaining social distancing protocols and (6) prevention of the disease where availability of timely data IoT enables real-time information for prevention of the virus [61]. Meanwhile, in Australia, the risk of contraction is increasing by the day, forcing health professional to accept the use of virtual care models to monitor and track patients [6]. The applications of IoT can potentially boost the research in development of vaccines, prevention of infections, managing of infected people as well as prevention of virus contractions through development of smart devices such as smart social distancing glasses, cloths, etc. Therefore, given the current pandemic stage of the virus, this paper highlights the role of IoT in overcoming the previously mentioned problems. It is also projected that demand for IoT devices will climb to 10 percent in 2020 placing the number of devices at 718 million as Australia confronts COVID-19 [43]. In this regard, the current research has been carried out to explore

- the potential usage of IoT in combating the Pandemic
- various challenges in implementing IoT-enabled services
- the success stories of IoT in various areas

2 Problem Statement

With the continuous increase in the infection rate of the virus, there is also a desperate need to develop solutions towards containing the pandemic in a manner that can be easily managed. A need of technologies such as real-time information and data exchange is essential for getting a reliable vaccine more achievable in the shortest possible time. The practical implementation of technology-based remote patient monitoring system is also essential to safeguard the frontline medical personnel engaged in treating asymptomatic COVID-19. Further, monitoring and tracking exercise are required to furnish throat swabs, personal and contact particulars and travel details so that health authorities can track and trace those who have made contact with travellers displaying symptoms of COVID-19. This can only be achieved using bug data, cloud technology and Internet of Things for the purpose of faster data processing as well as for determining efficient and effective methodology of managing the current situation. The insertion of sensors and connection of care devices like smart watches, heart rate monitors and glucose monitoring devices and other life-saving devices emanating remote signals to technological gadgets of healthcare professionals to act speedily to save lives has transitioned health care from being a responsive medical-based system into a more hands-on wellness-based system [60].

3 Literature Review

Internet of Things (IoT) has revolutionized the way in which different organizations have conducted their activities. Consequently, their dependency on IoT has increased over time as industries look for better ways to cut costs and increase profits. Therefore, the end of 2020 will witness installation of more than 24 million IoT devices with estimated revenue of more than 300 billion dollars [60] for businesses globally. Particularly, infusing sensors into everyday usable items like TV remote controls and more sophisticated devices like smart offices designed to save energy costs by controlling the electricity or temperature has increased the level of dependency of businesses on IoT. The use of sensors has gained industry-wide popularity guaranteeing its wide deployment and adoption because of the sensors' significant qualities such as (1) low prices, (2) capability and (3) size.

The vast range applications of IoT are forecasted to transform ordinary everyday objects to valuable intelligent devices that disseminate vital data for effective decision making. IoT is deployed in various forms and devices solely aimed at offering users with real-time information that is crucial for all aspects of life. Therefore, IoT's deployment is evident in (1) wearables such as heart rate monitors for people with acute cardiac difficulties, smart watches for sports people and others to monitor their body's functionality and glucose monitoring devices for patients with diabetes, (2) smart home applications whereby energy usage and temperature levels are controlled to save energy costs, (3) healthcare application where patients' care devices are connected to a network that send signals to care givers' mobile devices, (4) smart cities applications where IoT is used in traffic lights controls, providing important data on cleanliness of water and air in the periphery of the city, for instance, in Palo Alto, California, USA, in order to reduce traffic congestion, sensors are installed in parking areas that alert motorist of occupancy status, (5) agricultural applications where smart greenhouse allows bountiful harvests by managing environmental factors like climate and plant requirements, (6) industrial application like automation where IoT enables faster developments of products and yields high returns on investment as factories become more digitalized, effective monitoring of product flows, better inventory management, assuring safety and security, efficient quality control, packaging optimization, and logistics and supply chain optimization [60]. Therefore, making industries more efficient supersedes production costs which lead to profits because organizations target cost reduction in all areas while maximizing profits. Hence, different industries have jumped on the IoT bandwagon because it provides (1) manufacturing efficiency by eliminating bottlenecks and minimizing wastage and time, (2) energy efficiency whereby sensors observe lightings, temperature and energy usage in a plant, (3) agricultural efficiency by scientifically detecting soil humidity, anticipating weather changes, ensuring smart irrigation systems to water plants and eliminating water wastages, (4) inventory efficiency by assigning radio frequency identification (RFID) tags on every single product designed to track its location and movements within the inventory management system right up to purchase by customers, (5) improving healthcare systems' efficiency by intensifying surveillance, monitoring, tracking, detection and prevention of viruses and diseases.

IoT defines a system of interconnected devices inclusive of hardware and software such as computers, digital and mechanical devices which are connected together through the Internet for the purpose of monitoring, control, data transmission, etc., which is meant to achieve certain set of objectives without necessary interference from humans at various levels. This has been successfully integrated in our everyday life that most people make use of IoT unknowingly. A simple task as controlling home security, lighting and cooling system amongst other is a very typical application of IoT [28, 42]. It is therefore needless to say that such interesting technology can be used in times of crisis as in the case of the present COVID-19. Several researches earnestly kicked off in the early stage of the pandemic and so much money was invested by government agencies as well as private organization in a bid to handle the situation. This led to some interesting applications of IoT from researchers of engineering and physical science background in an attempt to tackle the challenge by exploring untapped areas and developing new theories and concepts which adapted IoT.

One of the aforementioned developments in the application of IoT is readily seen in India. A group of researchers developed "ArogyaSetu" [51] which is a smartphone-based application to connect the people to the relevant medical services required; similarly, several applications have been developed to tell the proximity of a user to a positively identified patient [29]. Furthermore, Beark et al. [7] have previously developed a smart chair for patients' health monitoring. Their design involved IoT application in remote transfer of data relating to heart rate and blood pressure monitoring. Similar mobile-based applications have been used in scheduling hospital visits [27], fall detection of elderly [36], etc. Majorly, this IoT drive seems to be caused by the expenditure cost relating to medical services, need for solutions to cover up lapses, expanding growth of wireless connectivity and the overall requirement of precise and efficient healthcare targeted to take care of all individuals without bias [46].

Of recent, the expansion of cloud-based computing and analytics of big data have also seen an increased application in the fight against the pandemic. Nowadays, a massive amount of data could be collected with the application of smart technologies such as phones to gather data in real time. Mostly, this have rendered the older methodologies used extremely obsolete to the point that countries with limited knowledge and capacities have a hard time keeping up. This application of cloud computing and data analysis has been practically implemented by Taiwan. Their adopted model was based on the militarization of case identification as well as suppression and provision of resources. Their methodology was based on the use of big data in acquiring records of personnel immigration history which analysed the data to predict and provide real-time alerts to hospitals when anyone went for clinical visits based on the travel history of the individual and symptoms identified. This was aided by the use of QR codes and connected reports, etc. [63].

In addition to these recent applications, wireless in the medical field is not a new phenomenon. This technology has always been used in fatal electrocardiograms [58], monitoring of patients' breathing parameters [15] and condition in real time, monitoring of ICU patients [9], assistive identification of apnea, stethoscope applications, to mention a few [57]. Similarly, the use of wireless or cloud-based computing has been applied in data storage such as taking record of physiological conditions such as breathing rate, temperature, heart rate, blood pressure, etc. [35]. This has further been enhanced with the increasing development of nanotechnology which has been applied in the development of smart shirts [23] which incorporates an array of sensor networks for monitoring conditions such as heart rate, lung functions as well as other body parameters.

Going by further review of more recent works, we see that smart technologies, which are meant to be less intrusive, interconnected, dynamic and intelligent, are now moderately being developed by less technologically advanced countries as can be seen in smartphones, watches, devices, etc., which is capable of meeting the minimum requirement for effective usability. This therefore explains the predominant adoption of mobiles and smart watches in medical monitoring which in effect boosts the increasing interest in these smart devices. It is now easy to integrate these smart networks in medical applications such as interconnected healthcare services using sensors such as GPS receiver to provide real-time data on the location of users, vision systems for remote connections between medical services and users, and mobile sensors such as accelerometer, gyroscope, etc. could also provide valid data relating to a multi-faceted surrounding situation. More often, these mobile smartphones work in tandem with smart watches [16], smart glasses or other smart devices which can account for the lapses in gathering sufficient data by a single device (Table 1).

As seen in the table above, medical care mobile applications could be designed to support a wide range of functions for varying medical conditions. Of these applications, the major classifications are the apps which provide functions based on embedded sensors to monitor and keep track of health conditions such as heart rate, pulse, oxygen level, etc.; secondly, other apps are based on recommending activities to the users with respect to but not limited to health, nutrition, available stores, etc.; thirdly, other apps provide medical-related tips, advice and other related services in contrast to medical care applications which can be used to schedule clinical appointments or place order for drugs prescription. The potential application of IoT which is categorized under 3 sections as shown in Table 1 could be helpful for supporting disjointed branches of the medical system [59]. However, recent

Table 1 Sample functions of medical care application

Users function	Medical services function	System admin function
Medication compliance	Receive alerts	Resource and communication
Users condition monitoring	Connected medical reference	Sensors data acquisition
Remote care	Remote diagnostic	Health information extraction
Managing conditions history	Health record access	Cloud storage
Wellness/fitness tracking	Time management and scheduling	Adaptive visualization
	Communications and consultation	System management
	Real-time management and monitoring	Optional services (intelligent alerts, recommendations and guidelines, etc.)
	Related education and training	
[17, 19, 30]	[34, 40]	[5, 10]

occurrence has spurred the urgent need as a peer collaborative system working hand in hand to offer a robust model of medical care.

Looking at past emergency preparedness and response, the adoption of smart devices and big data analytics is now being used to interconnect laboratories working on possible vaccines as medical devices. Advancements also include the capability of these wearable sensors to measure cardiac rhythms, biometric data, etc., which increase the ability of these devices to predict or diagnose some existing conditions which may be present as well as for remote patient monitoring in the case of severe conditions which needs to be monitored in real time [54–56]. This potentially maximises the hopes of quickly developing and improving technologies [2]. There is also an overwhelming multitude of open source articles, codes and designs available on the Internet as part of individual's contribution. This has seen the use of thermal imaging, IoT and even ventilators development to assist in rural communities with limited manpower and resources. Although the implementation of thermal imaging is not self-sufficient in adequately detecting possible infection, the combination of this technology with artificial intelligence (AI) and IoT offers endless possibilities [8] provide that there are more accurate and comprehensive real-time information on outbreaks as well as the infection dispersion tracking. This will definitely make the risk management more effective and manageable (Table 2).

Table 2 Current IoT applications in fighting COVID-19 [51]

Methodology	Remark
Interconnected hospitals	This integrates various functions within the hospital facility
Emergency alerts	Allows for quick medical response when required
Treatment management	Unbiased treatment, method selections, etc.
Assignment scheduling	Assists medical personnel in management of duties and work functions
Remote monitoring	This enables remote patient in getting the possible best medical assistance through connected communications service as well as remote medical equipment control
Wireless identification	Application of mobile applications in identity authentication for secure access to facilities, that is, barcodes, etc.
Contact tracing	Applications of internet and GPS services in tracing patients.
Information sharing	Connected real-time and validated information sharing channels.
Rapid screening	Efficient and effective diagnosis through applications of smart devices.
Research	Application in vaccines research, monitoring, etc.
Connected hardware	Connections of medical equipment and other device for real-time information management.
Prediction	Uses statistical analysis to identify and predict future conditions for medical staffs, government agencies, data analyst as well as researchers.

4   Methodology

A subject advances when prior studies are synthesized logically based on the findings of prior studies [31]. Hence, this research is conducted for extracting and analysing literature pertinent to IoT application from various journals and other published articles. This could be helpful for both the practitioners and academicians in formulating a response tailored to the societal needs. Literature reviews, as a research methodology [53], contribute significantly for conceptual, methodological, and thematic development of different domains [24, 39].

In the preparation of this paper, the main source of information was collected from published journals from the database of SCOPUS, Google Scholar and Research Gate coupled with other sources from credible organizations' reports and blogs such as the world health organization, etc. The major key words used in the search were "Internet of Things (IoT)", "Coronavirus", "COVID-19", "Artificial Intelligence (AI)", "Pandemic", etc. Finally, a comparison of the methods applied were made based on information obtained from the journals.

5 Practical Applications of IoT in Combating COVID-19

Given that we are in a digital age surrounded by advancing technology, IoT has a lot of advantages to offer in the way of combating the pandemic. There is an increasing amount of data that needs to be analysed all ranging from immigration records, medical symptoms, transport, entertainment, health sector, etc., and in the case of COVID-19, such analysis becomes critical and needs to be executed without compromising efforts put in place to contain the virus spread. With a large number of IoT devices now in circulation, there is a possibility of a large number of interconnected devices which can be used to form proper management systems in handling the virus effect as well as for making proper decision. Some of the possible applications are briefly summarized below.

5.1 Prediction and Spread Prevention

Early enough in the initial phase of the virus spread from Wuhan, China, smart hospitals were integrated such that they provided additional functions such as the combination of AI and IoT to assist in monitoring incoming patients' conditions such as temperature, oxygen levels, etc. [12]. Other robots were designed to prevent contact between the infected patients and the medical personnel. These robots were tasked with essential services such as in medicine and food delivery, etc. This function highlights the enhanced accessibility and availability towards information related to healthcare system management such that patient's medical information and images, etc. were possible to be stored in the cloud for unobstructed access by designated medical personnel attending to the patient from mobile smart platforms such as phone, etc.

Furthermore, the use of contact tracing applications has seen huge adaptation in most western countries which are mostly hit with the pandemic such as the United States. Contact tracing applications in this case refers to mobile applications that are developed with the sole purpose of tracing and identifying individuals who have been exposed to an infected individual at any point in time while taking into consideration the virus incubation period. This app aims to reduce the data processing cycle and identify connectivity between apps supporting healthcare system. They are also capable of warning, tracing and quickly identifying all persons who have come in contact with an infected patient at one point of another. In less developed countries, most of the tracing is done manually and a major issue with this system is that most of the time, the infected patient may not remember everyone they have been in contact with over a period of time. This application is therefore useful in identifying infection hotspots, isolation setup and diagnosis. The process involved in tracing contacts is initial identification of close contact with and infected person, detailed information of people in contact and finally setting up testing of those individuals as quickly as possible.

Additionally, geo-fencing, which is an existing technology [62], has been previously used in monitoring prisoners [38]. This system combines the use of IoT with wireless network as well as a geographical fence to reduce the number of agents monitoring a group of isolated people. In practical relation to COVID-19 control, when abnormal traits are noticed within an isolated group, it becomes easy to mobilize for transfer and immediate treatment; these are additional steps that can be implemented to prevent virus spread.

Recent data on coronavirus show that there are some cases where patients who are infected might be asymptomatic. This makes early detection more difficult. Moreover, almost eighty percent of infected individuals recover from the virus without any form of medical intervention or specific treatment [64]. However, for the older population, there is a higher risk of the occurrence of complications. To prevent this, it is imperative that alternative and more efficient and effective systems for early detection are implemented. IoT can be combined together with other equipment such as thermal imaging and optical cameras which gathers data for remote post processing which can help in rapid identification of infected patients.

5.2 Treatment

Due to the successful reduction in the total cases in China, other countries have now adopted the use of IoT, thereby bringing it to the forefront of the pandemic fight. According to Forrester's analysis [25], the application of IoT in the Asian healthcare system prior to the virus was estimated to be 7%, and this number has sharply risen with the mandatory implementation of smart systems for measuring people's vital signals such as temperatures in most open spaces and business facilities such as restaurants, train stations, airports, etc. Most of this setup are often analysed in real time.

Given the increasing rate of infection in other regions, prevention and treatment become more strained. This is where IoT comes to play. Technologies such as real-time information and data exchange make the prospect of getting a reliable vaccine more achievable in the shortest possible time. The practical implementation of remote patient monitoring has also assisted frontline medical personnel in treating asymptomatic COVID-19 patients remotely. This has tremendously helped in decongesting medical facilities for more critical cases. This technology implements IoT in monitoring patient's health conditions using wireless sensors [50].

In the process of developing suitable vaccines in fighting the pandemic, AI and IoT are combined to aid the research effort such that possible outcomes can be predicted before trial commences. Similarly, this technology is applied in designing efficient drug delivery system which exponentially increases the drugs testing phase in real time as compared with the standard methodologies previously implemented [13, 22]. This study also studies the reaction to certain drugs based on patient's behaviour and body response [52]. Finally, once a vaccine has been discovered, there will be significant pressure due to the large demand for the vaccine. This will present a problem at that stage, and hence, there will be need for accelerated mass production of vaccines. Thankfully we are closer to the reality of a commercial vaccine due to advancement in physical, biological and chemical modelling of data toward computer-aided research for potential drugs, and for efficient delivery, AI and machine learning techniques can be deployed for a predictive and prescriptive analysis of the vaccine production. Now scientists and engineers can monitor production parameters through specific location of sensors throughout the whole production process for real-time analysis in ensuring constant research and development, quality control as well as an optimized product [44].

Similarly, in COVID-19 treatment, IoT is currently playing a big role by utilizing AI algorithms as well as interconnecting computing power in others to increase research ability. This can be practically seen in Googles DeepMind which is a platform where findings related to treatments of COVID-19 are constantly being published. Similarly, Benevolent AI is an artificial intelligence system which is purposely set up for the treatment of severe cases of diseases. It is used in building suitable drugs that aid the ongoing search for a vaccine [47]. Furthermore, robots are now being deployed in hospitals which assist medical personnel in the treatment of infected patients. This relieves the overworked staffs as the cases continue to surge.

The use of IoT along with Cloud and AI are leading the way to combat the crisis especially in the United States, Britain, Germany and elsewhere. With the help of global information systems (GIS) on IoT-assisted epidemiologist to track and monitor high-risk infected patients on their mobile data usage which helped in the process of identifying people they have come into contact with and allowing healthcare professionals to adequately act to contain the spread of the deadly virus [20]. IoT was infused into wearables, electrical gadgets and such as smart suits, IoT buttons and electrical thermometers. Smart suits are designed to monitor the human body and signal the patient of changes in body temperature exceeding 37.5 degrees while also analysing the movements of the body. IoT buttons provide advice of any cleaning or maintenance needs that pose grave danger of people contracting the virus as they come into contact with surfaces handled by patients. Electrical thermometers are connected to energy sources which are employed in checking and screening patients and staff for any deviations in temperature as they enter and exit quarantine and hospital facilities [20].

5.3 Direction and Prospect of IoT

Continuous advancements in technology relating to wireless technology, Internet of Things and artificial intelligence have led to the emergence of an hyper-connected world which have attracted practitioners and researchers to develop ways for the practical implementation of these technologies in healthcare services [55, 56]. As of now, the current IoT implementation in the medical field is still at a young stage with promising future prospect if fully developed. The use of this technology however should be based on the main aim of preventing, treating and research towards human health. This would require more adaptation to other technologies such as machine learning, computer vision for a broader application as well as the development of more secure communication and consultation platform. Finally, future prospect should yield an effective system with a robust quality control across all the nodes of the health sector. Future solutions should also address the issue of interoperability of devices as well as security of user's data.

The future of IoT in health care looks more promising with potentially new frontiers to be discovered in the interaction and connectivity of medical professionals, technology and devices purposed to make the healthcare system smarter and even more successful by ensuring that medical devices are all connected to a network, tracking and monitoring of patients progress, video- and ID-enabled security systems, tagging and tracking of vital assets and organs and effective maintenance of essential equipment done without hampering medical care are the trends in healthcare attributed to IoT. However, certain setbacks have thrown caution in the wind where healthcare administrators are working round the clock to ensure that the extraordinary depth of data available, hurdles of managing and operating network infrastructures and heighted security problems associated with the unprecedented charter of information delay effective implementation of IoT [26].

In the treatment of COVID-19, there have been extensive development and experience gained from the application of Augmented Reality (AR) technology in the study and management of the lung's nodules. This is the integration of virtual information with IoT and can utilize a variety of data streams such as three-dimensional models, registration details, real-time tracking, etc. in the diagnosis and treatment models which are applied towards the treatment of pandemic-infected patients [32, 33]. Additionally, since the COVID-19 is a viral infection affecting the lung and leading to Acute Respiratory Distress Syndrome (ARDS), this means that addressing the symptoms as quickly as possible increases survival rate. Therefore, IoT systems can be tasked with early support for treatments and monitoring of patients which would relieve pressure of medical staffs.

6 IoT—Challenges and Opportunities

While the use of data is crucial to the success of the IoT system, there are a lot of concerns about data management such as the way these data are being collected and managed in terms of storage, processing and accessibility by selected users [2–4]. Needless to say, these concerns are not baseless and must be adequately addressed going by previous experiences with security breach and selling of user's data to government organizations without permission. One major issue regarding data access is the political factor. This is due to the fact that data have now become a gold rush given its potential in influencing geopolitical standings in both government and private corporations. Hence, there is a rush to control this flow of data. This is very obvious in the recent fallout between the United States, United Kingdom and China in the planned implementation of 5G networks in the country's infrastructure which have generated huge pushback pioneered by the United States [14].

Similarly, this has pitted the famous mobile application TikTok in the middle of a battle between countries due to political reason. Although the Internet services such as the 5G network offers a vast opportunity presented in the form of increased Internet speed, this therefore means more handling of user's data. This political quest to control data will clearly affect the adaptation of interconnected IoT device at an international level. Furthermore, some countries tend to be more critical of others in terms of approach and methodology of handling the pandemic rather than focussing on the productive application of technology in containing the spread. A practical example of this is the obvious disparity in the COVID-19 data of both the United States and China. The virus originated from China with a population of 1.34 billion residents. Common logic would expect that the infection and death rate should be extremely high given the condition of living of its residents as well as it being an epicentre of a novel virus which means it took some time to identify and formally declare emergency. On the other hand, the United States with a mere fractional population of 311 million when compared with China's population accounts for the highest infection and death rate in the world despite the fact that they had enough time to put in place preventive measures. Of course, there might be other factors which accounts for these disparities, we simply cannot ignore the fact that the adoption of these aforementioned technologies had an important role to play in containing the pandemic in China. Due to this geopolitical fight for data control, it is obvious that certain economies will push for their dominant control of critical information which will in turn slow down the international effort geared at controlling the pandemic and developing vaccines if critical information regarding COVID-19 is not freely shared.

Additionally, in implementing a smart interconnected mobile health system, this will require the use of some custom sensors which might be embedded for adaptability. These systems must remain interconnected at all time to ensure a robust and intelligently functioning system [59]. The main issue here is the fact that mobile sensors and applications can only provide limited amount of information. For more critical cases as in remote monitoring of patients, the use of mobile applications to measure parameters such as heart rate, pressure, etc. might not be feasible or entirely inadequate. Also, different healthcare facilities are very much against the idea of data sharing or interconnected scheduling. Furthermore, it becomes a burden for physically disabled individuals to utilize some of these IoT systems set up for monitoring; also, given that some health monitoring applications employ the use of short surveys in determining the well-being of users, this is subject to misinformation being that sometimes, it is not convenient or the users are not in a right mind frame when filling those details [41].

Developed countries such as New Zealand and Australia are also facing challenges that thwarted their best efforts in mitigating the impacts of COVID-19 as the number of new cases surges. COVID-19 has introduced a new trend to business vocabulary which is "stay-at-home-economy" where more businesses and people are taking advantage of social distancing limitations by conducting business while at home. This new trend has caused a surge in the shipment of IoT devices to 718 million units in 2020 alone from China despite the anxiety and threat of COVID-19. Other similar technologies on high demand are smart personal audio devices which rose to 15.5 percent and wearable bands to 3.8 percent in 2020 alone [43]. These surges in the demand for IoT gadgets will stress the current technological capability and burdening the existing infrastructure as more people are using IoT devices. This will certainly pose monumental challenges like (1) unreliability of the network as more people are connected using more than one IoT devices, (2) huge number of data flowing at any one time making analysis and interpretation difficult reducing real-time application, (3) use of multiple devices creates confusion in tracking, detecting and preventing the rapid spread of COVID-19 and (4) elimination of the human component of the healthcare system affects patients who have the virus.

Irrespective of these issues, there are lots of opportunities which can be leveraged on if the adaptation of IoT is successfully implemented in the fight against COVID-19. Technologies such as cloud computing where data can be processed and stored online offers a unique advantage. This is because it does not require a lot of effort to set up or maintain and IoT enabled health services could be hosted on platforms like this. Also, with the introduction of 5G networks, this greatly increases the speed at which information can be processed online and improves the efficiency of remote patient monitoring as well as other medical services such as contact tracing, testing and even treatment.

7 Conclusion

The advantages of IoT in saving time and money make IoT unavoidable as organizations prioritize their efforts to ensure that effective adoption of IoT is vital for organizations. Therefore, IoT cannot be ignored nor dispensed by organizations but incorporated into their activities to minimize costs and maximize revenues. The long awaited overhaul of the healthcare system through the integration of IoT has taken place as healthcare professionals seek new ways of incorporating IoT in their practices by building infrastructure that can enable implementation of IoT and health care. IoT facilitates connection of medical equipment with Internet and aids in collection and sharing of important information regarding health condition of patients, their life style and treatment process [56]. As a result, vital patients' needs are communicated through a network to the devices of health professionals, which allows them to offer customized services to patients' special needs. IoT has become pivotal in the fight against COVID-19 as governments and health professionals attempt to curb the steady increase of the deadly virus in the efforts to detect, control and predict the spread of COVID-19. By providing an integrated network for healthcare professionals by maintaining connectivity with devices during the critical situation with the virus, vital information is passed on to health professionals which enables them to act adequately in their pursuit of provided relief and solution to this deadly enemy.

Although smart technologies and IoT have been introduced to the medical sector, there are still several ways that this methodology has not been adequately explored for better services. Drawing from the contents of the preceding sections, it is obvious that IoT is capable of delivering an extensive interconnected system which can be optimally used in tackling the pandemic. This application ranges from infection identification, tracing of infected individuals, management of hospital and healthcare services as well as inter-health service communications for management and research purpose. Though major cities around the world were prepared for potential outbreak of COVID-19, it was observed that these cities had varying policies pertaining to the pandemic control, which led to eventual collapse of the approved protocol that could potentially be avoided if there was a global and unified approach towards research and treatment of the virus. The sharing of data in critical situations will enable a proactive approach in identifying and setting up better measures to combat health-related issues.

This study has highlighted the potential applications as well as the challenges in the implementation of IoT. It is important to note that an unobstructed service is key for the successful implementation of IoT in the fight to contain the virus. Additionally, due to the successful applications in western countries, this model of service can be applied to obtain similar improvement.

8   Limitations and Scope of Future Work

Various challenges of IoT adoption in healthcare practices have been highlighted in this study. The current work doesn't reveal whether the challenges encountered during IoT implementation have equal impact on it or not. In future studies, multi-criteria decision making (MCDM) tools such as analytic hierarchy process (AHP) could be employed to rank these challenges. This would be helpful for organizations interested in IoT integration in their operational practices to rationale their resources [37].

References

1. Alberti, P., Beretta, S., Piatti, M., Karantzoulis, A., Piatti, M. L., Santoro, P., ... Ferrarese, C. (2020). Guillain-Barré syndrome related to COVID-19 infection. *Neurology: Neuroimmunology and NeuroInflammation, 7*(4), e741. https://doi.org/10.1212/NXI.0000000000000741.
2. Allam, Z. (2019a). Cities and the digital revolution: Aligning technology and humanity. In *Cities and the digital revolution: Aligning technology and humanity*. https://doi.org/10.1007/978-3-030-29800-5.
3. Allam, Z. (2019b). The emergence of anti-privacy and control at the nexus between the concepts of safe city and smart city. *Smart Cities*. https://doi.org/10.3390/smartcities2010007.

4. Allam, Z., & Dhunny, Z. A. (2019). On big data, artificial intelligence and smart cities. *Cities*. https://doi.org/10.1016/j.cities.2019.01.032.

5. Amin, M. B., Banos, O., Khan, W. A., Bilal, H. S. M., Gong, J., Bui, D. M., & Lee, S. (2016). On curating multimodal sensory data for health and wellness platforms. *Sensors (Switzerland)*. https://doi.org/10.3390/s16070980.

6. Angus, D., Connolly, M., & Salita, M. (2020). The shift to virtual care in response to COVID 19. *PwC Report*.

7. Baek, H. J., Chung, G. S., Kim, K. K., & Park, K. S. (2012). A smart health monitoring chair for nonintrusive measurement of biological signals. *IEEE Transactions on Information Technology in Biomedicine*. https://doi.org/10.1109/TITB.2011.2175742.

8. Banerjee, A., Chakraborty, C., Kumar, A., & Biswas, D. (2020). Emerging trends in IoT and big data analytics for biomedical and health care technologies. In *Handbook of data science approaches for biomedical engineering*. https://doi.org/10.1016/b978-0-12-818318-2.00005-2.

9. Basilotta, F., Riario, S., Stradolini, F., Taurino, I., Demarchi, D., De Micheli, G., & Carrara, S. (2015). Wireless monitoring in intensive care units by a 3D-printed system with embedded electronic. In *IEEE biomedical circuits and systems conference: engineering for healthy minds and able bodies, BioCAS 2015 – Proceedings*. https://doi.org/10.1109/BioCAS.2015.7348311.

10. Bassi, A., Arfin, S., John, O., & Jha, V. (2020). An overview of mobile applications (apps) to support the coronavirus disease-2019 response in India. *Indian Journal of Medical Research, 151*(5), 468–473. https://doi.org/10.4103/ijmr.ijmr_1200_20.

11. Boopathi, S., Poma, A. B., & Kolandaivel, P. (2020). Novel 2019 coronavirus structure, mechanism of action, antiviral drug promises and rule out against its treatment. *Journal of Biomolecular Structure & Dynamics, 39*(9), 3409–3418. https://doi.org/10.1080/07391102.2020.1758788.

12. Capolongo, S., Rebecchi, A., Buffoli, M., Appolloni, L., Signorelli, C., Fara, G. M., & D'Alessandro, D. (2020). COVID-19 and cities: From urban health strategies to the pandemic challenge. A Decalogue of public health opportunities. *Acta Bio-Medica: Atenei Parmensis, 91*(2), 13–22. https://doi.org/10.23750/abm.v91i2.9615.

13. Cascella, M., Rajnik, M., Cuomo, A., Dulebohn, S. C., & Di Napoli, R. (2020). *Features, evaluation and treatment coronavirus (COVID-19)*. StatPearls.

14. Cassell, B.-L., & Packham, C. (2019). *How Australia led the US in its global war against Huawei*. The Sydney Morning Herald.

15. Cesareo, A., Previtali, Y., Biffi, E., & Aliverti, A. (2019). Assessment of breathing parameters using an inertial measurement unit (IMU)-based system. *Sensors (Switzerland), 19*(1), 88. https://doi.org/10.3390/s19010088.

16. Chen, T. (2014). A fuzzy parallel processing scheme for enhancing the effectiveness of a dynamic just-in-time location-aware service system. *Entropy, 16*(4), 2001–2022. https://doi.org/10.3390/e16042001.

17. Chen, T. (2016). Ubiquitous multicriteria clinic recommendation system. *Journal of Medical Systems, 40*(5), 113. https://doi.org/10.1007/s10916-016-0469-6.

18. Cheval, S., Adamescu, C. M., Georgiadis, T., Herrnegger, M., Piticar, A., & Legates, D. R. (2020). Observed and potential impacts of the covid-19 pandemic on the environment. *International Journal of Environmental Research and Public Health, 17*(11), 4140. https://doi.org/10.3390/ijerph17114140.

19. Chiou, W. K., Kao, C. Y., Lo, L. M., Huang, D. H., Wang, M. H., & Chen, B. H. (2017). Feasibility of utilizing E-mental health with mobile app interface for social support enhancement: A conceptional solution for postpartum depression in Taiwan. *Lecture Notes in Computer Science (Including Subseries Lecture Notes in Artificial Intelligence and Lecture Notes in Bioinformatics)*. https://doi.org/10.1007/978-3-319-58637-3_15.

20. Choudhary, M. (2020). *How IoT can help fight the COVID 19 battle.* https://www.geospatialworld.net/blogs/how-iot-can-help-fight

21. Dash, M., Shadangi, P. Y., Muduli, K., Luhach, A. K., & Mohamed, A. (2021). Predicting the motivators of telemedicine acceptance in COVID-19 pandemic using multiple regression and ANN approach. *Journal of Statistics and Management Systems, 24*(2), 319–339.

22. Haleem, A., Vaishya, R., Javaid, M., & Khan, I. H. (2020). Artificial intelligence (AI) applications in orthopaedics: An innovative technology to embrace. *Journal of Clinical Orthopaedics and Trauma, 11*, S80–S81. https://doi.org/10.1016/j.jcot.2019.06.012.

23. Harifi, T., & Montazer, M. (2017). Application of nanotechnology in sports clothing and flooring for enhanced sport activities, performance, efficiency and comfort: A review. *Journal of Industrial Textiles, 46*(5), 1147–1169. https://doi.org/10.1177/1528083715601512.

24. Hulland, J., & Houston, M. B. (2020). Why systematic review papers and meta-analyses matter: An introduction to the special issue on generalizations in marketing. *Journal of the Academy of Marketing Science, 48*, 351–359.

25. Home · Forrester. (n.d.). Retrieved July 13, 2020, from https://go.forrester.com/

26. IoT in Healthcare. (2018). *Build a secure foundation to leverage IoT solutions and optimize the care pathways.* Alcatel Lucent: ALE.

27. Kaewkungwal, J., Singhasivanon, P., Khamsiriwatchara, A., Sawang, S., Meankaew, P., & Wechsart, A. (2010). Application of smart phone in "better border healthcare program": A module for mother and child care. *BMC Medical Informatics and Decision Making, 10*(1), 1–2.

28. Jia, M., Komeily, A., Wang, Y., & Srinivasan, R. S. (2019). Adopting internet of things for the development of smart buildings: A review of enabling technologies and applications. *Automation in Construction, 101*, 111–126. https://doi.org/10.1016/j.autcon.2019.01.023.

29. Keeling, M. J., Hollingsworth, T. D., & Read, J. M. (2020). Efficacy of contact tracing for the containment of the 2019 novel coronavirus (COVID-19). *Journal of Epidemiology and Community Health, 74*(10), 861–866. https://doi.org/10.1136/jech-2020-214051.

30. Kumar, D., Hemmige, V., Kallen, M. A., Giordano, T. P., & Arya, M. (2019a). Mobile phones may not bridge the digital divide: A look at mobile phone literacy in an underserved patient population. *Cureus, 11*(2), e4104. https://doi.org/10.7759/cureus.4104.

31. Kumar, A., Paul, J., & Unnithan, A. B. (2019b). 'Masstige' marketing: A review, synthesis and research agenda. *Journal of Business Research, 113*, 384–398. https://doi.org/10.1016/j.jbusres.2019.09.030.

32. Le, V., Yang, D., Zhu, Y., Zheng, B., Bai, C., Nguyen, Q., & Shi, H. (2017). Automated classification of pulmonary nodules for lung adenocarcinomas risk evaluation: An effective CT analysis by clustering density distribution algorithm. *Journal of Medical Imaging and Health Informatics, 7*(8), 1753–1758. https://doi.org/10.1166/jmihi.2017.2259.

33. Le, V., Yang, D., Zhu, Y., Zheng, B., Bai, C., Shi, H., . . . Lu, S. (2018). Quantitative CT analysis of pulmonary nodules for lung adenocarcinoma risk classification based on an exponential weighted grey scale angular density distribution feature. *Computer Methods and Programs in Biomedicine, 160*, 141–151. https://doi.org/10.1016/j.cmpb.2018.04.001.

34. Moazzami, B., Razavi-Khorasani, N., Dooghaie Moghadam, A., Farokhi, E., & Rezaei, N. (2020). COVID-19 and telemedicine: Immediate action required for maintaining healthcare providers well-being. *Journal of Clinical Virology, 126*, 104345. https://doi.org/10.1016/j.jcv.2020.104345.

35. Mora, H., Gil, D., Terol, R. M., Azorín, J., & Szymanski, J. (2017). An IoT-based computational framework for healthcare monitoring in mobile environments. *Sensors (Switzerland), 17*(10), 2302. https://doi.org/10.3390/s17102302.

36. Mutiara, G. A., Hapsari, G. I., & Rijalul, R. (2016). Smart guide extension for blind cane. In *2016 4th international conference on information and communication technology, ICoICT 2016.* https://doi.org/10.1109/ICoICT.2016.7571896.

37. Muduli, K., & Barve, A. (2013). Developing a framework for study of GSCM criteria in Indian mining industries. *APCBEE Procedia, 5*, 22–26.

38. Nandyal, S., & Mugali, S. (2018). An automated system for identification of fire and behaviour of prisoners through posture using iot. In *Proceedings of the 2nd international conference on green computing and internet of things, ICGCIoT 2018.* https://doi.org/10.1109/ICGCIoT.2018.8753041.

39. Palmatier, R. W., Houston, M. B., & Hulland, J. (2018). Review articles: Purpose, process, and structure. *Journal of Academy of Marketing Science, 46,* 1–5. https://doi.org/10.1007/s11747-017-0563-4.

40. Parker, M. J., Fraser, C., Abeler-Dörner, L., & Bonsall, D. (2020). Ethics of instantaneous contract tracing using mobile phone apps in the control of the COVID-19 pandemic. *Journal of Medical Ethics, 46*(7), 427–431. https://doi.org/10.1136/medethics-2020-106314.

41. Porzi, L., Messelodi, S., Modena, C. M., & Ricci, E. (2013). A smart watch-based gesture recognition system for assisting people with visual impairments. In *IMMPD 2013 – Proceedings of the 3rd ACM international workshop on interactive multimedia on mobile and portable devices, co-located with ACM multimedia 2013.* https://doi.org/10.1145/2505483.2505487.

42. Pramanik, P. K. D., Pal, S., & Choudhury, P. (2018). *Beyond.* Automation: The Cognitive IoT. Artificial Intelligence Brings Sense to the Internet of Things. https://doi.org/10.1007/978-3-319-70688-7_1.

43. Remote Worker Tech. (2020). IoT device shipments to grow 9.8% despite COVID-19 uncertainty. *Remote Worker Tech.*

44. Ren, S., Zhang, Y., Liu, Y., Sakao, T., Huisingh, D., & Almeida, C. M. V. B. (2019). A comprehensive review of big data analytics throughout product lifecycle to support sustainable smart manufacturing: A framework, challenges and future research directions. *Journal of Cleaner Production, 210,* 1343–1365. https://doi.org/10.1016/j.jclepro.2018.11.025.

45. Sahoo, K. K., Muduli, K. K., Luhach, A. K., & Poonia, R. C. (2021). Pandemic COVID-19: An empirical analysis of impact on Indian higher education system. *Journal of Statistics and Management Systems, 24*(2), 341–355.

46. Santos-Peyret, A., Durón, R. M., Sebastián-Díaz, M. A., Crail-Meléndez, D., Gómez-Ventura, S., Briceño-González, E., ... Martínez-Juárez, I. E. (2020). E-health tools to overcome the gap in epilepsy care before, during and after COVID-19 pandemics. *Revista de Neurologia, 70*(9), 323–328. https://doi.org/10.33588/RN.7009.2020173.

47. Santosh, K. C. (2020). AI-driven tools for coronavirus outbreak: Need of active learning and cross-population train/test models on multitudinal/multimodal data. *Journal of Medical Systems, 44*(5), 1–5. https://doi.org/10.1007/s10916-020-01562-1.

48. Scott, B. K., Miller, G. T., Fonda, S. J., Yeaw, R. E., Gaudaen, J. C., Pavliscsak, H. H., ... Pamplin, J. C. (2020). Advanced digital health technologies for COVID-19 and future emergencies. *Telemedicine and E-Health, 26*(10), 1226–1233. https://doi.org/10.1089/tmj.2020.0140.

49. Shanker, S., Barve, A., Muduli, K., Kumar, A., Garza-Reyes, J. A., & Joshi, S. (2021). Enhancing resiliency of perishable product supply chains in the context of the COVID-19 outbreak. *International Journal of Logistics Research and Applications, 25,* 1–25.

50. Seshadri, D. R., Davies, E. V., Harlow, E. R., Hsu, J. J., Knighton, S. C., Walker, T. A., ... Drummond, C. K. (2020). Wearable sensors for COVID-19: A call to action to harness our digital infrastructure for remote patient monitoring and virtual assessments. *Frontiers in Digital Health, 2,* 8. https://doi.org/10.3389/fdgth.2020.00008.

51. Singh, R. P., Javaid, M., Haleem, A., & Suman, R. (2020). Internet of things (IoT) applications to fight against COVID-19 pandemic. *Diabetes and Metabolic Syndrome: Clinical Research and Reviews, 14*(4), 521–524. https://doi.org/10.1016/j.dsx.2020.04.041.

52. Sohrabi, C., Alsafi, Z., O'Neill, N., Khan, M., Kerwan, A., Al-Jabir, A., ... Agha, R. (2020). World Health Organization declares global emergency: A review of the 2019 novel coronavirus (COVID-19). *International Journal of Surgery, 76,* 71–76. https://doi.org/10.1016/j.ijsu.2020.02.034.

53. Snyder, H. (2019). Literature review as a research methodology: An overview and guidelines. *Journal of Business Research, 104,* 333–339.

54. Swain, S., Peter, O., Adimuthu, R., & Muduli, K. (2021a). 9 Blockchain Technology for Limiting the Impact of Pandemic. Computational Modeling and Data Analysis in COVID-19 Research, 165.
55. Swain, S., Peter, O., & Muduli, K. (2021b). Intelligent technologies for excellency in sustainable operational performance in health care sector. *International Journal of Social Ecology and Sustainable Development, 14*(6), 64–77.
56. Swain, S., Muduli, K., Kommula, V. P., & Sahoo, K. K. (2021c). Innovations in internet of medical things, artificial intelligence and readiness of health care sector towards health 4.0 adoption. *International Journal of Social Ecology and Sustainable Development, 15*(1).
57. Szot, S., Levin, A., Ragazzi, A., & Ning, T. (2019). A wireless digital stethoscope design. *International Conference on Signal Processing Proceedings, ICSP.* https://doi.org/10.1109/ICSP.2018.8652475.
58. Traikov, L., Sabit, Z., Bogdanov, T., Bakalov, D., Hadjiolova, R., Petrova, J., & Ushiama, A. (2019). Animal models based on telemetry for investigation of mechanisms of neurological and cardiovascular diseases (in vivo). *Physica Medica, 68,* 155–156. https://doi.org/10.1016/j.ejmp.2019.10.005.
59. Trescot, A. M. (2016). Peripheral nerve entrapments: Clinical diagnosis and management. In *Peripheral nerve entrapments: Clinical diagnosis and management.* https://doi.org/10.1007/978-3-319-27482-9.
60. Upsana. (2020). *Real World IoT application in different domains.* http://IoT Applications | Internet Of Things Examples | Real World IoT Examples | Edureka.
61. Vaishyu, R., Javaid, M., Khan, I. H., & Haleem, A. (2020). Artificial intelligence applications of COVID 19 pandemic. *Diabetes and Metabolic Syndrome: Clinical Research and Reviews, 14*(2020), 337–339.
62. Victor, F., & Zickau, S. (2019). Geofences on the blockchain: Enabling decentralized location-based services. In *IEEE International Conference on Data Mining Workshops, ICDMW* (pp. 97–104). IEEE. https://doi.org/10.1109/ICDMW.2018.00021.
63. Wang, C. J., Ng, C. Y., & Brook, R. H. (2020a). Response to COVID-19 in Taiwan: Big data analytics, new technology, and proactive testing. *JAMA – Journal of the American Medical Association, 323*(14), 1341–1342. https://doi.org/10.1001/jama.2020.3151.
64. Wang, Y., Wang, Y., Chen, Y., & Qin, Q. (2020b). Unique epidemiological and clinical features of the emerging 2019 novel coronavirus pneumonia (COVID-19) implicate special control measures. *Journal of Medical Virology, 92*(6), 568–576. https://doi.org/10.1002/jmv.25748.
65. WHO Coronavirus Disease (COVID-19) Dashboard | WHO Coronavirus Disease (COVID-19) Dashboard. (2020). Retrieved July 11, 2020, from https://covid19.who.int/

Machine Learning Approaches for COVID-19 Pandemic

Charles Oluwaseun Adetunji ⓘ, Olugbemi Tope Olaniyan,
Olorunsola Adeyomoye, Ayobami Dare, Mayowa J. Adeniyi, Enoch Alex,
Maksim Rebezov, Ekaterina Petukhova, and Mohammad Ali Shariati

1 Introduction

The outbreak of SARs-CoV-2 virus that caused COVID-19 disease, in 2019, has continued to exert a heavy financial toll and socioeconomic challenge across the

C. O. Adetunji (✉)
Applied Microbiology, Biotechnology and Nanotechnology Laboratory, Department of
Microbiology, Edo University Iyamho, Auchi, Edo State, Nigeria
e-mail: adetunji.charles@edouniversity.edu.ng

O. T. Olaniyan
Laboratory for Reproductive Biology and Developmental Programming, Department of
Physiology, Edo University Iyamho, Iyamho, Nigeria

O. Adeyomoye
Department of Physiology, University of Medical Sciences, Ondo City, Nigeria

A. Dare
Department of Physiology, School of Laboratory Medicine and Medical Sciences, College of
Health Sciences, Westville Campus, University of KwaZulu-Natal, Durban, South Africa

M. J. Adeniyi
Department of Physiology, Edo State University Uzairue, Iyamho, Edo State, Nigeria

E. Alex
Department of Human Physiology, Ahmadu Bello University Zaria, Kaduna State, Nigeria

M. Rebezov
Prokhorov General Physics Institute, Russian Academy of Sciences, Moscow, Russia

K.G. Razumovsky Moscow State University of Technologies and Management (the First Cossack
University), Moscow, Russia

E. Petukhova · M. A. Shariati
K.G. Razumovsky Moscow State University of Technologies and Management (the First Cossack
University), Moscow, Russia

S. K. Pani et al. (eds.), *Assessing COVID-19 and Other Pandemics and Epidemics using
Computational Modelling and Data Analysis*,
https://doi.org/10.1007/978-3-030-79753-9_8

133

globe. This disease has spread to across many countries and responsible for substantial number of morbidity and mortality, especially in patients with comorbidities [1]. This pandemic is progressively causing serious setback to the healthcare system in different regards including rapid increase in in-hospital patients, exceeding the available medical facilities, as well as increased exposure and risk among healthcare providers [2]. Also, adequate diagnosis of this disease, especially in developing countries, using reverse transcriptase polymerase chain reaction (RT-PCR), is limited due to shortage of equipment, contributing to poor diagnosis, high prevalence, with poor preventive and treatment guidelines. These setbacks facilitate the need to reduce prevalence and formulation of policies that promote effective treatment in order to reduce the menace of this disease. Thus, there is a need for proper screening to promote rapid and adequate diagnosis of COVID-19 disease, which will help to decrease the harmful influence of this disease on the healthcare system. The routine diagnosis with RT-PCR test is time-consuming, sometimes requiring up to 2 days to complete; there is likelihood of false-negative result, thereby requiring serial testing for confirmation as well as insufficient supply of PCR test kits, suggesting the imperative need to incorporate substitute procedures for quick and precise identification of this disease. Computed tomography (CT) scan of the chest has been used to assess patients with mild infection. However, this procedure alone may not provide confirmatory result especially in the early stage of infection, where patients present with normal radiological findings. To improve diagnosis, methods such as machine learning that can utilize the large amount of data that has been generated from several laboratory tests, clinical symptoms, and radiological scans will be of great benefits by improving screening and facilitating rapid diagnosis on a large scale which may not be achieved via physical testing [3]. Machine learning technique can utilize the available quantitative and qualitative data to predict infected patients [4, 5]. The result obtained from this technique can assist the physician in taking adequate decision during diagnosis and treatment with subsequent improvement in combating the pandemic.

General Overview of COVID-19

Since the first outbreak of SARs-CoV-2 in 2019, the virus has spread all around the world and has been regarded as a global issue of public health concern. Therefore, World Health Organization declared COVID-19 disease caused by SAR-CoV-2 as a pandemic [6, 7]. The major transmission of this virus is via direct contact with infected person or through respiratory droplets [8], while the likelihood of airborne transmission of the disease has been reported under certain conditions like continuous exposure in an air-tight space with poor ventilation [9, 10]. The virus spread rapidly across the world with >72 million active cases and over 1 million mortalities as on December 2020. This data may have underestimated the exact infection cases because some countries have decided not to report their statistics, while in those countries that reported, asymptomatic patients may not be tested and reduced efficiency of the testing procedures and reporting protocols [11].

While SARs-CoV-2 shares similar genetic sequence with SARs-CoV, it is less deadly. However, SARs-CoV-2 spread easily and rapidly, with long latent period and characterized by no or moderate symptoms, thereby causing unpredicted difficulty in adequate diagnosis, contact tracing, and treatment of the disease [12]. The long latent period of about 2–14 days after first exposure varies between age and in patients with comorbidities [13].

COVID-19 disease affects majorly the respiratory system, but other systems have also been implicated such as the reproductive [14], nervous, and gastrointestinal systems with clinical symptoms ranging from mild respiratory disease to severe acute respiratory distress syndrome, which subsequently cause cellular dysfunction in at-risk patients and eventually leads to mortality [15]. In practice, the most observed symptoms of COVID-19 include dry cough, difficulty in breathing, fatigue, high fever, pneumonia, and infection [16]. Several experimental and clinical studies have proposed different therapeutic strategies to manage this disease; none of these pharmacological agents has been able to provide an overall treatment [17, 18]. The available therapeutic regimen and preventive protocols against COVID-19 include monoclonal antibodies, convalescent plasma, repurposing of drugs used in the hospitals, and series of vaccines which are currently been explored [19].

Vaccination against the causative virus has been suggested as an effective therapeutic strategy, and many nations of the world have been looking forward to such medical breakthrough. However, major setbacks in vaccine development includes, but not limited to, inability to validate and target the most suitable vaccine platform technologies, failure of developing chronic immunity, and difficulty in alleviating the cytokine storm as well as mutation of the SARs-CoV-2 virus [20]. These setbacks have made many countries and healthcare providers to doubt the long-term safety and efficacy of these vaccines, suggesting that many of these agents need to be redesigned and reevaluated. The numerous application of machine learning for the management of COVID-19 diseases is shown in Fig. 1:

2 Machine Learning and Artificial Intelligence

Before COVID-19 outbreak in 2019, artificial intelligence has been used especially in disease diagnosis, prognosis, and forecast. Areas where artificial intelligence can prove useful are

1. Contact tracing

Together with mobile health packages, artificial intelligence can be used in monitoring of population at high COVID-19 risk. A14COVID-19 is an example application that can be based on audio recording [21].

2. Early diagnosis

It has been discovered that Multi-Objective Differential Evolution and convolutional neural networks could be applied for early identification of COVID-19. Scientists have developed a new deep learning using chest Computerized Tomography [22]. COVID-19 detection neural network (COVNet) is a deep learning

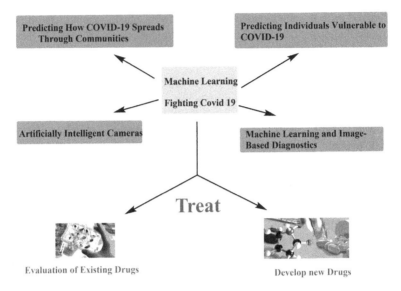

Fig. 1 Shows the numerous application of machine learning for the management of COVID-19 diseases

technology that has the tendency of differentiating between community-acquired pneumonia and COVID-19. This model was developed on the basis of visual 2D and 3D parameters gotten from volumetric chest Computerized Tomography scan [23]. With the use of automatic and progressing resizing, a model known as COVID-ResNet has been devised in COVID-19 diagnosis [24]. Artificial intelligence-related models for predicting real time polymerase chain reaction results of COVID-19 cases has also been designed [25].

3. Disease management

In the diagnosis of COVID-19, identification of mortality rate, studying of the trend of disease, and forecast of future disease outbreaks, artificial intelligence has been utilized.

4. Disease monitoring

Prediction of mortality and recovery chances from COVID-19 revolves around artificial intelligence. Evidence from data obtained from clinical parameters, statistics and artificial intelligence demonstrates the tendency of being used to avail requisite information for decision-making for prioritization of the need for respiratory supports especially in Intensive Care Unit [26]. Monitoring of COVID-19 patients and disease prognosis also revolves around the use of artificial intelligence.

5. Disease prediction and tracking

With the help of artificial intelligence, warning signals will serve as a medium for predicting the possible COVID-19 transmission. Using machine learning technique, a technique known as Bluedot can recognize pneumonia cluster.

6. Prediction of protein structure

ResNets has been utilized for protein structure prediction such as nsp6, protein 3a, nsp4, and COVID-19 papain-like-C-terminal domain [27]. A program that is based on deep convolutional neural network has been utilized to obtain protein structure of COVID-19 [28].

7. Development of pharmacological agents

Vaxign reverse vaccinology is a machine learning mechanism that can possibly be used in making prediction on the possible COVID-19 vaccine [29]. Artificial intelligence will play key roles in drug and vaccine development. An integrative network adopted by Zhou et al. [30] is being used for exploring COVID-19 drugs from the existing drug repertoire. Polypharm DB, incl Project, and many more are artificial intelligence-oriented mechanisms that are being used in the identification of potential drugs against COVID-19. Concerning drug discovery, machine learning-based Benevolent AI has been deployed and with this baricitinib has been identified as a potential pharmacological candidate against COVID-19 [30].

8. Alleviation of pressure on health personnel

With anthropomorphic and service robots, delivery of routine functions and primary healthcare services can be achieved using artificial intelligence. Such activities include general cleaning and disinfection among others [31, 32].

9. Classification of viral genomes

With the use of artificial intelligence, classification of COVID-19 is much more achievable with high levels of accuracy [33].

3 Machine Learning (ML)/Artificial Intelligence (AL) and COVID-19

The current global COVID-19 pandemic has been a treat to millions of persons, stability of infrastructure, and world's economies [34]. The disease has put a great burden on healthcare system because of the unavailability of adequate facilities to respond to the challenges posed by the disease [34]. Moreover, it has been discovered that the usage of AI plays several crucial roles in healthcare system so that many lives could be saved as well as improves the quality of healthcare service delivery [35]. Also, unlike many chronic diseases, that put the health of individuals at risk, COVID-19 is a global pandemic which jeopardized the health of the world's populace. Furthermore, with the improvement in the simulation of human intelligence, it is important to be able to use AL model and machine-based learning methods to forecast the menace of experiencing adverse events in SARS-CoV-2-infected persons [36]. Some of these events include mortality, need for healthcare resources, and the timing of these needs. Some of the risk factors that can predispose an infected person to these adverse events include comorbidities with other diseases like diabetes mellitus, hypertension, immunosuppressive conditions, cystic fibrosis, and cancer [37]. In addition, a person's age could also be a predisposing factor. The use of machine learning method could therefore be very efficient, using existing data to predict the risk of exposure or possible adverse events that may occur in SARS-CoV-2-infected persons. Since artificial intelligence is based on data gathering, it may therefore be useful in identifying the paramount technique to manage SARS-CoV-2-infected persons using the existing information or characteristics of such individuals including the presence of comorbidities [38]. Furthermore, it may also help to develop treatment plans and even predict the outcomes of treatments. Artificial intelligence model, for instance, can predict the outcome of what will happen if the use of mechanical ventilator is delayed in chronically infected COVID-19 patients. Also, effective collaborations between private and public sectors and with data obtained from airlines, social media, traffic places, and other COVID-19 prone areas, it will be easy to use AL models and ML to predict the menace of being infected, death rate, and other adverse event outcomes.

4   Computer Communication-Assisted Diagnosis of COVID-19

The computer communication network has a role to play in the diagnosis of COVID-19 disease. They have large capacity which surpasses those of the physicians to analyze, interpret data, and predict the outcomes of exposure to COVID-19 infections [39]. The computer communication-assisted network includes the use of smartphones which have data sensing capabilities to communicate between devices. With vast intelligence gathering, computer communication network is finding their way into the detection and prevention of COVID-19 disease. The current COVID-19 pandemic has placed face-to-face clinical restrictions between patients and the physicians; hence, much of the hospital health care are being delivered through the use of smartphone technology which is an extension of telemedicine [40]. The use of smartphones between the patients in isolation centers and the health professionals has greatly enhanced communication and effective management of COVID-19 infections. The use of smartphone technologies has also modified medical care protocols, as patients are not required to queue to consult the healthcare providers [41]. This limits the risk of contamination and curtails the spread of the disease. Smartphone technologies are important in contact tracing of infected persons. With the development of contact tracing applications, infected persons can easily be located; however, the major limitation is the issue of privacy as some of the information that are provided on the application include the location of the person with a suspected case of the virus [42]. Smartphones has provided people with opportunities to get information in the social media which greatly helps to avoid going to COVID-19-prone areas and the likelihood of contacting the infection. Smartphone technologies also provide information on global mortality rate and sensitize the populace on the danger of being infected. In addition to the use of smartphones, other media centers like the television stations, radio stations, and newspapers also provide update on the spread of the disease [43]. Even though the use of these media services is relevant, the populace is more accustomed and still relies on the use of their smartphones that serve as the most reliable and best way to communicate with the people and educate them on what to do when they notice the symptoms associated with COVID-19 in them or people in their environment.

5 Specific Authors Who Have Applied Machine Learning, Artificial Intelligence, and Smart Sensing for COVID-19

Jamshidi et al. [44] stated that the rapid spread and devastating effects of COVID-19 have resulted in serious global crisis with death tolls amounting to 6 million. The authors noted that to contest the pandemic, AL and deep learning approaches such as Generative Adversarial Networks, Long/Short-Term Memory, and Extreme Learning Machine can be adopted. This entails integrated bioinformatics approach by combining various information dataset in user-friendly platforms for healthcare workers and investigators. These approaches facilitate rapid and efficient way of treatment, diagnostics, and study of transmission of COVID-19 with the utilization of medical imaging and clinical data.

Gupta et al. [45] revealed that the rapid development and growth of Internet of Things has continued to offer a wide range of services and support to medical research, treatment, and diagnosis. Internet, actuators, connected sensors, networking, artificial intelligence, communication technology, and smart cyberphysical systems have impacted positively on human health particularly with the new occurrence of COVID-19. The authors reported that smart connected technologies and Internet of Things with the application of data-driven can perform a critical role in stoppage, prompt enforcement of guidelines, checking of patients, mitigation, diagnosis, administrative orders, and rules to contain such epidemics in the future. Numerous smart infrastructures are E-Health, supply chain management, smart city and communities, smart cyberphysical systems, Amazon Web Services, smart home, cloud infrastructures and ecosystem, and smart transportation, which are generally harnessed together to mitigate pandemic situation. Khan et al. [46] described that the spread of COVID-19 has caused serious negative impact on the economy and human safety, thus reducing the adoption and implementation of various smart technologies to fight against the pandemic situation. The authors revealed that novel implementation and adoption of smart sensing technology will help to curb the spread of COVID-19. These smart sensing technologies include drone, optical sensing, artificial intelligence, robotics, sensor technology, 3D printing, and mask. Many of these smart technologies can be implemented in many healthcare facilities to assist in healthcare surveillance, disinfection process, and diagnosis. These offer a robust opportunity for bulk data computing, providing medical assistance, predicting infection threats, analyzing diagnosis results, and generation of medical imaging.

Naseem et al. [47] reported that the current global COVID-19 pandemic has caused serious limitations to the capacity of the healthcare systems generating novel approaches of different methods on how to tackle the global crisis. Machine learning and artificial intelligence have been suggested by the authors to possess enhanced potential to optimize exponentially the healthcare practice, treatment protocol, strategies, and research. Adopting AI-driven tools in many developing nations can assist in mitigating against healthcare burdens and inequalities. The authors suggested from their findings that smart technologies can increase the accuracy and speed of identifying different cases, enhance data mining, efficient way of diagnosis and contact tracing, virus detection, increased smart city terminal tracking system, outbreak predictions, boost drug development and patient monitoring, data generation linked with COVID-19, vaccine formulation, reduce workload, and analysis of diverse medical.

Ana et al. (2020) reported that urban growth requires adequate urban management and planning for robust urban safe, inclusive, sustainable, and resilient. The authors discovered that COVID-19 has compromised the management of smart cities and communities; hence, the novel solution is to rapid address the issue of health and environment. The implementation of ML methodology has been reported to assist in the optimization and sustainability of smart cities and communities. Tarik et al. (2021) reported that deep learning AI and ML can be useful in the mitigation of COVID-19 through effective diagnosis and treatment strategies.

6 Conclusion and Future Perspectives

This chapter has given detailed facts on the relevance of machine learning for the detection of diagnosis, prognosis, and forecasting of diseases. Moreover, specific information was also provided on the utilization of ML in contact tracing, early diagnosis, disease management, disease monitoring, disease prediction and tracking, prediction of protein structure, development of pharmacological agents, alleviation of pressure on health personnel, and classification of viral genomes. Also, relevant facts on the usage of ML/AL and COVID-19, Computer communication-assisted diagnosis of COVID-19 were highlighted.

References

1. Sanyaolu, A., Okorie, C., Marinkovic, A., Patidar, R., Younis, K., Desai, P., Hosein, Z., Padda, I., Mangat, J., & Altaf, M. (2020). Comorbidity and its Impact on Patients with COVID-19. *SN Comprehensive Clinical Medicine, 2*(8), 1069–1076. https://doi.org/10.1007/s42399-020-00363-4

2. Olaniyan, O. T., Adetunji, C. O., Okotie, G. E., Adeyomoye, O., Anani, O. A., & Mali, P. C. (2021). Impact of COVID-19 on assisted reproductive technologies and its multifacet influence on global bioeconomy. *Journal of Reproductive Healthcare and Medicine, 2*(Suppl_1), 92–104. https://doi.org/10.25259/JRHM_44_2020

3. Mei, X., Lee, H. C., Diao, K. Y., Huang, M., Lin, B., Liu, C., Xie, Z., Ma, Y., Robson, P. M., Chung, M., Bernheim, A., Mani, V., Calcagno, C., Li, K., Li, S., Shan, H., Lv, J., Zhao, T., Xia, J., ... Yang, Y. (2020). Artificial intelligence-enabled rapid diagnosis of patients with COVID-19. *Nature Medicine, 26*(8), 1224–1228. https://doi.org/10.1038/s41591-020-0931-3. PMID: 32427924; PMCID: PMC7446729.

4. Panch, T., Szolovits, P., & Atun, R. (2018). Artificial intelligence, machine learning and health systems. *Journal of Global Health, 8*(2), 020303.

5. Sear, R. F., Velasquez, N., Leahy, R., Restrepo, N. J., Oud, S. E., Gabriel, N., Lupu, Y., & Johnson, N. F. (2020). Quantifying COVID-19 content in the online health opinion war using machine learning. *IEEE Access, 8*, 91886–91893. https://doi.org/10.1109/access.2020.2993967

6. Cohen, J., & Normile, D. (2020). New SARS-like virus in China triggers alarm. *Science, 367*(6475), 234–235.

7. Lupia, T., Scabini, S., Mornese Pinna, S., Di Perri, G., De Rosa, F. G., & Corcione, S. (2020). 2019 novel coronavirus (2019-nCoV) outbreak: A new challenge. *Journal of Global Antimicrobial Resistance, 21*, 22–27.

8. Mürbe, D., Kriegel, M., Lange, J., Schumann, L., Hartmann, A., & Fleischer, M. (2021). Aerosol emission of adolescents voices during speaking, singing and shouting. *PLoS One, 16*(2), e0246819. https://doi.org/10.1371/journal.pone.0246819

9. Morawska, L., & Milton, D. K. (2020). It is time to address airborne transmission of COVID-19 Clin. *Infectious Diseases, 6*, 2311–2313.

10. Yu, I. T., Li, Y., Wong, T. W., Tam, W., Chan, A. T., Lee, J. H., Leung, D. Y., & Ho, T. (2004). Evidence of airborne transmission of the severe acute respiratory syndrome virus. *New England Journal of Medicine, 350*, 1731–1739.

11. Oran, D. P., & Topol, E. J. (2020). Prevalence of asymptomatic SARS-CoV-2 infection: A narrative review Ann. *Internal Medicine, 173*, 362–367.

12. Lauer, S. A., Grantz, K. H., Bi, Q., Jones, F. K., Zheng, Q., Meredith, H. R., Azman, A. S., Reich, N. G., & Lessler, J. (2020). The incubation period of coronavirus disease 2019 (COVID-19) from publicly reported confirmed cases: Estimation and application. *Annals of Internal Medicine, 172*, 577–582.

13. Florindo, H. F., Kleiner, R., Vaskovich-Koubi, D., Acurcio, R. C., Carreira, B., Yeini, E., Tiram, G., Liubomirski, Y., & Satchi-Fainaro, R. (2020). Immune-mediated approaches against COVID-19. *Nature Nanotechnology, 15*, 630–645.

14. Olaniyan, O. T., Dare, A., Okotie, G. E., Adetunji, C. O., Ibitoye, B. O., Bamidele, O. J., & Eweoya, O. O. (2020). Testis and blood-testis barrier in Covid-19 infestation: Role of angiotensin converting enzyme 2 in male infertility. *Journal of Basic and Clinical Physiology and Pharmacology, 31*(6), 1–13. https://doi.org/10.1515/jbcpp-2020-0156

15. Tay, M. Z., Poh, C. M., Renia, L., MacAry, P. A., & Ng, L. F. P. (2020). The trinity of COVID-19: Immunity, inflammation and intervention. *Nature Reviews. Immunology, 20*, 363.

16. Jiang, F., Deng, L., Zhang, L., Cai, Y., Cheung, C. W., & Xia, Z. (2020). Review of the clinical characteristics of coronavirus disease 2019 (COVID-19). *Journal of General Internal Medicine, 35*, 1545.

17. Alvi, M. M., Sivasankaran, S., & Singh, M. (2020). Pharmacological and non-pharmacological efforts at prevention, mitigation, and treatment for COVID-19. *Journal of Drug Targeting, 28*(7–8), 742–754. https://doi.org/10.1080/1061186X.2020.1793990

18. Onwudiwe, O. A., Weli, H., Shaanu, T. A., Akata, N. M., & Ebong, I. L. (2020). Pharmacological treatment of COVID-19: An update. *Journal of Global Health Reports, 4*, e2020090. https://doi.org/10.29392/001c.17372

19. Tang, Z., Zhang, X., Shu, Y., Guo, M., Zhang, H., & Tao, W. (2021). Insights from nanotechnology in COVID-19 treatment. *Nano Today, 36*, 101019. https://doi.org/10.1016/j.nantod.2020.101019

20. Zhang, L., Jackson, C. B., Mou, H., Ojha, A., Rangarajan, E. S., Izard, T., Farzan, M., & Choe, H. (2020a). The D614G mutation in the SARS-CoV-2 spike protein reduces S1 shedding and increases infectivity. *Nature Communications, 11*, 6013.

21. Imran, A., Posokhova, I., Qureshi, H. N., Masood, U., Riaz, M. S., Ali, K., John, C. N., Hussain, M. I., & Nabeel, M. (2020). AI4COVID-19: AI enabled preliminary diagnosis for COVID-19 from cough samples via an app. *Informatics in Medicine Unlocked, 20*, 100378. https://doi.org/10.1016/j.imu.2020.100378. PMID: 32839734; PMCID: PMC7318970.

22. Singh, D., Kumar, V., & Vaishali, M. K. (2020). Classification of COVID-19 patients from chest CT images using multi-objective differential evolution-based convolutional neural networks. *European Journal of Clinical Microbiology & Infectious Diseases, 39*, 1379–1389.

23. Li, L., Qin, L., Xu, Z., Yin, Y., Wang, X., Kong, B., Bai, J., Lu, Y., Fang, Z., Song, Q., Cao, K., Liu, D., Wang, G., Xu, Q., Fang, X., Zhang, S., Xia, J., & Xia, J. (2020). Using artificial intelligence to detect COVID-19 and community-acquired pneumonia based on pulmonary CT: Evaluation of the diagnostic accuracy. *Radiology, 296*(2), E65–E71. https://doi.org/10.1148/radiol.2020200905

24. Farooq, M., & Hafeez, A. (2020). COVID-ResNet: a deep learning framework for screening of COVID-19 from radiographs. *arXiv arXiv:2003.14395*, 1–5.

25. Soarez, F. (2020). A novel specific artificial intelligence-based method to identify COVID-19 cases using simple blood exams. *MedRxiv.* https://doi.org/10.1101/2020.04.10.20061036

26. Rahmatizadeh, S., Valizadeh-Haghi, S., & Dabbagh, A. (2020). The role of artificial intelligence in management of critical COVID-19 patients. *Journal of Cellular & Molecular Anesthesia, 5*(1), 16–22.

27. Senior, A. W., Evans, R., Jumper, J., Kirkpatrick, J., Sifre, L., Green, T., Qin, C., Žídek, A., Nelson, A. W. R., Bridgland, A., Penedones, H., Petersen, S., Simonyan, K., Crossan, S., Kohli, P., Jones, D. T., Silver, D., Kavukcuoglu, K., & Hassabis, D. (2020). Improved protein structure prediction using potentials from deep learning. *Nature, 577*(7792), 706–710. https://doi.org/10.1038/s41586-019-1923-7. PMID: 31942072.

28. Pfab, J., Phan, N. M., & Si, D. (2020). DeepTracer: Fast cryo-EM protein structure modeling and special studies on CoV-related complexes. *bioRxiv.* https://doi.org/10.1101/2020.07.21.214064

29. Ong, E., Wong, M. U., Huffman, A., & He, Y. (2020). COVID-19 coronavirus vaccine design using reverse vaccinology and machine learning. *Frontiers in Immunology, 11*, 1581.

30. Zhou, Y., Hou, Y., Shen, J., Huang, Y., Martin, W., & Cheng, F. (2020). Network-based drug repurposing for novel coronavirus 2019-nCoV/SARS-CoV-2. *Cell Discovery, 6*, 14. https://doi.org/10.1038/s41421-020-0153-3. PMID: 32194980; PMCID: PMC7073332.

31. Yang, G. Z., Nelson, B. J., Murphy, R. R., Choset, H., Christensen, H., Collins, S. H., Dario, P., Goldberg, K., Ikuta, K., Jacobstein, N., & Kragic, D. (2020). Combating COVID-19-the role of robotics in managing public health and infectious diseases. *Science Robotics, 5*(40), eabb5589. https://doi.org/10.1126/scirobotics.abb5589. PMID: 33022599.

32. Zeng, Z., Chen, P. J., & Lew, A. A. (2020). From high-touch to high-tech: COVID-19 drives robotics adoption. *Tourism Geographies, 22*(3), 1–11.

33. Randhawa, G. S., Soltysiak, M. P. M., El Roz, H., de Souza, C. P. E., Hill, K. A., & Kari, L. (2020). Machine learning using intrinsic genomic signatures for rapid classification of novel pathogens: COVID-19 case study. *PLoS One, 15*(4), e0232391. https://doi.org/10.1371/journal.pone.0232391. Erratum in: PLoS One. 2021 Jan 27;16(1):e0246465. PMID: 32330208; PMCID: PMC7182198.
34. Sharma, O., Sultan, A. A., Ding, H., & Triggle, C. R. (2020). A review of the Progress and challenges of developing a vaccine for COVID-19. *Frontiers in Immunology, 11*, 585354.
35. Alimadadi, A., Aryal, S., Manandhar, I., Munroe, P. B., Joe, B., & Cheng, X. (2020). Artificial intelligence and machine learning to fight COVID-19. *Physiological Genomics, 52*(4), 200–202.
36. van der Schaar, M., Alaa, A. M., Floto, A., Gimson, A., Scholtes, S., Wood, A., McKinney, E., Jarrett, D., Lio, P., & Ercole, A. (2020). How artificial intelligence and machine learning can help healthcare systems respond to COVID-19. *Machine Learning, 110*(1), 1–14.
37. Leyfman, Y., Erick, T. K., Reddy, S. S., Galwankar, S., Nanayakkara, P., Di Somma, S., Sharma, P., Stawicki, S. P., & Chaudry, I. H. (2020). Potential immunotherapeutic targets for hypoxia due to COVI-flu. *Shock (Augusta, Ga.), 54*(4), 438–450.
38. Sodhi, G. K., Kaur, S., Gaba, G. S., Kansal, L., Sharma, A., & Dhiman, G. (2021). COVID-19: Role of robotics, artificial intelligence, and machine learning during pandemic. *Current Medical Imaging.* https://doi.org/10.2174/1573405617666210224115722
39. Ting, D., Carin, L., Dzau, V., & Wong, T. Y. (2020). Digital technology and COVID-19. *Nature Medicine, 26*(4), 459–461.
40. Whitelaw, S., Mamas, M. A., Topol, E., & Van Spall, H. (2020). Applications of digital technology in COVID-19 pandemic planning and response. *The Lancet Digital Health, 2*(8), e435–e440.
41. Baumgart, D. C. (2020). Digital advantage in the COVID-19 response: Perspective from Canada's largest integrated digitalized healthcare system. *NPJ Digital Medicine, 3*, 114.
42. Zhang, M., Chow, A., & Smith, H. (2020b). COVID-19 contact-tracing apps: Analysis of the readability of privacy policies. *Journal of Medical Internet Research, 22*(12), e21572.
43. González-Padilla, D. A., & Tortolero-Blanco, L. (2020). Social media influence in the COVID-19 pandemic. *International Brazilian Journal of Urology, 46*(suppl.1), 120–124.
44. Jamshidi, M., Lalbakhsh, A., Talla, J., Peroutka, Z., Hadjilooei, F., Lalbakhsh, P., Jamshidi, M., La Spada, L., Mirmozafari, M., Dehghani, M., & Sabet, A. (2020). Artificial intelligence and COVID-19: Deep learning approaches for diagnosis and treatment. *IEEE Access, 8*, 109581–109595. https://doi.org/10.1109/ACCESS.2020.3001973
45. Gupta, D., Bhatt, S., Gupta, M., & Tosun, A. S. (2020). Future smart connected communities to fight COVID-19 outbreak. *Internet of Things, 13*(100342), 1–26.
46. Khan, H., Kushwah, K. K., Singh, S., Urkude, H., Maurya, M. R., & Sadasivuni, K. K. (2021). Smart technologies driven approaches to tackle COVID-19 pandemic: A review. *3 Biotech, 11*, 50. https://doi.org/10.1007/s13205-020-02581-y
47. Naseem, M., Akhund, R., Arshad, H., & Ibrahim, M. T. (2020). Exploring the potential of artificial intelligence and machine learning to combat COVID-19 and existing opportunities for LMIC: A scoping review. *Journal of Primary Care & Community Health, 11*, 1–11.

Smart Sensing for COVID-19 Pandemic

Charles Oluwaseun Adetunji ⓘ **, Olugbemi Tope Olaniyan,**
Olorunsola Adeyomoye, Ayobami Dare, Mayowa J. Adeniyi, Enoch Alex,
Maksim Rebezov, Olga Isabekova, and Mohammad Ali Shariati

1 Introduction

COVID-19 is a severe respiratory disease of public health concern with detrimental effects on human health [1, 2]. Moreover, owing to its rapid and widespread

C. O. Adetunji (✉)
Applied Microbiology, Biotechnology and Nanotechnology Laboratory, Department of Microbiology, Edo University Iyamho, Auchi, Edo State, Nigeria
e-mail: adetunji.charles@edouniversity.edu.ng

O. T. Olaniyan
Laboratory for Reproductive Biology and Developmental Programming, Department of Physiology, Edo University Iyamho, Iyamho, Nigeria

O. Adeyomoye
Department of Physiology, University of Medical Sciences, Ondo City, Nigeria

A. Dare
Department of Physiology, School of Laboratory Medicine and Medical Sciences, College of Health Sciences, Westville Campus, University of KwaZulu-Natal, Durban, South Africa

M. J. Adeniyi
Department of Physiology, Edo State University Uzairue, Iyamho, Edo State, Nigeria

E. Alex
Department of Human Physiology, Ahmadu Bello University Zaria, Kaduna State, Nigeria

M. Rebezov
Prokhorov General Physics Institute, Russian Academy of Sciences, Moscow, Russia

K.G. Razumovsky Moscow State University of Technologies and Management (the First Cossack University), Moscow, Russia

O. Isabekova · M. A. Shariati
K.G. Razumovsky Moscow State University of Technologies and Management (the First Cossack University), Moscow, Russia

transmission, smart sensing has been considered as global strategies that could be applied in reducing the rate of transmission at the early stage. This has been shown to be very important technique that could be utilized in prevention of this pandemic situation [3].

It has been discovered that different healthcare system across different countries have incorporated massive testing devices, sensors and biosensors that help in the identification of COVID-19 on infected patients, as well as provision of potent therapeutic intervention for these patients. Also, several countries have adopted other protocols which entail lockdown regulations most especially in endemic areas that could help in reducing the spread of infection during COVID-19 pandemic [4]. Therefore, to sustain the healthcare system and effectively implement the restriction protocol across the countries, potent approaches to shape the society is highly important. Typical examples of these approaches entail the application of smart technologies in different sectors including the healthcare sector [5]. This paradigm shift toward technology-based approach has increased the urge for novel techniques, thereby increasing the significance of technologies in the health sector [6]. Smart technologies include drones, telehealth, robotics, and artificial intelligence which are used during COVID-19 pandemic to disinfection, promote consultation between patients and healthcare providers, enhance tracking and contact tracing of infected individuals, deliver medical equipment, and survey and screen people for early diagnosis of SARs-CoV-2 infection [7]. The synergistic contributions from several health sectors, incorporation of technology-based approach, and adequate government regulation have provided a potent strategy to alleviate COVID-19 pandemic. Undoubtedly, the significant impact of technology-based approach in curtailing COVID-19 infection requires proper understanding. This will enable scientist to decipher the mode of operation of these innovative techniques during COVID-19 pandemic. The article highlights the various smart technologies and their potential application to curtail COVID-19 pandemic.

Therefore, this chapter intends to provide detailed information on the utilization of smart technology in reducing the spread of COVID-19 diseases. Figure 1 shows the various applications of biosensors in the detection of COVID-19 diseases.

2 Application of Sensors and Biosensors for Monitoring and Detection of COVID-19

It has been discovered that there is a need to innovate a diagnostic test that could be applied in rapid recognition of affected people with outbreak of the coronavirus disease (COVID-19) in order to prevent the spreading of this virus. Mojsoska et al. [8] stated the role of a proof-of-concept label-free electrochemical immunoassay for swift identification of SARS-CoV-2 virus through the spike surface protein. The assay entails a grapheme that works with electrode functionalized with anti-spike antibodies. The immunosensor works on the principles which entails identification

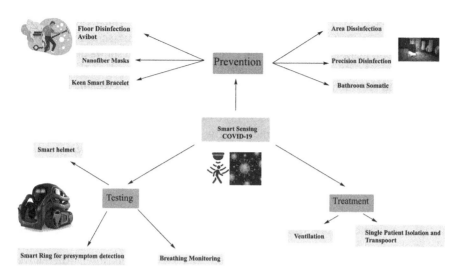

Fig. 1 Shows the various applications of biosensors in the detection of COVID-19 diseases

of signal perturbation derived from ferri/ferrocyanide determination after joining with the antigen during 45 min of incubation with a sample. The total alteration in the $[Fe(CN)_6]^{3-}/^{4-}$ current upon enhancing the level of antigen concentrations on the immunosensor surface was applied in evaluating the identification range of the spike protein. The sensor portent the capability to identify a specific signal above 260 nM (20 μg/mL) of subunit 1 of recombinant spike protein. Moreover, it could identify SARS-CoV-2 at a concentration of 5.5×10^5 PFU/mL, which falls within physiologically applicable concentration values. It was discovered that novel immunosensor has a meaningfully faster investigation time than the standard quantitative polymerase chain reaction (qPCR) coupled with a portable device that can enable on-site evaluation of infection.

Quick and timely identification of COVID-19 diseases has been identified as a measure that could be applied in the prevention of this pandemic outbreak. It has become mandatory to enhance healthcare quality in airports, markets, public places, schools so as to generate necessary insight into the technological environment and assist researchers recognize the selections and gaps existing in this field. In view of this, Taha et al. [9] wrote a comprehensive report on the application of biosensors that works based on the application of laser detection technology.

The authors provide three useful points such as means of assessment of hazard of pandemic COVID-19 transmission styles in comparison with Middle East Respiratory Syndrome (MERS) so as to establish the main factor responsible for the spreading of the virus, a life-threatening evaluation that can be applied in identification of coronavirus disease 2019 (COVID-19) that depends on artificial intelligence through *chest x-ray* (CXR) images, computerized tomography (CT) scans, and types of biosensors.

Hussein et al. [10] pointed out five different techniques which entails micro-neutralization evaluation and cell-culture recognition, genetic screening through Real-time polymerase chain reaction (RT-PCR) coupled which shows the gold standard for virus identification most especially in the nasopharyngeal swabs. Moreover, it was stated that immunoassays could be developed either by screening or antigen identification of the whole virus, available in the blood, sera, and the whole virus. Also, proteomic mass-spectrometry (MS)-based technique was identified for the evaluation of the swab samples. Interestingly, virus-biosensing devices have been developed, while eye-based technologies and electrochemical immunosensors could also be used for the identification of virus present in swab. It was also stated that lateral flow point-of-care immunoassays are based on swab-based techniques for the detection of virus in the blood samples. They have merit which includes their portability for on-site evaluation in airports and hotspots, virtually without any sample treatment. Finally, virus-biosensing devices were efficiently designed. Both electrochemical immunosensors and eye-based technologies have been described, showing detection times lower than 10 min after swab introduction. Alternative to swab-based techniques, lateral flow point-of-care immunoassays are already commercially available for the analysis of blood samples. Such biosensing devices hold the advantage of being portable for on-site testing in hospitals, airports, and hotspots, virtually without complicated lab precautions or any sample treatment.

It has been discovered that due to the lack of prophylactic measure for effective management of viral infection, quick identification, as well as their quarantining of asymptomatic infected individuals most especially during this pandemic situation is very difficult. It has been discovered that the current techniques are time consuming and very expensive. Therefore, there is a need to search for more sustainable solutions that could be applied for quick identification of this virus. Vadlamani et al. [11] stated the application of a more cost-effective, and sensitive, cobalt-functionalized TiO_2 nanotubes (Co-TNTs)-based electrochemical sensor that could be applied for rapid identification of the spike (receptor binding domain (RBD)) available on the surface of the virus. The authors applied a cost-effective and simple one-step electrochemical anodization route for the fabrication of TNTs, together with wetting techniques for cobalt functionalization of the TNTs platform, coupled with a potentiostat for data collection. The sensor portends the capacity to identify the S-RBD protein of SARS-CoV-2 even when available in a lower concentration which varies from 14 to 1400 nM (nano molar). The authors also demonstrated a linear rejoinder in the identification of viral protein in a lot of varying concentrations. Therefore, the application of Co-TNT sensor has been discovered to be more efficient in the identification of SARS-CoV-2 S-RBD protein within 30 s, which shows that it can be utilized for development of a point-of-care diagnostics for quick identification of SARS-CoV-2 in saliva samples and nasal secretions.

Mavrikou et al. [12] stated the proof-of-concept development of a biosensor that could be applied in the identification of spike protein present on the surface of the virus. The biosensor works on the principle of membrane-engineered mammalian cells having the human chimeric spike S1 antibody. Their study indicated that the joining of the protein to the membrane-bound antibodies led to discriminatory and considerable alteration in the cellular bioelectric features determined through the help of Bioelectric Recognition Assay. The novel biosensor could detect a semi-linear range of response between 10 fg and 1 μg/mL and a detection limit of 1 fg/mL, while there was no cross-reactivity against the SARS-CoV-2 nucleocapsid protein. Moreover, the biosensor was constructed as a ready-to-use platform, with a portable read-out device that works through a tablet or a smartphone. Their study shows that this novel biosensor could be utilized for mass screening of SARS-CoV-2 surface antigens without going through the process of sample processing. This indicated that this biosensor could be available as a sustainable technique that could be applied for rapid identification and monitoring of this global coronavirus.

Khan et al. [13] reported that the current COVID-19 pandemic can be reduced by adopting the use of smart technology, thus the implementation of these novel innovations will ensure therapeutic advantage together with vaccine development. The authors pointed out that in the implementation of smart technology for the mitigation of global COVID-19 pandemic, specific attention must be given to the utilization of drone, artificial intelligence, robotics, 3D printing, optical sensing, and sensor technology and mask technology in healthcare services through disinfection process, medical assistance, surveillance, prediction of infection diseases, evaluation, data analysis and computing, diagnostics, and controlled monitoring of infected patients.

Erdem et al. [14] demonstrated that COVID-19 has caused a serious of devastating effects on global health system, which needs rapid and aggressive solutions to quickly curtail the spread. The authors noted that effective clinical management and implement strategies through prompt diagnostic techniques are a mainstay. Many of the diagnostic techniques for the screening of this disease include real-time reverse transcriptase (RT)-PCR test, serological assays which are very slow, expensive, and require skilled personnel resulting in greater impediments especially in the low-resource settings. Today, sensor systems provide huge opportunity particularly in the clinical diagnostics settings as complementary options or alternative to the present diagnostic approaches in point-of-care settings, time-effective, and economic perspective way. Sensor technology accommodate the utilization of smart materials like photosensitive materials, nanomaterials, electrically sensitive materials, wearable tools with tremendous analytical performances and physiochemical properties such as linear dynamic range, specificity, and sensitivity for COVID-19 diagnosis.

Gupta et al. [15] showed that in the past few years, rapid development has been seen in the adoption of Internet of Things for healthcare support through different connection with networking, sensors, communication technology, actuators, smart cyberphysical systems, and artificial intelligence globally. The authors noted that the implementation of Internet of Things to mitigate the spread of COVID-19 is paramount. Hence, smart connected technologies and Internet of Things which combine data-driven applications and algorithms are crucial to the continuous monitoring, prevention, evaluation, mitigation, analysis, prediction, and diagnosis of patients and future outbreaks. Recently, different smart technologies like smart home, E-Health, smart transportation, supply chain management, and smart communities have been implemented through cloud-enabled Internet of Things and Amazon Web Services as novel strategy to combat COVID-19.

Damin et al. [16] reported that reverse transcription-quantitative polymerase chain reaction with other specificity and sensitivity analytical techniques have been utilized for the diagnosis of COVID-19. The authors showed that these techniques are less sensitive particularly in asymptomatic individuals with lot of time-wasting efforts. Hence, the introduction of microarray-based assay which is a solid-phase hybridization involving fluorescently labeled amplicons with reverse transcription and RNA extraction. This technique is very sensitive due to the optical-physical properties of the silicon substrate to the viral genes.

3 Application of Drone Technology-Driven Technology and Robotics in Supporting Disinfection Process, Surveillance, and Health System

COVID-19 is a transmissible disease that has claimed responsibility for the death of over 500,000 people worldwide since its onset. The transmissibility of coronavirus necessitates introduction of technology-aided measures to curb the spread of the diseases. Among the technology-aided measures are drone and robotics.

4 Application of Drone-Driven Technology in Supporting Disinfectant Process, Surveillance, and Health System

Drones (Unmanned Aerial Vehicles) are aircrafts that are devoid of pilot. The first unmanned aircraft surfaced during the First World War, but earlier, unmanned aerovehicle known as balloon carrier had been utilized in fighting war in 1849. Drones with different potential benefits as far COVID-19 prevention is concerned have been developed. For instance, drone-based healthcare services for patients with chronic ailments useful in data collection, medication, and healthcare were developed by Kim et al. [17]. Drone-based rescuers and disasters scenario platform created by Camara et al. [18] are useful in monitoring and data collection. Another system for medical services which was useful in monitoring, data collection, medication, and healthcare was created by Graboyes and Skorup [19]. With respect to services such as medication and healthcare, a drone-based system in medical drug supply has been developed [20]. Sethuraman et al. [21] developed a drone-based healthcare platform useful in monitoring and healthcare. Kumar et al. [22] developed a drone-based smart healthcare platform with artificial intelligence which among others may be useful in monitoring of COVID-19, social distancing, data analysis, and statistics.

5   Application in Data Collection

There are many potential drone-based platforms for data collection and statistics. Typical examples of this is found in the drone -based healthcare services which is applicable to patients that are suffering from chronic disease that could help in data collection as described by Kim et al. [17]. Drone-based platform created by Camara et al. [18] was useful in monitoring and data collection. Likewise, the one developed by Graboyes and Skorup [19] is useful in data collection.

6   Application in Aerial Disinfection

Drones have been extensively used in agriculture for spraying. These drones are also applicable in spraying the environment in order to rid the environment of microorganisms [23]. Achieving optimum disinfection is vital for minimizing active or passive COVID-19 transmission. This is much more required in view of the report emanating from New England Journal of Medicine that claims that coronavirus can survive for about 3 h, 4 h, 24 h, and 3 days in the air, copper, cardboard, and steel, respectively [24]. In developing countries like Nigeria with ebbed technology, disinfections were largely achieved through the use of manpower (www.premiumtimesng.com/news/top-news) exposing more lives to COVID-19 risks.

7   Application in Transportation of Medical Materials

Drones have been used in the transport of fragile medical items and pharmaceutical agents. With drones, medical kits and items were delivered to locations and this has contributed to curtaining the spread of the virus. The use of drones for the delivery of medical items occurred in countries like India, China, Chile, Estonia, Canada, Australia, United States, and many more [22].

8   Policy Monitoring and Surveillance

One of the World Health Organization strategies for the control of COVID-19 spread is social distancing. Drones are being used to monitor citizens' compliance to social distancing regulation, crowd tracking, and remote measurements [23].

9   Dissemination of Information During COVID-19

The application of telecommunication was applied for effective dissemination of information to the people in remote locations where communication means are not available. The application of several communication gadgets were deployed for successful dissemination of information majorly on how to mitigate against COVID-19 diseases, application of face mask, constant cleaning of hoods, social distancing, and application of alcohol-based sanitizers. Most people in the rural area in most developing county get access to important information through the help of their telephone and other telecommunication gadgets.

10   Application of Smart Technology in Medical Assistance, Forecasting Infection Threats, Investigating Diagnosis

Robotics are programmable machines that could be applied to perform numerous tasks in different areas in the health sector. The integration of machine learning algorithms and artificial intelligence play major roles most especially in the medical services.

11   Application in the Diagnosis and Rehabilitation

Before the outbreak of COVID-19 in 2019, tele-robots, a mechanism that involves the use of wireless communication network to control robot from a distance, have been used in clinics and hospitals for diagnosis, management, and rehabilitation [25].

12   Collection of COVID-19 Sample

Collaborative robotics, a technology that is operative in proximity with human being, shows promising roles especially with the development of a swab-collecting robot [26]. This robotics is capable of not only collecting specimen from the throats but also submitting it in a sample box [27].

13 Sanitation, Safety, and Management of COVID-19 Situation

It has been documented that it is possible to disinfect public places during COVID-19 era with the help of autonomous robotics. Robotics are also capable of interacting with patients without human interventions, recording health information, and data. These robotics are also capable of giving medical treatment and supplying personal protective equipment and food [28].

14 Application of Robotics in Protecting People During Pandemic Period

The application of robots was shown to be more effective in protecting people from COVID-19 infection. In the COVID-19 pandemic, social robots are used for social interaction and guiding peoples in the hospital. This was also used in fumigation of the environment [29].

15 Measurement of Vital Signs

The application of wearable robotics was normally placed on the body of numerous people in order to measure their body vital signs. Mohammed et al. [30] proposed smart helmet for adequate measurement of body temperature and facial recognition. This was established to be more effective for adequate recording, monitoring, and quick detection of COVID-19 diseases on patients most especially in most crowded areas.

16 Conclusion and Future Directions

This chapter has provided detailed information on the applications of smart sensing as well as biosensors and sensors during COVID-19 pandemic. The application of biosensors and different sensors was established as a major benchmark that played a crucial role in ravaging prevalence of COVID-19 pandemic which actually play a significant role in the prevention of this pandemic disease. Moreover, the application of drones, telehealth, robotics, and artificial intelligence used during COVID-19 pandemic for disinfection, promote consultation between patients and healthcare providers as well as their role in the tracking and contact tracing of infected individuals were highlighted, deliver medical equipment, survey, and screen people for early diagnosis of SARs-CoV-2 infection were highlighted during this study. Also it has been established that the application of technology-driven approaches could play a significant role in promoting social distancing measures, predict possible infections, and optimize the delivery of essential services and resources in a swift and efficient manner. However, large-scale implementation of smart sensing in many societies is usually confronted with numerous difficulties like government regulation and policies, security, and confidentiality, which need to be properly addressed to promote acceptability of this innovative concept. Also, incorporation of smart sensing will facilitate collaborative research that cuts across various disciplines with the development of appropriate technologies to prevent subsequent outbreak. Also, the application of technology-driven approaches with little or no human interference in smart sensing could promote the implementation of preventative strategies that controls the transmission of infection in the society.

References

1. Olaniyan, O. T., Dare, A., Okotie, G. E., Adetunji, C. O., Ibitoye, B. O., Bamidele, O. J., & Eweoya, O. O. (2020). Testis and blood-testis barrier in Covid-19 infestation: Role of angiotensin converting enzyme 2 in male infertility. *Journal of Basic and Clinical Physiology and Pharmacology, 31*(6), 1–13. https://doi.org/10.1515/jbcpp-2020-0156
2. World Health Organization. (2020). *Clinical management of severe acute respiratory infection (SARI) when covid-19 disease is suspected.* Interim Guidance. World Health Organization
3. Lai, C. C., Shih, T. P., Ko, W. C., Tang, H. J., & Hsueh, P. R. (2020). Severe acute respiratory syndrome coronavirus 2 (SARS-CoV-2) and corona virus disease-2019 (COVID-19): The epidemic and the challenges. *International Journal of Antimicrobial Agents, 55*, 105924–105932.
4. Olaniyan, O. T., Adetunji, C. O., Okotie, G. E., Adeyomoye, O., Anani, O. A., & Mali, P. C. (2021). Impact of COVID-19 on assisted reproductive technologies and its multifacet influence on global bioeconomy. *Journal of Reproductive Healthcare and Medicine, 2*(Suppl_1), 92–104. https://doi.org/10.25259/JRHM_44_2020.
5. Elavarasan, R. M., & Pugazhendhi, R. (2020). Restructured society and environment: A review on potential technological strategies to control the COVID-19 pandemic. *Science of the Total Environment, 725*, 138858–138875.
6. Javaid, M., Haleem, A., Vaishya, R., Bahl, S., Suman, R., & Vaish, A. (2020). Industry 4.0 technologies and their applications in fighting COVID-19 pandemic. *Diabetes and Metabolic Syndrome: Clinical Research and Reviews, 14*, 419–422.
7. European Parliament. (2020). *Ten technologies to fight coronavirus.* https://www.europarl.europa.eu/RegData/etudes/IDAN/2020/641543/EPRS_IDA(2020)641543_EN.pdf. Accessed 27 Jun 2020.
8. Mojsoska, B., Larsen, S., Olsen, D. A., Madsen, J. S., Brandslund, I., & Alatraktchi, F. A. (2021). Rapid SARS-CoV-2 detection using electrochemical immunosensor. *Sensors, 21*(2), 390. https://doi.org/10.3390/s21020390
9. Taha, B. A., Al Mashhadany, Y., Hafiz Mokhtar, M. H., Dzulkefly Bin Zan, M. S., & Arsad, N. (2020). An analysis review of detection coronavirus disease 2019 (COVID-19) based on biosensor application. *Sensors, 20*(23), 6764. https://doi.org/10.3390/s20236764
10. Hussein, H. A., Hassan, R. Y. A., Chino, M., & Febbraio, F. (2020). Point-of-care diagnostics of COVID-19: From current work to future perspectives. *Sensors, 20*(15), 4289. https://doi.org/10.3390/s20154289

11. Vadlamani, B. S., Uppal, T., Verma, S. C., & Misra, M. (2020). Functionalized TiO_2 nanotube-based electrochemical biosensor for rapid detection of SARS-CoV-2. *Sensors, 20*(20), 5871. https://doi.org/10.3390/s20205871

12. Mavrikou, S., Moschopoulou, G., Tsekouras, V., & Kintzios, S. (2020). Development of a portable, ultra-rapid and ultra-sensitive cell-based biosensor for the direct detection of the SARS-CoV-2 S1 spike protein antigen. *Sensors, 20*(11), 3121. https://doi.org/10.3390/s20113121

13. Khan, H., Kushwah, K. K., Singh, S., Urkude, H., Maurya, M. R., & Sadasivuni, K. K. (2021). Smart technologies driven approaches to tackle COVID-19 pandemic: A review. *3 Biotech, 11*, 50. https://doi.org/10.1007/s13205-020-02581-y

14. Erdem, Ö., Derin, E., Sagdic, K., Eylul Gulsen Yilmaz and Fatih Inci (2021)Smart materials-integrated sensor technologies for COVID-19 diagnosis. Emergent Materials 4, 169–185. https://doi.org/10.1007/s42247-020-00150-w.

15. Gupta, D., Bhatt, S., Gupta, M., & Tosun, A. S. (2021). Future smart connected communities to fight COVID-19 outbreak. *Internet of Things, 13*, 100342. https://doi.org/10.1016/j.iot.2020.100342

16. Damin, F., Galbiati, S., Gagliardi, S., Cereda, C., Dragoni, F., Fenizia, C., Savasi, V., Sola, L., & CovidArray, C. M. (2021). A microarray-based assay with high sensitivity for the detection of SARS-CoV-2 in nasopharyngeal swabs. *Sensors, 21*(7), 2490. https://doi.org/10.3390/s21072490

17. Kim, S. J., Lim, G. J., Cho, J., & Cote, M. J. (2017). Drone-aided healthcare services for patients with chronic diseases in rural areas. *Journal of Intelligent & Robotic Systems, 88*(1), 163–180.

18. Camara, D. (2014). Cavalry to the rescue: Drones fleet to help rescuers operations over disasters scenarios. In *2014 IEEE Conference on Antenna Measurements and Applications, CAMA 2014* (pp. 1–4). IEEE.

19. Graboyes, R. F., & Skorup, B. (2020). Medical drones in the United States and a survey of technical and policy challenges. *Mercatus Center Policy Brief.*

20. Jones, R. W., & Despotou, G. (2019). Unmanned aerial systems and healthcare: Possibilities and challenges. In *2019 14th IEEE Conference on Industrial Electronics and Applications (ICIEA), 2019* (pp. 189–194). IEEE.

21. Sethuraman, S. C., Vijayakumar, V., & Walczak, S. (2020). Cyber-attacks on healthcare devices using unmanned aerial vehicles. *Journal of Medical Systems, 44*(1), 29.

22. Kumar, A., Sharma, K., Singh, H., Naugriya, S. G., Gill, S. S., & Buyya, R. (2021). A drone-based networked system and methods for combating coronavirus disease (COVID-19) pandemic. *Future generations computer systems, 115*, 1–19.

23. Kramar, V. (2020). UAS (drone) in response to coronavirus. In S. Balandin, V. Turchet, & T. Tuytina (Eds.), *Proceedings of the FRUCT'27, Trento, Italy, 7–9* (pp. 90–100). FRUCT.

24. Van Doremalen, N., Bushmaker, T., Morris, D. H., Holbrook, M. G., Gamble, A., Williamson, B. N., Tamin, A., Harcourt, J. L., Thornburg, N. J., Gerber, S. I., & Lloyd-Smith, J. O. (2020). Aerosol and surface stability of SARS-CoV-2 as compared with SARS-CoV-1. *The New England Journal of Medicine, 382*, 1564–1567. https://doi.org/10.1056/NEJMc2004973

25. Agostini, M., Moja, L., Banzi, R., Pistotti, V., Tonin, P., Venneri, A., & Turolla, A. (2015). Telerehabilitation and recovery of motor function: A systematic review and meta-analysis. *Journal of Telemedicine and Telecare, 21*, 202–213.

26. Evening Standard. (2020). *Robots offer a contact-free way of getting swabbed for coronavirus.* https://www.standard.co.uk/tech/robots-offer-new-coronavirus-swab-technique-a4477396.html. Accessed 2 Jul 2020.

27. Healthcare Packaging. (2020). *New robots perform COVID-19 swab tests.* https://www.healthcarepackaging.com/covid-19/article/21136422/quick-hits-new-robots-perform-covid19-swab-tests. Accessed 30 Jun 2020.

28. Nikkei Asian Review. (2020). *JD.com makes drone deliveries as coronavirus cuts off usual modes.* https://asia.nikkei.com/Spotlight/Coronavirus/JD.com-makes-drone-deliveries-as-coronavirus-cuts-off-usual-modes. Accessed 26 Jun 2020.

29. The Peninsula. (2020). *Robots may become heroes in war on coronavirus.* https://www.thepeninsulaqatar.com/article/09/04/2020/Robots-may-become-heroes-in-war-on-coronavirus. Accessed 30 Jun 2020.
30. Mohammed, M. N., Syamsudin, H., Al-Zubaidi, S., & Yusuf, E. (2020). Novel COVID-19 detection and diagnosis system using IOT based smart helmet. *International Journal of Psychosocial Rehabilitation, 24*, 2296–2303. https://doi.org/10.37200/IJPR/V24I7/PR270221

eHealth, mHealth, and Telemedicine for COVID-19 Pandemic

Charles Oluwaseun Adetunji ⓘ**, Olugbemi Tope Olaniyan,
Olorunsola Adeyomoye, Ayobami Dare, Mayowa J. Adeniyi, Enoch Alex,
Maksim Rebezov, Larisa Garipova, and Mohammad Ali Shariati**

1 Introduction

One of the strategies developed by many countries worldwide was to enforce movement restriction or lockdown during COVID-19 pandemic situation. The aim

C. O. Adetunji (✉)
Applied Microbiology, Biotechnology and Nanotechnology Laboratory, Department of Microbiology, Edo University Iyamho, Auchi, Edo State, Nigeria
e-mail: adetunji.charles@edouniversity.edu.ng

O. T. Olaniyan
Laboratory for Reproductive Biology and Developmental Programming, Department of Physiology, Edo University Iyamho, Iyamho, Nigeria

O. Adeyomoye
Department of Physiology, University of Medical Sciences, Ondo City, Nigeria

A. Dare
Department of Physiology, School of Laboratory Medicine and Medical Sciences, College of Health Sciences, Westville Campus, University of KwaZulu-Natal, Durban, South Africa

M. J. Adeniyi
Department of Physiology, Edo State University Uzairue, Iyamho, Edo State, Nigeria

E. Alex
Department of Human Physiology, Ahmadu Bello University Zaria, Kaduna State, Nigeria

M. Rebezov
Prokhorov General Physics Institute, Russian Academy of Sciences, Moscow, Russia

K.G. Razumovsky Moscow State University of Technologies and Management (the First Cossack University), Moscow, Russia

L. Garipova · M. A. Shariati
K.G. Razumovsky Moscow State University of Technologies and Management (the First Cossack University), Moscow, Russia

© The Author(s), under exclusive license to Springer Nature Switzerland AG 2022   157
S. K. Pani et al. (eds.), *Assessing COVID-19 and Other Pandemics and Epidemics using Computational Modelling and Data Analysis*,
https://doi.org/10.1007/978-3-030-79753-9_10

of restricting peoples' movement is to control the spread of this infectious disease [1]. As a result of this, attention has shifted to other mechanisms of achieving effective healthcare delivery without physical movements. eHealth, telemedicine, and mHealth are technology-aided media through which healthcare can be accessed. Moreover, eHealth and telemedicine center on the use of electronic mechanisms and communication for healthcare delivery most especially through the use of telecommunication technology and electronic information in healthcare delivery [1–3]; mHealth deals with the use of mobile communication gadgets such as tablet computers, mobile phones, wearable devices (such as watch), and many more for the delivery of health services.

Many countries of the world have placed emphasis on the use of mobile technology for the adequate sensitization of healthcare personnel during COVID-19 era in order to provide adequate to healthcare services to mostly affected areas. For instance, in Iran, with the use of telemedicine and social media platform such as WhatsApp, email, and messaging software was reported by Davarpanah et al. [4]. It was discovered that this led to drastic reduction in need to send patients to congested hospitals, provision of consultation in areas where people had restricted access to radiology experts and brought experts from all countries of the globes together. In China, Zhai et al. [5] showed that the application of live video conferencing and mobile helped patients to have adequate access to guidelines regarding inhibition and management of COVID-19. It also reduced exposure risk to respiratory discharges, foiled potential healthcare provider-patient transmission. In the USA, with the use of video using a Health Insurance Portability and Accountability (HIPAA)-compliant platform, electronic health record (EHR), digital photography, telephone, patient portal messaging, electronic medical record, and website phone calls, patient's health concerns were addressed [6]. Besides, COVID-19 tracking and monitoring were achieved, updating of travel and symptom screening was possible, and training sessions were conducted.

Moreover, the major barrier to the general use of telemedicine in most African countries is ebbed level of development and infrastructures. Most cities in Africa are situated in such a way that around 65% of the populace live in rural poorly infrastructure [7]. However, the application of mobile technology played a crucial role in quick mitigation of COVID-19 diseases. For instance, District Health Information Software 2, an open health management software mechanism, was utilized in 67 countries. In Uganda and Rwanda, WelTelvirtual care system was utilized for remote monitoring of COVID-19. In Nigeria and many other African countries, text messages were sent on daily, weekly, or monthly basis to citizens regarding disease awareness and adherence to COVID-19 protocol. The various applications of telemedicine in the management of COVID-19 diseases is illustrated in Fig. 1.

Therefore, this chapter intends to provide a detailed information on the application of eHealth, mHealth, and telemedicine as a preemptive measure to increase clinical care for effective management of COVID-19 infection.

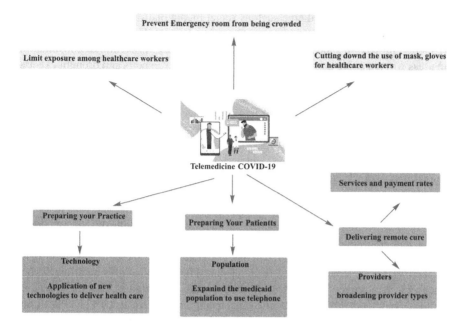

Fig. 1 The various applications of telemedicine in the detection of COVID-19 diseases

2 eHealth, mHealth, and Telemedicine Applications and COVID-19 outside of Africa

COVID-19 is a highly infectious disease. One of the strategies developed by many countries worldwide was enforcement of movement restriction or lockdown [8]. The aim of restricting peoples' movement is to control the spread of the disease [9]. As a result of this, there was attention shift to other mechanisms of achieving effective healthcare delivery without physical movements. One of these mechanisms is the use of mobile technology platforms such as eHealth, mHealth, and telemedicine.

eHealth entails the application of electronic mechanisms and communication for healthcare delivery. It encompasses mHealth, the use of mobile communication gadgets such as tablet computers, mobile phones, wearable devices (such as watch), and many more for the delivery of health services and telemedicine. Telemedicine is an acronym formed by two words – tele (distant) and medicine (health). It is the use of telecommunication technology and electronic information in healthcare delivery [1–3]. For quite a while, telemedicine has found its place in the qualitative and quantitative delivery of healthcare, diagnosis, and prognosis in rural, urban, and remote areas. With telemedicine, health services can be delivered without transport, mobility, funding, or staff restriction barriers [1]. Where physical distance exists, telemedicine has been used to provide diagnosis, monitoring, consultation, and many more.

With the use of telemedicine such as social media platform such as WhatsApp, email, and messaging software in Iran, Davarpanah et al. [4] reported that there was reduction in need to send patients to congested hospitals, provision of consultation in areas where people had restricted access to radiology experts and brought experts from all countries of the globe together. In China, Zhai et al. [5] showed that the application of mobile phones and live video conferencing could help patients to have access to guidelines regarding hindrance and management of COVID-19. It also reduced exposure risk to respiratory discharges and foiled potential healthcare provider-patient transmission. In the USA, with the use of electronic health record (EHR), telephone, patient portal messaging, electronic medical record, digital photography, video using a HIPAA-compliant platform, and website phone calls, patient's health concerns were addressed [6]. Besides, COVID-19 tracking and monitoring were achieved, updating of travel and symptom screening was possible, and training sessions were conducted. In many countries, with the use of video, telephone, and laptops, there was a massive reduction in COVID-19 transmission risk during radiotherapy, minimized risks of brain drain and infection, and fast-tracked access to treatment and hospital data [10]. Applications such as Skype, Facebook Messenger video chat, Mobile Health Technology, Apple FaceTime reduced healthcare provider/patient and hastened access to treatment [11]. Platforms such as Weibo, TikTok, and WeChat have also enabled healthcare specialists to provide mental health facilities during COVID-19 pandemic [12].

3 eHealth, mHealth, and Telemedicine Applications and COVID-19 within Africa

As far as African countries are concerned, the major barrier to the general use of telemedicine is ebbed level of development and infrastructures. Africa is situated in such a way that around 65% of the populace live in rural poorly infrastructure [7]. However, mobile technology for COVID-19 was devised. For instance, District Health Information Software 2, an open health management software mechanism, was utilized in 67 countries (District Health Information Software) [13]. In Uganda and Rwanda, WelTelvirtual care system was utilized for remote monitoring of COVID-19. In Nigeria and many other Africa countries, text messages were sent on daily, weekly, or monthly basis to citizens regarding disease awareness and adherence to COVID-19 protocol. With an Unstructured Supplementary Service Data (USSD) code, (*920*222#), 080-CORONA, and specific phone numbers, citizens of Ghana and Nigeria had the opportunity of communicating with health agencies in charge of COVID-19 prevention and control [14]. Good hygiene practices messages were broadcast through SMS services in Senegal [15].

4 Application of eHealth, mHealth, Telemedicine, and Pandemic

Application of eHealth, mHealth, and telemedicine has great benefits to reduce the spread of pandemic by using high-speed telecommunication platform and software to provide, manage, and monitor healthcare services. These platforms help to reduce physical contact thereby protecting healthcare providers and the general populace against COVID-19 infections [16]. Applying these platforms can provide quick access to information, avoiding the stress of travelling long distance to the clinic for mild health services as well as preventing the risk of more infection [17].

Telemedicine has been used to improve communication via camera-enabled computers, smartphones, and other internet-enabled devices, which has helped to promote real-time consultation with the clinicians through virtual meetings and teleconferencing for adequate health advice [18, 19]. Virtual consultation has helped to avoid congestion at the clinic, reduce wastage of resources such as personal protective equipment (PPEs), and improve access to health services [20]. Virtual primary health enables safe handling of insignificant cases of COVID-19 thereby increasing the availability of personnel and equipment to severely ill COVID-19 patients. Telemedicine can upkeep the handling of COVID-19 and mitigation of its transmission between patients and healthcare providers [21].

Telemedicine and eHealth strategies can assist clinicians to remotely recognize patients who are severely affected with SARs-Cov-2 and may need extensive hospital care [22, 23]. eHealth can aid investigation systems for trace contacts, quick regulation of pandemics diseases [24]. mHealth enhances easy sharing of information between patients and healthcare providers via mobile phones with dependable software uses, which are vital for adequate orientation of the masses about this disease and effective management of COVID-19 pandemic [25].

Undeniably, telemedicine is a vital platform with numerous benefits during COVID-19 pandemic, where transmission is via direct physical contact. It has been reported that there are several benefits for health facilities applying mHealth, eHealth, and telemedicine to support patient care. These benefits include reduced workload on the healthcare providers, reduced exposure of the clinicians to infection, sustainable staff strength, and decrease wastage of personal protective equipment (PPEs) [26].

5 Specific Authors That Worked on COVID-19 Pandemic Using eHealth, mHealth, and Telemedicine

Babalola et al. [27] reported that in remote healthcare approach, telehealth is a fast growing technology that has been proven to be very useful in addressing many pandemic diseases. Even though in the Sub-Saharan African states the technology has not been fully implemented and adopted in the public health settings, the technology tends to offer great opportunity in increasing the efficiency of healthcare resources, inhibition of medical personnel infection, forward triaging, participation, supporting medical students' clinical surveillance, and guarantee of social support for patients. The authors noted that many factors may impede the implementation of this technology such as low political will, absence of policy frame work and regulations, price of maintenance of telehealth services, inadequate funding, willingness to pay, and patient and healthcare personnel bias on telehealth.

Elham and Alireza [28] reported that the public health emergency due to the COVID-19 pandemic has caused a lot of international concerns. The authors suggested that telehealth is a valuable tool for curbing the spread of the disease and improving health services through effective diagnosing, preventing, control, and treatment regime. Bokolo [29] revealed that information and communication technology, eHealth, and telemedicine embedded with different software programs and algorithms provide a platform for high-speed telecommunications systems in health system for effective healthcare management, delivery, monitoring, and diagnostic approaches. The authors noted that the application of telemedicine has provided a lot of timely intervention particularly during the recent global COVID-19 pandemic situation. These platforms offer great opportunity to protect patients and medical practitioners from contaminations or infections, thus curbing the spread of the disease. This technology is safe, convenient, effective, and scalable for various clinical care in health sector.

Okereafor et al. [16] demonstrated that there is urgent need for rapid development of novel strategies to expand and promote innovative measures in the detection, monitoring, and control of COVID-19 through the adoption and implementation of Telehealth, non-contact technologies, telemedicine, use of high-speed telecommunications systems, for improved service delivery, monitoring, generation of dataset, analysis of data, and management of healthcare system. Betancourt et al. [30] reported the devastating effects of COVID-19 on the US healthcare sector, thus suggested the adoption and implementation of telemedicine. This technology assists in facilitating the continuity, maintenance of standard practice and accessibility patient care, and strengthens the healthcare. Williams et al. [31] showed that in order to effectively restrain the extent and effect of the current global pandemic, concerted and drastic efforts must be put in place by health workers, researchers, and governments to combat the virus. The authors suggested that mobile health technology could offer great opportunity as a tool to create more awareness, information dissemination, smart tracking of affected persons, remote monitoring of patients, and virtual screening. The authors noted that many factors may hinder the adoption and implementation such as inadequate level of knowledge in mHealth, telehealth, accountability, eHealth and telemedicine, information privacy, transparency, and cost of maintenance.

Yamin and Alyoubi [32] stated the relevance of telemedicine and wireless sensor network applications particularly during COVID-19 pandemic and the way they influence implementation and utility of telemedicine in sustainable management of the COVID-19 pandemic around the globe as well as their relevance in health services. Doraiswamy et al. [33] reported that COVID-19 has resulted in economic and societal upheaval of unprecedented magnitude bringing about positive solutions to healthcare system across the globe. This attempt is aimed at mitigating the influence of COVID-19 through the application of telehealth. The authors noted that application of telehealth particularly in developing countries is still very vague, thus more effort must be put in place to boost the present technological competences in these regions.

Bhaskar et al. [34] reported that the adoption of telemedicine globally is on the rise; thus, many nations have encountered different issues with regard to implementation. The authors revealed that telemedicine is basically an integral aspect of public health for safety-net particularly during COVID-19 pandemic.

6 Role of Telemedicine in Early Detection and Control of COVID-19 Disease

Telehealth including mHealth, eHealth, and telemedicine can integrate numerous organizations and healthcare conditions into one virtual system, controlled by the central clinic and centers of rehabilitation centers as well as all registered patients in different locations. Telemedicine has played a significant function in the prevention of COVID-19 [35]. Telehealth can utilize all phases of the healthcare system to reduce the spread of COVID-19 infection, ensure people conform to good health practices, enhance safety protocol by providing online health services, protect patients, healthcare providers, and the entire population against exposure to SARs-CoV-2 infection, and ultimately reduce the burden on the clinicians and health system. Adequate application of these platforms has been associated with great benefits during COVID-19 pandemic with significant benefits on the patients, healthcare provider, and the general population. Using these platforms, the general population can have quick and convenient access to medical information thereby reducing the menace of contact to the infection [36]. Telemedicine can assist in early detection and control of COVID-19 disease, thereby reducing the broadcast of the causative virus, and decrease the burden of healthcare providers in different health centers.

7 High-Speed Telecommunications Systems and COVID-19

The high-speed telecommunication system includes the local and wide area networks, wired and wireless networks, and the soft and hardware components which help a system to connect with other devices [37]. Most of the connections are provided by network service providers which ensures or enable information sharing using different types of technologies which include wire, radio, optical, and other electromagnetic systems [38]. In the twenty-first century, most long-distance communication occurs through electrical and electromagnetic technologies, some of which include smartphones, television, radio, satellite, telegraph, networks, and many others. With the outbreak of COVID-19 in the year 2020, the use of high-speed telecommunication system has gained significant relevance. The COVID-19 pandemic caused lockdown restrictions which result in unprecedented health and socioeconomic crisis in the world at large [39]. The lockdown causes surge or increased use of network services and income generation for the network service providers [40]. The use of communication networks enhanced connectivity in many organizations and even the education sector is not left out as many institutions switched to virtual learning using different online learning platforms [41]. In the health sector, most patients with COVID-19 and those with other illnesses were restricted from meeting healthcare providers and therefore most ailments and even COVID-19 were treated or managed using the high-speed telecommunication system [42]. With the introduction of 5G high frequency powered network, communication will be very fast and efficient; however, it is important to assess the health safety of this network in order to preserve lives. Most institutions and organizations have continued to utilize the high-speed telecommunication system for effective service delivery even after the COVID-19 lockdown restriction order.

8 Application of eHealth, mHealth, Telemedicine, and Clinical Practice

eHealth is a health practice used particularly by patients to assess their health status. Most smart phones' eHealth applications can deliver good and quality health services and reduce the burden on healthcare professionals [43]. For instance, a pregnant woman can use eHealth applications to assess her health condition and that of the fetus [44]. In addition, can be used in calculating the time of delivery. In non-pregnant women, eHealth applications are used to monitor their menstrual cycle and the time of ovulation [45]. In healthcare research, several eHealth applications are available to calculate indices such as the body mass index, homeostasis model assessment insulin resistance (HOMA-IR), insulin sensitivity, lipid profile indices, and many others [46]. Definitely, the rapid progress in development of eHealth application will have great impact on healthcare services in the nearest future. The mobile health (mHealth) is a branch of the eHealth which uses electronic devices with software applications to provide healthcare services. mHealth enhances the healthcare providers and in most developing countries where there is shortage of health professionals, mHealth application will go a long way in delivering quality healthcare and improving patient satisfaction. In addition, some mobile health devices are now available to individuals for use and can be attached to different body parts to monitor vital signs in individuals with some clinical conditions [47]. With telemedicine, patients can access the medical services from the comfort of their homes. Those persons living in rural communities and who always have difficulty in assessing health professionals can now do that virtually, and even readings on medical devices can easily be captured and used for diagnosis.

9 Conclusion and Future Perspectives

This chapter provides adequate information on the application of eHealth, mHealth, and telemedicine applications and COVID-19. Relevant information are also provided application of telemedicine as an effective and safe applied by numerous healthcare experts for the detection of symptoms and prevention of COVID-19 infection. There is a need for the government to lay more emphasis on the significance of eHealth, mHealth, and telemedicine especially in the developing country. There is a need for policy makers, government, and numerous funding bodies to support many poor people in the developing countries.

References

1. Mashima, P. A., & Doarn, C. R. (2008). Overview of telehealth activities in speech-language pathology. *Telemedicine Journal and E-Health, 14*(10), 1101–1117.
2. Shaw, D. K. (2009). Overview of telehealth and its application to cardiopulmonary physical therapy. *Cardiopulmonary Physical Therapy Journal, 20*(2), 13–18.
3. Masson, M. (2014). Benefits of TED talks. *Canadian Family Physician, 60*(12), 1080.
4. Davarpanah, A. H., Mahdavi, A., Sabri, A., Langroudi, T. F., Kahkouee, S., Haseli, S., Kazemi, M. A., Mehrian, P., Mahdavi, A., Falahati, F., Tuchayi, A. M., Bakhshayeshkaram, M., & Taheri, M. S. (2020). Novel screening and triage strategy in Iran during deadly coronavirus disease 2019 (COVID-19) epidemic: Value of humanitarian teleconsultation service. *Journal of the American College of Radiology, 6*, 734–738. https://doi.org/10.1016/j.jacr.2020.03.015. Epub 2020 Mar 24. PMID: 32208138; PMCID: PMC7118529.
5. Zhai, Y., Wang, Y., Zhang, M., Gittell, J. H., Jiang, S., Chen, B., Cui, F., He, X., Zhao, J., & Wang, X. (2020). From isolation to coordination: how can telemedicine help combat the COVID-19 outbreak? *medRxiv, 2020.* https://doi.org/10.1101/2020.02.20.20025957.
6. Reeves, J. J., Hollandsworth, H. M., Torriani, F. J., Taplitz, R., Abeles, S., Tai-Seale, M., Millen, M., Clay, B. J., & Longhurst, C. A. (2020). Rapid response to COVID-19: Health informatics support for outbreak management in an academic health system. *Journal of the American Medical Informatics Association, 27*(6), 853–859. https://doi.org/10.1093/jamia/ocaa037. PMID: 32208481; PMCID: PMC7184393.
7. Ibekwe, T. S., & Fasunla, A. J. (2020). Telemedicine in otorhinolaryngological practice during COVID-19 pandemic. *Nigerian Medical Journal, 61*, 111–113.
8. Olaniyan, O. T., Adetunji, C. O., Okotie, G. E., Adeyomoye, O., Anani, O. A., & Mali, P. C. (2021). Impact of COVID-19 on assisted reproductive technologies and its multifacet influence on global bioeconomy. *Journal of Reproductive Healthcare and Medicine, 2*(Suppl_1), 92–104. https://doi.org/10.25259/JRHM_44_2020.
9. Smith, A. C., Thomas, E., Snoswell, C. L., Haydon, H., Mehrotra, A., Clemensen, J., & Caffery, L. J. (2020). Telehealth for global emergencies: Implications for coronavirus disease 2019 (COVID-19). *Journal of Telemedicine and Telecare, 26*(5), 309–313. https://doi.org/10.1177/1357633X20916567. Epub. PMID: 32196391; PMCID: PMC7140977.

10. Simcock, R., Thomas, T. V., Estes, C., Filippi, A. R., Katz, M. A., Pereira, I. J., & Saeed, H. (2020). COVID-19: Global radiation oncology's targeted response for pandemic preparedness. *Clinical and Translational Radiation Oncology, 22*, 55–68. https://doi.org/10.1016/j.ctro.2020.03.009. PMID: 32274425; PMCID: PMC7102593.

11. Monaghesh, E., & Hajizadeh, A. (2020). The role of telehealth during COVID-19 outbreak: A systematic review based on current evidence. *BMC Public Health, 20*(1193), 1–9. https://doi.org/10.1186/s12889-020-09301-4.

12. Liu, S., Yang, L., Zhang, C., Xiang, Y. T., Liu, Z., Hu, S., & Zhang, B. (2020). Online mental health services in China during the COVID-19 outbreak. *Lancet Psychiatry, 4*, e17–e18. https://doi.org/10.1016/S2215-0366(20)30077-8. Epub 2020 Feb 19. PMID: 32085841; PMCID: PMC7129099.

13. Khobi, J. A. M., Mtebe, J. S., & Mbelwa, J. T. (2020). Factors influencing district health information system usage in Sierra Leone: A study using the technology-organization environment framework. *Electronic Journal of Information Systems in Developing Countries, e12140*, 1–15. https://doi.org/10.1002/isd2.12140.

14. The Washington Post. (2020). South Africa Is Hunting DownCoronavirus with Thousands of Health Workers. Available at:https://www.washingtonpost.com/world/africa/south-africa-is-hunting-down-coronavirus-with-tens-of-thousands-of-health-workers/2020/04/21/6511307a-8306-11ea-81a3-9690c9881111_story.html. Accessed May 20, 2020.

15. United Purpose. (2020). COVID-19: Our Response in Senegal. Available at: https://united-purpose.org/covid19-senegal.Accessed May 20, 2020.

16. Okereafor, K., Olajide, A., & Rania, D. (2020). Exploring the potentials of telemedicine and other non-contact electronic health technologies in controlling the spread of the novel coronavirus disease (COVID-19). *International Journal in IT and Engineering (IJITE), 8*(4), 1–13.

17. Kapoor, A., Guha, S., Das, M. K., Goswami, K. C., & Yadav, R. (2020). Digital healthcare: The only solution for better healthcare during COVID-19 pandemic? *Indian Heart Journal, 72*(2), 61–64. https://doi.org/10.1016/j.ihj.2020.04.001.

18. Chauhan, V., Galwankar, S., Arquilla, B., Garg, M., Di Somma, S., El-Menyar, A., & Stawicki, S. P. (2020). Novel coronavirus (COVID-19): Leveraging telemedicine to optimize care while minimizing exposures and viral transmission. *Journal of Emergencies, Trauma, and Shock, 13*(1), 20.

19. Hollander, J. E., & Carr, B. G. (2020). Virtually perfect? Telemedicine for Covid-19. *The New England Journal of Medicine, 382*(18), 1679–1681. https://doi.org/10.1056/NEJMp2003539. Epub. PMID: 32160451.

20. Pecchia, L., Piaggio, D., Maccaro, A., Formisano, C., & Iadanza, E. (2020). The inadequacy of regulatory frameworks in time of crisis and in low-resource settings: Personal protective equipment and COVID-19. *Health and Technology*, 1–9. Advance online publication. https://doi.org/10.1007/s12553-020-00429-2.

21. Hong, Y. R., Lawrence, J., Williams, D., Jr., & Mainous, A., III. (2020). Population-level interest and telehealth capacity of US hospitals in response to COVID-19: Cross-sectional analysis of Google search and National Hospital Survey Data. *JMIR Public Health and Surveillance, 6*(2), e18961. https://doi.org/10.2196/18961. PMID: 32250963; PMCID: PMC7141249.

22. Hutchings, O. R., Dearing, C., Jagers, D., Shaw, M. J., Raffan, F., Jones, A., Taggart, R., Sinclair, T., Anderson, T., & Ritchie, A. G. (2021). Virtual health care for community management of patients with COVID-19 in Australia: Observational cohort study. *Journal of Medical Internet Research, 23*(3), e21064. https://doi.org/10.2196/21064. PMID: 33687341; PMCID: PMC7945978.

23. Mann, D. M., Chen, J., Chunara, R., Testa, P. A., & Nov, O. (2020). COVID-19 transforms health care through telemedicine: Evidence from the field. *Journal of the American Medical Informatics Association, 27*(7), 1132–1135. https://doi.org/10.1093/jamia/ocaa072. PMID: 32324855; PMCID: PMC7188161.

24. Gong, M., Liu, L., Sun, X., Yang, Y., Wang, S., & Zhu, H. (2020). Cloud-based system for effective surveillance and control of COVID-19: Useful experiences from Hubei, China. *Journal of Medical Internet Research, 22*, e18948.

25. Yee, T. S., Seong, L. C., & Chin, W. S. (2019). Patient's intention to use mobile health app. *Journal of Management Research, 11*, 18.

26. Doshi, A., Platt, Y., Dressen, J. R., Mathews, B. K., & Siy, J. C. (2020). Keep calm and log on: Telemedicine for COVID-19 pandemic response. *Journal of Hospital Medicine, 15*(5), 302–304.

27. Babalola, D., Anayo, M., & Itoya, D. A. (2021). Telehealth during COVID-19: Why Sub-Saharan Africa is yet to log-in to virtual healthcare? *AIMS Medical Science, 8*(1), 46–55. https://doi.org/10.3934/medsci.2021006.

28. Elham, M., & Alireza, H. (2020). The role of telehealth during COVID-19 outbreak: A systematic review based on current evidence. *BMC Public Health, 20*, 1193. https://doi.org/10.1186/s12889-020-09301-4.

29. Bokolo, A. J. (2021). Application of telemedicine and eHealth technology for clinical services in response to COVID-19 pandemic. *Health and Technology (Springer)*. https://doi.org/10.1007/s12553-020-00516-4.

30. Betancourt, J. A., Rosenberg, M. A., Zevallos, A., Brown, J. R., & Mileski, M. (2020). The impact of COVID-19 on telemedicine utilization across multiple service lines in the United States. *Healthcare, 8*(380), 1–21. https://doi.org/10.3390/healthcare8040380.

31. Williams, S. Y., Adeyemi, S. O., Eyitayo, J. O., Odeyemi, O. E., Dada, O. E., Adesina, M. A., & Akintayo, A. D. (2020). Mobile health technology (Mhealth) in combating COVID-19 pandemic: Use, challenges and recommendations. *European Journal of Medical and Educational Technologies, 13*(4), em2018. https://doi.org/10.30935/ejmets/8572.

32. Yamin, M. A. Y., & Alyoubi, B. A. (2020). Adoption of telemedicine applications among Saudi citizens during COVID-19 pandemic: An alternative health delivery system. *Journal of Infection and Public Health, 13*, 1845–1855. https://doi.org/10.1016/j.jiph.2020.10.017.

33. Doraiswamy, S., Abraham, A., Mamtani, R., & Cheema, S. (2020). Use of telehealth during the COVID-19 pandemic: Scoping review. *Journal of Medical Internet Research, 22*(12), e24087. 1–15.

34. Bhaskar, S., Bradley, S., Chattu, V. K., Adisesh, A., Nurtazina, A., Kyrykbayeva, S., Sakhamuri, S., Yaya, S., Sunil, T., Thomas, P., Mucci, V., Moguilner, S., Israel-Korn, S., Alacapa, J., Mishra, A., Pandya, S., Schroeder, S., Atreja, A., Banach, M., & Ray, D. (2020). Telemedicine across the globe-position paper from the COVID-19 pandemic health system resilience PROGRAM (REPROGRAM) international consortium (part 1). *Frontiers in Public Health, 8*, 556720. https://doi.org/10.3389/fpubh.2020.556720.

35. Pinzon-Perez, H., & Zelinski, C. (2016). The role of teleconferences in global public health education. *Global Health Promotion, 23*, 38–44.

36. Gao, Y., Liu, R., Zhou, Q., Wang, X., Huang, L., Shi, Q., Wang, Z., Lu, S., Li, W., Ma, Y., Luo, X., Fukuoka, T., Ahn, H. S., Lee, M. S., Luo, Z., Liu, E., Chen, Y., Shu, C., & Tian, D. (2020). Application of telemedicine during the coronavirus disease epidemics: A rapid review and meta-analysis. *Annals of Translational Medicine, 8*(10), 626. https://doi.org/10.21037/atm-20-3315.

37. Higuchi, K., Nakazawa, Y., Sakata, N., Takizawa, M., Ohso, K., Tanaka, M., Yanagisawa, R., & Koike, K. (2011). Telecommunication system for children undergoing stem cell transplantation. *Pediatrics International: Official Journal of the Japan Pediatric Society, 53*(6), 1002–1009.

38. Kvedar, J., Coye, M. J., & Everett, W. (2014). Connected health: A review of technologies and strategies to improve patient care with telemedicine and telehealth. *Health Affairs (Project Hope), 33*(2), 194–199.

39. Webb, R. (2020). COVID-19 and lockdown: Living in 'interesting times'. *Journal of Wound Care, 29*(5), 243.

40. Wosik, J., Fudim, M., Cameron, B., Gellad, Z. F., Cho, A., Phinney, D., Curtis, S., Roman, M., Poon, E. G., Ferranti, J., Katz, J. N., & Tcheng, J. (2020). Telehealth transformation: COVID-19 and the rise of virtual care. *Journal of the American Medical Informatics Association: JAMIA, 27*(6), 957–962.

41. Odriozola-González, P., Planchuelo-Gómez, Á., Irurtia, M. J., & de Luis-García, R. (2020). Psychological effects of the COVID-19 outbreak and lockdown among students and workers of a Spanish university. *Psychiatry Research, 290*, 113108.

42. Ghosh, A., Gupta, R., & Misra, A. (2020). Telemedicine for diabetes care in India during COVID19 pandemic and national lockdown period: Guidelines for physicians. *Diabetes & Metabolic Syndrome, 14*(4), 273–276.

43. Guo, U., Chen, L., & Mehta, P. H. (2017). Electronic health record innovations: Helping physicians – One less click at a time. *Health Information Management: Journal of the Health Information Management Association of Australia, 46*(3), 140–144.

44. van den Heuvel, J. F., Groenhof, T. K., Veerbeek, J. H., van Solinge, W. W., Lely, A. T., Franx, A., & Bekker, M. N. (2018). eHealth as the next-generation perinatal care: An overview of the literature. *Journal of Medical Internet Research, 20*(6), e202.

45. DeNicola, N., & Marko, K. (2020). Connected health and Mobile apps in obstetrics and gynecology. *Obstetrics and Gynecology Clinics of North America, 47*(2), 317–331.

46. Fagherazzi, G., & Ravaud, P. (2019). Digital diabetes: Perspectives for diabetes prevention, management and research. *Diabetes & Metabolism, 45*(4), 322–329.

47. Majumder, S., Mondal, T., & Deen, M. J. (2017). Wearable sensors for remote health monitoring. *Sensors (Basel, Switzerland), 17*(1), 130.

Prediction of Care for Patients in a COVID-19 Pandemic Situation Based on Hematological Parameters

Arianne Sarmento Torcate, Flávio Secco Fonseca, Antônio Ravely T. Lima, Flaviano Palmeira Santos, Tássia D. Muniz S. Oliveira, Maíra Araújo de Santana, Juliana Carneiro Gomes, Clarisse Lins de Lima, Valter Augusto de Freitas Barbosa, Ricardo Emmanuel de Souza, and Wellington P. dos Santos (iD)

1 Introduction

In December 2019, the World Health Organization (WHO) received notifications from China regarding cases of pneumonia and severe respiratory syndrome among workers at the seafood market in Wuhan, China, having unknown cause [1]. In March 2020, WHO classified the state of this disease as pandemic, known as COVID-19, which has a high rate of transmissibility, with the ability to spread fast throughout the world [2].

After a year of pandemic, the clinical manifestations of COVID-19 are not fully understood, as there is still limited information to characterize the clinical picture of the disease [3]. However, at the beginning of the pandemic, Brazilian Ministry of Health established the flu syndrome as the most common manifestation of this disease, which is defined as an acute respiratory condition, characterized by a feverish sensation or fever, accompanied by cough, sore throat, runny nose, or difficulty breathing.

The World Health Organization clarifies that the initial signs and symptoms of the disease resemble a common flu-like condition, but can vary from person to person, and may manifest through pneumonia, severe pneumonia, and severe acute

A. S. Torcate · F. S. Fonseca · M. A. de Santana · J. C. Gomes · C. L. de Lima
Graduate Program in Computer Engineering, Polytechnique School of the University of Pernambuco, Recife, Brazil

A. R. T. Lima · F. P. Santos · T. D. M. S. Oliveira · R. E. de Souza · W. P. dos Santos (✉)
Department of Biomedical Engineering, Federal University of Pernambuco, Recife, Pernambuco, Brazil
e-mail: wellington.santos@ufpe.br

V. A. de Freitas Barbosa
Academic Unit of Serra Talhada, Rural Federal University of Pernambuco, Serra Talhada, Brazil

© The Author(s), under exclusive license to Springer Nature Switzerland AG 2022
S. K. Pani et al. (eds.), *Assessing COVID-19 and Other Pandemics and Epidemics using Computational Modelling and Data Analysis*,
https://doi.org/10.1007/978-3-030-79753-9_11

respiratory syndrome [4]. In the current context, people with comorbidities such as diabetes, cardiovascular diseases, obesity, hypertension, tuberculosis, and others are at higher risk of rapid worsening of the disease, which can lead to death.

With that in mind, the early detection and management of this disease is essential, especially in the Brazilian context, a country with continental dimensions and diverse territorial realities, with great social inequality and limitations in access to health services by the population. Thus, it is necessary to know and check information about local realities (states and cities) to make decisions in this scenario.

It is worth mentioning that, based on this problem, organizations and researchers from different areas of knowledge have sought answers to questions related to health problems caused by COVID-19. Scientific investigations aim to provide immediate actions that collaborate to control the pandemic, that is, that contribute to assist in clinical, social, or political decision-making, all based on scientific evidence, to maximize the benefits and minimize injuries and costs.

From the existing initiatives to assist the health professional in decision-making, we highlighted in this research the Artificial Intelligence (AI) and biostatistics, which together can predict, for instance, the survivor or risk of patients from their physiological parameters. These predictions allow treatment individualization, and greater chances of complete recovery. Among the subareas of AI, we highlight machine learning, whose algorithms have become one of the most used forms of classification and prediction of patterns in large data nowadays [5].

Thereby, the main goal of this research is to make predictions about the type of hospitalization and severity assessments of patients with and without COVID-19. For this, we used hematological data from patients of the units of the Unified Health System (SUS) in the city of Paudalho, Brazil. The aim is to analyze algorithms that are capable of making hospitalization predictions into one of the three possible choices: regular ward, semi-intensive care unit, and intensive care unit, corresponding to mild (non-critical), moderate, and severe cases. In order to perform this study, seven classic classifiers were applied to the data set, such as Support Vector Machine (SVM), Multilayer Perceptron (MLP), Random Tree, Random Forest, Bayesian Networks, and J48 Decision Trees, all with well-defined parameter settings tested along the experiments.

This chapter is structured as follows. After this introduction, Sect. 2 and its respective subsections highlight the theoretical background used in this project. Section 3 presents the works related to the developed research. Section 4 sets out the methodological path adopted, while Sect. 5 describes in detail the obtained results. Section 6 presents a general analysis, building discussions of the results. Final considerations and perspectives for future work are presented in Sect. 7.

2 Theoretical Foundation

In this section, we present the theoretical approaches related to the main topics of this research, with emphasis on COVID-19 and machine learning.

2.1 COVID-19

Characterized as a contagious respiratory disease, COVID-19 is associated with high mortality rates since its emergence in December 2019 [6]. According to the World Health Organization, the coronavirus pandemic is putting even the best health systems worldwide under tremendous pressure [7]. The study by Nemati et al. [6] states that on March 24, 2020, the virus had spread to more than 170 countries, with more than 422,613 confirmed cases and 18,891 deaths. In addition, mortality rates may vary between countries due to demographic differences, age distribution, and healthcare infrastructure.

The symptoms of COVID-19 are similar to a common flu condition, such as malaise, fever, fatigue, cough, mild dyspnea, anorexia, sore throat, body pain, headache, and nasal congestion [3]. However, as this disease progresses, patients feel shortness of breath, nausea culminating in pneumonia, and multiple organ failure. Because of this, the best way so far to prevent the disease is individual protection, such as hand washing, correct use of masks, and social isolation. These protective measures must be aligned with government actions, such as providing more beds in intensive care units, hiring and qualifying frontline professionals, providing basic sanitation, acquiring specialized medical equipment such as mechanical fans, and increasing and decentralizing the performance of rapid tests and vaccination of the population.

The best accepted test to diagnose COVID-19 is the molecular test, such as reverse transcription followed by polymerase chain reaction (RT-PCR), which identifies the SARS-CoV-2 viral RNA. In this type of test, secretions from the nasopharynx are collected. However, according to Iser et al. [3], citing the work developed by Woelfel et al. [8], Tolia et al. [9], and Hadaya et al. [10], these tests should be performed between the third and seventh day of infection. This period guarantees greater precision of the method and reduction of false-negative results. The problem is that in many cases, it is difficult to identify when the patient became infected. In addition, RT-PCR tests are normally performed on health professionals and on symptomatic patients who have been hospitalized, due to the high cost and the scarcity of certified laboratories for its performance. Serologic tests by immunochromatography, known as rapid tests, have become an option for: (1) people with mild to moderate symptoms, without the need for hospitalization; (2) public health system, for tracking asymptomatic cases, epidemiological survey of confirmed cases, and estimating the population's immunization rate. In this case, they should be used 7 days after the onset of symptoms. Unfortunately, rapid tests are non-specific for the detection of virus presence directly.

Briefly, the coronavirus increased the need for immediate clinical decisions and the effective use of health resources, as record pressure was imposed on health systems worldwide. The aim of developing techniques that assist in decision-making is to contribute to the control not only of the pandemic, but of the factors associated with the problem of interest. As a result, scientific investigations, ethical commitment, and the ability to perceive important clinical gaps for characterizing and defining hypotheses have become essential.

2.2 *Machine Learning*

Machine learning (ML) is a subarea of Artificial Intelligence. From a more technical point of view, ML stems from the difficulty in manually handling a large volume of available data, proposing intelligent systems that can learn the patterns or regularities in the data [12]. One of the objectives of ML techniques is to perform pattern detection in databases [13]. This is possible through data that provides machines with the ability to learn and, from that, recognize patterns and create relationships between variables.

The use of machine learning has become very valuable in several intelligent applications, solving most data-related problems [14]. With the algorithms present in machine learning it is possible to work with hundreds of attributes, either in the detection/use of the interactions between the attributes and thus favor the support to the diagnosis in the health area [15]. Machine learning techniques have been used in activities that mainly involve the identification of patterns [13, 15–20]. These techniques are significantly contributing to the resolution of real problems such as prediction, diagnosis, and recognition of health problems.

In this COVID-19 scenario, prediction has become a priority for public health. Thus, the use of ML algorithms has been useful to health professionals, and can help in several factors, such as for pneumonia detection by analyzing chest X-ray images [16], support for diagnosis through blood tests [13], predictive model to help doctors choose the best therapeutic strategies for patients with COVID-19 [21], prediction of mortality from blood tests [22], and others.

3 Related Works

To contribute in the context of forecasts that estimate the outbreak of COVID-19, the study by Ardabili et al. [23] presents a thorough and comparative analysis of machine learning and soft computing models, both in alternative to the Susceptible-Infected-Recovered (SIR) and Susceptible-Exposed-Infectious-Removed (SEIR) models. The methods were applied to data from five countries (Italy, Germany, Iran, the USA, and China) for the total cases obtained in 30 days. Among a range of investigated ML models, the Multilayer Perceptron (MLP) and the Adaptive Neuro-Fuzzy Inference System (ANFIS) stood out and showed promising results. The authors reinforce the importance of researchers dedicating themselves to investigating predictions to also estimate the number of infected patients, as well as the number of deaths. Thus, ML models must be analyzed and take into account the individual context of each country.

The research carried out by Nemati et al. [6] also contributes in this context by developing predictive models with the ability to predict the length of stay of patients in the hospital. For this, survival characteristics of 1182 patients were considered, where the time of discharge was chosen as a variable of interest, along with survival analysis techniques, including statistical analysis and seven algorithms machine learning. The results obtained indicate that being male or belonging to older age groups is associated with lower probabilities of hospital discharge. In addition, the Gradient Boosting (GB) survival model surpasses other models for predicting patient survival in the researched context. Stagewise GB, on the other hand, offers the most accurate download time prediction compared to other algorithms. But, Kaplan-Meier Estimator and Cox regression methods suggest that the gender and age of hospitalized patients have a direct effect on recovery time. Finally, all findings are of great relevance to help healthcare professionals make more assertive decisions during the outbreak.

Still analyzing the various possibilities of developing predictive models to contribute in the context of COVID-19, we also highlight the work carried out by Batista et al. [11] at the beginning of the pandemic in Brazil. The work aimed to predict the risk of a positive diagnosis of COVID-19 with machine learning, using data resulting from admission exams in the RT-PCR tests of 235 adult patients, in the emergency room of Hospital Israelita Albert Einstein in São Paulo, from March 17 to 30, 2020. In all, five ML algorithms (neural networks, Random Forests, gradient augmentation trees, logistic regression, and SVM) were used. The results show that the best predictive performance was obtained by the SVM algorithm (AUC: 0.85; Sensitivity: 0.68; Specificity: 0.85; BrierScore: 0.16). The three most important variables for the predictive performance of the algorithm were the number of lymphocytes, leukocytes, and eosinophils, respectively.

Researchers also appear in order to propose diagnoses that are faster and cheaper. An example of this is the work developed by Kumar et al. [7]. A classifier based on ML and Deep Learning using ResNet152 on chest X-ray images of patients with COVID-19 was proposed for prediction of the new coronavirus. This work focused on the prevention of spread of the virus by asymptomatic patients. The authors used the SMOTE technique to balance the data points and then apply the algorithms. The best results were obtained using the Random Forest model, which stood out in precision (0.973), sensitivity (0.974), specificity (0.986), F1 score (0.973), and AUC (0.997). About the predictive model, the XGBoost performed better, with precision (0.977), sensitivity (0.977), specificity (0.988), F1 score (0.977), and AUC (0.998). Both models contribute to the effective clinical prediction of COVID-19.

Gomes et al. [16] proposed an intelligent system to support the diagnosis of COVID-19, investigating radiographs from different databases. Radiographies of patients with viral pneumonia, bacterial pneumonia, and healthy patients were obtained from the Kaggle website. The X-rays of patients with COVID-19 were obtained from four different databases: open source GitHub repository shared by Dr. Joseph Cohen et al. [24]; COVID-19 database, made available online by Societa Italiana di Radiologia Medica e Interventistica [25], and Peshmerga Erbil Hospital database. The authors analyzed the classification performance, using five metrics: accuracy, sensitivity, precision, specificity, and kappa index. The machine learning methods used to classify X-ray images were the Multilayer Perceptron, Support Vector Machine, Decision Trees, Bayesian Network, and Naive Bayes, and all experiments were carried out with the Weka software. The work showed that the system can diagnose COVID-19 with an average accuracy of 89.78%. Its prototype is already developed, it was able to differentiate COVID-19 from viral and bacterial pneumonia and has low computational cost.

In the work of de Barbosa et al. [13], several experiments are also carried out with machine learning methods, such as MLP, SVM, Random Trees, Random Forest, Bayesian Networks, and Naive Bayes in order to propose an intelligent system with classic classifiers and low computational cost to support the diagnosis COVID-19 based on blood tests. Six metrics were chosen to analyze the classification performance: accuracy, precision, sensitivity, specificity, recall, and precision. The databases were made available by Hospital Israelita Albert Einstein located in São Paulo, Brazil, which are available on the Kaggle platform. Bayes Network was the best method that stood out, being able to achieve high diagnostic performance, with general precision of 95.159% $\pm$ 0.693 of general precision, kappa index of 0.903 $\pm$ 0.014, sensitivity of 0.968 $\pm$ 0.007, precision of 0.938 $\pm$ 0.010, and specificity of 0.936 $\pm$ 0.011. According to the authors, the availability of this software system combined with rapid and low-cost tests, based on blood tests, can be of great help in overcoming the testing challenges that are being faced worldwide.

With the works presented above, it is evident that the machine learning area can be applied to different types of data contributing to the pandemic scenario caused by COVID-19, considering different goals and objectives, either at the global or individual level of each country. For this, this research field has peculiarities, potentialities, and interdisciplinary alignment with other areas. For comparative effect and better understanding, each of the works cited in this section, they were summarized in Table 1, considering the main goal, methods used, and obtained results.

4 Methods

In this study, we evaluated 41 hematological data (blood tests) from patients treated at public healthcare units in the city of Paudalho, Brazil. These 41 tests (Table 2) were the features used as classification input. Patient records covered the period from December 2019 to August 2020. In order to predict hospitalization, we assessed data from three different hospitalization conditions: regular ward, semi-intensive care unit, and intensive care unit. Regular ward refers to regular service or non-critical cases, while semi-intensive care unit corresponds to moderate cases, and intensive care unit is related to severe cases. It is important to mention that all procedures involving human participants were performed in accordance with the 1964 Helsinki declaration and with the ethical standards of the institutional research committee from the Federal University of Pernambuco, registered under number 34932020.3.0000.5208.

The experiments were performed using the WEKA software, version 3.8 [26]. For a better understanding, Fig. 1 illustrates the methodological path used to carry out the experiments.

Table 1 Summary of related works

Authors/Year	Objective	Method	Results
Ardabili et al. [23]	Perform a comparative analysis of machine learning and soft computing models to predict the outbreak of COVID-19	Genetic algorithms (GA), particle swarm optimizer (PSO), gray wolf optimizer (GWO), and others	The MLP and ANFIS models stood out in the analysis due to the high generalization capacity for long-term forecasting
Nemati et al. [6]	Analyze the survival characteristics of patients with COVID-19 by computational techniques and predict the length of stay of these patients in the hospital	Statistical analysis techniques along with machine learning taking into account the survival characteristics of 1182 patients	The results obtained show that the gradient boosting model surpasses the other models for predicting patient survival, followed by the KM and cox regression methods
Batista et al. [11]	Predict the risk of a positive diagnosis of COVID-19 with machine learning, using the results of admission exams in the RT-PCR tests of 235 adult patients as predictors in the emergency room	Five machine learning algorithms were used in the experiments, namely, neural networks, random forests, gradient increase trees, logistic regression, and SVM	The best predictive performance was obtained by the SVM algorithm (with AUC: 0.85; SEN: 0.68; ESP: 0.85; BrierScore: 0.16). And, three variables were identified as most important for good predictive performance, namely: Number of lymphocytes, leukocytes, and eosinophils
Kumar et al. [7]	Propose a classifier capable of diagnosing patients with COVID-19 using chest X-ray images	ML-based classifier and deep learning using ResNet152 on chest X-ray images of patients with COVID-19 for early and non-invasive prediction of the new coronavirus	The best results were obtained using the random Forest model, which excelled in terms of accuracy (0.973), F1 score (0.973), AUC (0.997), SEN (0.974), and ESP (0.986)
Gomes et al. [16]	Development of an intelligent system to support the diagnosis of COVID-19, using radiographs from different databases	Multiclass classification, differentiating between multiple respiratory diseases, such as COVID-19, viral pneumonia, and bacterial pneumonia	The developed system can diagnose COVID-19 with up to 89.78% of average accuracy, also being able to differentiate COVID-19 from viral and bacterial pneumonia

(continued)

Table 1 (continued)

Authors/Year	Objective	Method	Results
de Barbosa et al. [13]	Propose an intelligent system capable of supporting the COVID-19 diagnosis based on blood tests	Experimenting with six classic machine learning models, namely, MLP, SVM, random trees, random Forest, Bayesian networks, and naive Bayes	The model that stood out most was the SVM, capable of achieving high diagnostic performance, with an overall accuracy of 95.159% ± 0.693 and low computational cost
This work	Analyze intelligent classifiers that are able to make hospitalization predictions considering three possible scenarios: Regular ward, semi-intensive care unit, and intensive care unit, corresponding to mild (non-critical), moderate, and serious cases	For this, we used hematological data from patients of the units of the unified health system in the city of Paudalho, Brazil. Where, seven classic classifiers were applied to the data set, such as SVM, MLP, random tree, random Forest, Bayesian networks, and J48 decision trees	The results obtained show the random Forest with 100 trees showed the best potential to perform the predictions for regular ward (ACC: 82%; KPP: 0.642; SEN: 0.730 and ESP: 0.913), as for the semi-intensive care unit (ACC: 81%; KPP: 0.633; SEN: 0.890 and ESP: 0.875), and intensive care unit (ACC: 82%; KPP: 0.640; SEN: 0.640 e ESP: 0.947)

It is valid to clarify that all steps were applied individually for each one of the three hospitalization conditions (regular, semi-intensive, and intensive). Class balancing was performed (Step 1), using SMOTE method (Synthetic Minority Oversampling Technique) [27], which aims to generate artificial instances based on existing samples to balance the classes. The settings used in this step can be seen in Table 3.

After class balancing, in Step 2 we pre-processed the data using MLPAutoencoder algorithm, which is an unsupervised learning method to select attributes and, consequently, decrease data dimensionality [28].

In Steps 3, 4, and 5 seven (7) classic classifiers were applied, both for balanced databases (Step 1) and for databases with selected attributes (Step 2). It is worth mentioning that in order to obtain individual statistical performance information for the analyses we tested each classifier 30 times, using the k-fold cross-validation method with the number of folds equal to 10. Table 4 shows which classifiers were used and their respective configuration.

Exceptionally, when using the original database, without pre-processing, (Step 5), only the classifiers that obtained the best results in Steps 3 and 4 were used, namely, Random Tree, Random Forest, and J48 Decision Tree, all configured with the parameters shown in Table 3. The choice of these algorithms was due to their performance in terms of the analyzed metrics, in addition to the consumption of time and memory involved in their processing. In Step 6, some evaluative metrics were used to perform a multimodal analysis of the obtained results. The metrics

Table 2 Features extracted from patients' blood tests

List of features				
Hematocrit	Leukocytes	Serum glucose	Metamyelocytes	Segmented
Hemoglobin	Basophils	Neutrophils	Myelocytes	HB saturation arterial blood gases
Platelets	Mean corpuscular hemoglobin Mch	Urea	Myeloblasts	Total CO_2 arterial blood gas analysis
Mean platelet volume	Eosinophils	C-reactive protein	Partial thromboplastin time PTT	Promyelocytes
Red blood cells	Mean corpuscular volume MCV	Creatinine	Lactic dehydrogenase	PCO_2 arterial blood gas analysis
Lymphocytes	Monocytes	Total bilirubin	Prothrombin time pt. activity	HCO_3 arterial blood gas analysis
Mean corpuscular hemoglobin concentration Mchc	Red blood cell distribution width rdw	Direct bilirubin	Lipase dosage	Indirect bilirubin
D-dimer	Base excess arterial blood gas analysis	PH arterial blood gas analysis	PO_2 arterial blood gas analysis	Arterial FIO$_2$
CTO$_2$ arterial blood gas analysis				

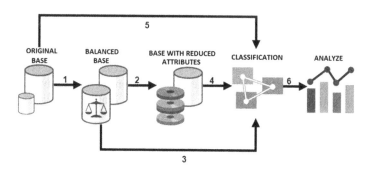

Fig. 1 Steps of the methodological path. In Step 1 the class balancing was carried out; in Step 2 we performed attribute selection; in Steps 3, 4, and 5, we trained and tested the classification models; and Step 6 consists in the metrics analyses

Table 3 Settings used for the SMOTE method

Care unit	Class value	Nearest neighbors	Percentage
Regular	0	2	95%
Semi-intensive	0	2	27.7%
Intensive	0	2	790%

Table 4 Classifiers configuration

Classifier	Parameters
Naive Bayes	Batch size: 100
Bayes net	Batch size: 100
Random tree	Batch size: 100 Seed: 1
J48 decision tree	Batch size: 100
Random Forest	Trees: 10, 20, 50 e 100 Batch size: 100
MLP	Neurons in the hidden layer: 20, 50 and 100 Learning rate: 0.3 Iterations: 500
SVM	Linear kernel ($P = 1$) Polynomial kernel ($P = 2$ and $P = 3$) RBF kernel (gamma: 0.01; 0.25 and 0.5)

were accuracy (ACC), kappa Statistics (KPP), sensitivity (SEN), specificity (ESP), ROC curve area (AUC), and training time (TT). In the results section, it is also reinforced that, for statistical purposes, the values presented refer to the averages and standard deviation of each metric, calculated from the 300 testing values found from the experiments.

5 Results

For better understanding, the results section was divided into three subsections, each referring to one of the care units. In each subsection we, respectively, assessed classifiers' performances in the prediction of regular ward hospitalization, semi-intensive care unit hospitalization, and intensive care unit hospitalization.

5.1 Regular Ward Hospitalization

In the data referring to the regular care wing, class 0 has 4110 instances and represents patients who did not need to be admitted to the regular ward. While class 1 is composed of 2105 instances and refers to patients who needed to be admitted to the regular ward. As a result, the imbalance between the two classes mentioned is visible, with a difference of 2005 instances. To solve this problem, we applied the SMOTE method considering the real instances as examples to generate synthetic data, based on two neighbors and using the percentage of 95%. After applying SMOTE, classes 0 and 1 were balanced, both with 4110 instances.

In Step 2, MLPAutoenconder was applied to select attributes and thus, reduce the dimensionality of the data. Then, from the 41 features of the original database (listed in Table 2), only 10 were identified by the model as relevant.

The results obtained after the application of the seven classifiers (Step 3) on the balanced database (Step 1) show that the model with the best performance in relation to accuracy (82.1%), kappa statistics (0.64), sensitivity (0.73), and specificity (0.91) was the Random Forest with 100 trees. On the other hand, the worst result achieved for this case was obtained by Naive Bayes algorithm, in relation to accuracy (65%), kappa (0.30), and sensitivity (0.38); however, the average specificity value of 0.98 was slightly higher than the other methods. These results can be seen in Fig. 2, and may be further analyzed in Table 5.

In Step 4, where we applied the MLPAutocoder to select attributes in the database pre-processed with SMOTE (Step 3), the Random Forest with 100 trees stood out in terms of accuracy (75%), kappa statistics (0.651), sensitivity (0.676), and specificity (0.824). Another notable point is the good results obtained by Random Forest with 50 trees, where kappa statistics (0.651) and sensitivity (0.676) are equal to the values of Random Forest with 100 trees, differing only in accuracy (74%) and specificity (0.800).

On the other hand, the SVM with linear kernel achieved a worse performance in terms of accuracy (67%) and sensitivity (0.320), but this same model showed better kappa statistics (0.489) and specificity (0.920) than the values obtained by Naive Bayes, with kappa of 0.298, and specificity of 0.880, but still with better accuracy (69%) and sensitivity (0.390).

Figure 3 presents the results obtained in Step 4. Briefly, it is clear that the algorithms with better performances were both Random Forest, using 50 and 100 trees, as already described. Naive Bayes, followed by SVM with linear kernel, is the model that stands out negatively when compared to the others to carry out the prediction of hospitalization in the regular ward.

Regarding Step 5, among the three classifiers, the accuracy (76.9%), kappa statistics (0.523), sensitivity (0.742), and specificity (0.822) of Random Forest with 100 trees stood out in relation to the other models. However, it is worth noting that Random Forest with 10, 20, and 50 trees also obtained good results, similar to those with 100 trees. Through Fig. 4 it is possible to view this information.

Even with good results, in this scenario, Random Tree continues to occupy the position of classifier with the worst performance in relation to the other models in terms of accuracy metrics (72%), kappa index (0.453), sensitivity (0.705), and specificity (0.798).

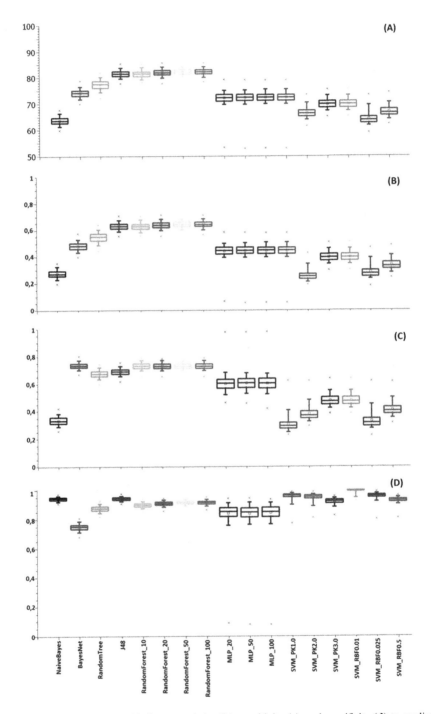

Fig. 2 Results of accuracy (**a**), kappa statistics (**b**), sensitivity (**c**), and specificity (**d**) to predict regular ward hospitalization using the balanced database (Step 3)

Table 5 Average and standard deviation of the evaluation metrics referring to the best models by type of care unit

Database	Type of care unit	Classifier	ACC (%)	KPP	SEN	SPE	AUC	TT (ms)
Original	Regular ward	RdmForest_100	76.936 ± 1.601	0.523 ± 0.034	0.742 ± 0.027	0.822 ± 0.045	0.854 ± 0.014	12.650 ± 5.309
	Semi-intensive	RdmForest_100	79.626 ± 1.568	0.594 ± 0.031	0.858 ± 0.022	0.861 ± 0.014	0.859 ± 0.056	74.831 ± 2.243
	Intensive	RdmForest_100	91.189 ± 0.674	0.335 ± 0.060	0.984 ± 0.004	0.262 ± 0.052	0.733 ± 0.034	3.327 ± 0.135
Balanced	Regular ward	RdmForest_100	82.175 ± 1.251	0.642 ± 0.030	0.730 ± 0.022	0.913 ± 0.013	0.884 ± 0.011	15.570 ± 4.101
	Semi-intensive	RdmForest_100	81.670 ± 1.710	0.633 ± 0.028	0.890 ± 0.017	0.742 ± 2.301	0.875 ± 0.012	8.217 ± 1.179
	Intensive	RdmForest_100	82.060 ± 1.040	0.640 ± 0.020	0.694 ± 0.018	0.947 ± 0.009	0.919 ± 0.007	7.261 ± 2.324
Selected features	Regular ward	RdmForest_100	75.033 ± 1.448	0.651 ± 0.029	0.676 ± 0.022	0.824 ± 0.018	0.808 ± 0.014	11.928 ± 0.510
	Semi-intensive	RdmForest_100	71.917 ± 1.615	0.438 ± 0.032	0.764 ± 0.022	0.674 ± 0.024	0.778 ± 0.016	5.059 ± 1.121
	Intensive	RdmForest_100	77.243 ± 4.205	0.545 ± 0.085	0.664 ± 0.111	0.882 ± 0.013	0.870 ± 0.054	16.758 ± 5.753

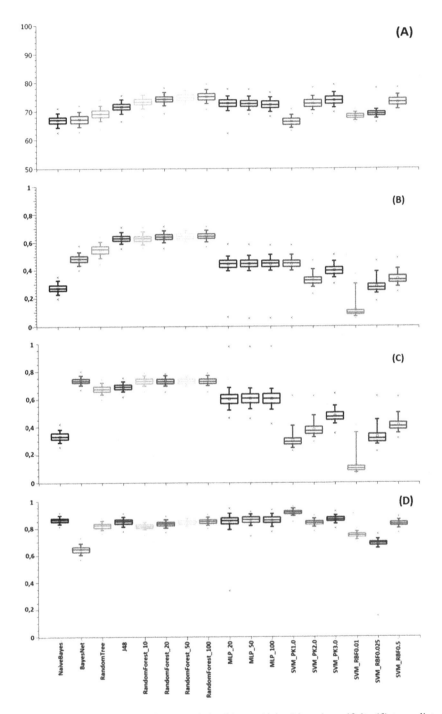

Fig. 3 Results of accuracy (**a**), kappa statistics (**b**), sensitivity (**c**), and specificity (**d**) to predict regular ward hospitalization using the database with features selected by MLPAutoencoder (Step 4)

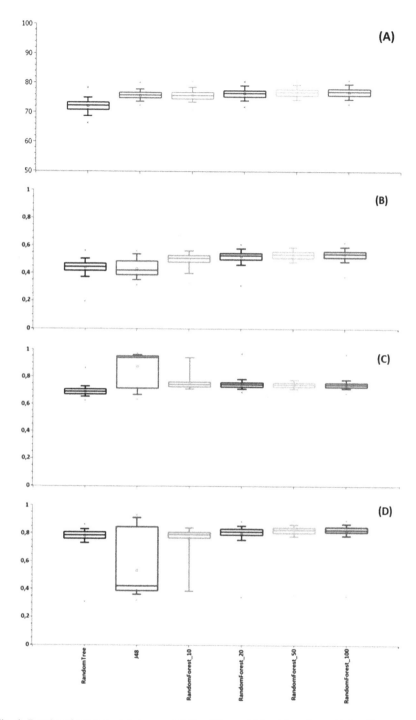

Fig. 4 Results of accuracy (**a**), kappa statistics (**b**), sensitivity (**c**), and specificity (**d**) to predict regular ward hospitalization using the original database, without pre-processing (Step 5)

5.2   Semi-Intensive Care Unit Hospitalization

In the data referring to semi-intensive care, class 0 is composed of 2728 instances and represents patients who did not need this type of care, while class 1 has 3487 instances and represents patients who needed care in the semi-intensive unit. There is clearly a difference of 759 instances between classes. In this unbalanced scenario, SMOTE was applied to these data in Step 1, considering two neighbors of the majority class and applying the percentage of 27.7% (as specified in Table 3). After this procedure, both classes were left with an equivalent number of instances of 3487.

Such as in the context of regular ward hospitalization, in Step 2, from the 41 original features, only 10 were selected by MLPAutoenconder. The prediction performance of the seven classifiers (Step 3) on the balanced database (Step 1) clearly demonstrate that the Random Forest with 100 trees stood out positively in terms of accuracy (81%), kappa statistics (0.633), sensitivity (0.890), and specificity (0.742). The Random Forest with 20 and 50 trees also performed well reaching similar results to that using 100 trees, as shown in Fig. 5.

On the other hand, Naive Bayes presents less satisfactory results regarding accuracy (61%), kappa statistics (0.230), and specificity (0.300); however, the sensitivity (0.980) of this classifier stands out in relation to the other models. The performance of this classifier is closely followed by SVM with RBF kernel and gamma of 0.01, with worse performances in the metrics of accuracy (60.8%), kappa statistics (0.240), and specificity (0.320). Similar to Naive Bayes, SVM also showed high sensitivity (0.870).

With the reduced number of attributes (Step 2), the results of Step 4 did not present major discrepancies between them. However, the best result was obtained by the Random Forest model with its respective configurations, highlighting the accuracy (71.9%) and the kappa statistics (0.438) of the model using 100 trees. In terms of sensitivity, the SVM with a 0.01 gamma and RBF kernel stood out, achieving the average value of 0.830. Regarding specificity, the Naive Bayes model performed better than the other methods, reaching 0.780. However, when comparing the training time (5 s) and the area under the ROC curve (0.778) of the 100 trees Random Forest with the other models, it presents the best results and is seen as a potential model for predicting hospitalization in semi-intensive care units in this scenario.

Analyzing from the perspective of the worst performing model, we highlight the SVM with linear kernel, which obtained the worst results in terms of accuracy (61%), kappa statistics (0.230), and specificity (0.380), improving only in sensitivity (0.850). In Fig. 6, it is possible to observe the performance of the aforementioned models.

In order to carry out a comparative analysis with the results obtained in Steps 3 and 4, in Step 5 experiments were performed to predict hospitalization in semi-intensive care units from the database without pre-processing. As shown in Fig. 7, it is clear that the Random Forest model with its respective tree configurations stands

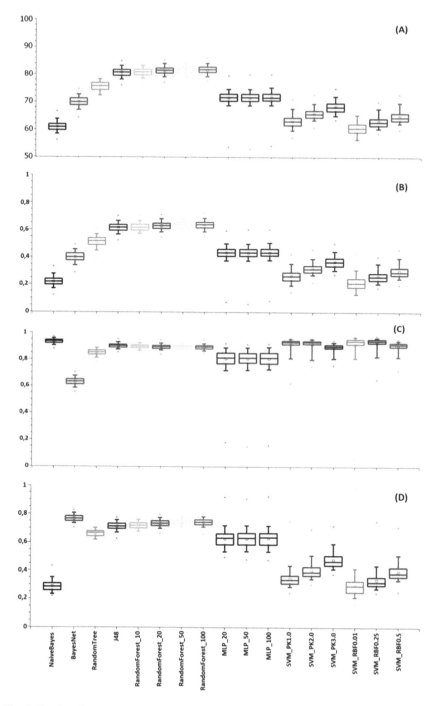

Fig. 5 Results of accuracy (**a**), kappa statistics (**b**), sensitivity (**c**), and specificity (**d**) to predict semi-intensive care unit hospitalization using the balanced database (Step 3)

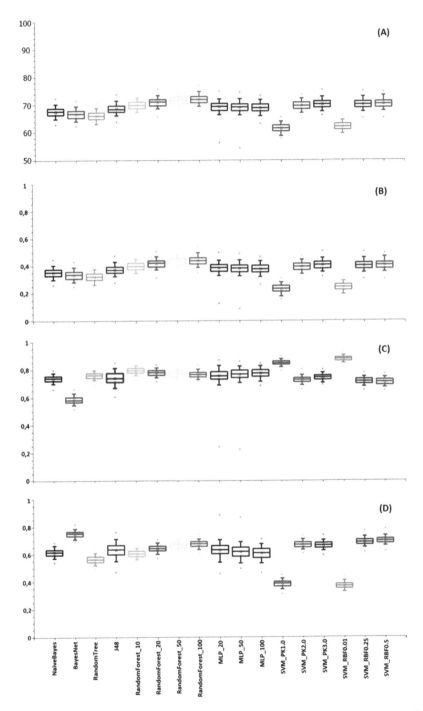

Fig. 6 Results of accuracy (**a**), kappa statistics (**b**), sensitivity (**c**), and specificity (**d**) to predict semi-intensive care unit hospitalization using the database with selected features (Step 4)

out, especially that of 100 trees regarding accuracy (79.6%), kappa statistics (0.594), sensitivity (0.858), and specificity (0.861).

Finally, analyzing the model with the worst performance in this scenario, we highlight the Random Tree model in terms of accuracy (73%), kappa statistics (0.512), sensitivity (0.710), and specificity (0.690). However, it is worth noting that in the general context, all three models (Random Tree, J48, and Random Forest) showed positive results and better performance than SVMs and MLPs. But among these three classifiers, Random Forest showed the best overall performance.

5.3 Intensive Care Unit Hospitalization

In the data regarding prediction of hospitalization in intensive care units, class 0 has 5592 instances and represents patients who did not need to be admitted to the intensive care unit. While class 1 is composed of 623 instances and refers to patients who needed intensive care. Clearly, it is possible to identify an imbalance between classes, with a difference of 4969 instances. In order to perform the class balancing, in Step 1 (SMOTE method), it was necessary to carry out an expansion of 790% of the smaller class based on two neighbors of the majority class, resulting in two balanced classes, each one with 5592 instances.

During the analysis, from the predictions of hospitalization in intensive care units using the database pre-processed with the SMOTE method (Step 3), the algorithm that obtained the best accuracy (82%) was Random Forest (as shown in Fig. 8a), configured with 50 and 100 trees, respectively. Bayes Net also stood out for accuracy, with 78.4%. For this same metric, the models that had the most outlier values were Naive Bayes (60%) and SVMs (ranging from 63% to 68%). This information can be seen in Fig. 8.

In Fig. 8, we also see the kappa statistics (Fig. 8b) obtained by the models in Step 3, which ended up following a pattern similar to the previous metric: with Random Forest of 100 trees achieving the best average kappa (0.65). Naive Bayes model with kappa of 0.22, followed by the SVMs with kappa statistics ranging from 0.23 to 0.43, showed less satisfactory results.

Regarding sensitivity, the SVM models performed better, reaching 0.98 in the identification of true positives. While Naive Bayes obtained only 0.26, it stands out, therefore, in three parameter analyses, as the worst model to be used in this context. The specificities of the SVMs models, on the other hand, were the lowest, thus disqualifying these models for the identification of true negatives.

In Step 4, after applying MLPAutocode, there was a reduction from 41 to 10 attributes in the database pre-processed with SMOTE (Step 3). By analyzing the accuracy of the models in detail, as shown in Fig. 9, it is clear that the Random Forest models of 100, 50, 20, and 10 trees stood out in comparison to the other algorithms, with accuracy between 76% and 78%. On the other hand, it was also found that SVM models did not obtain good results in this metric, specially SVM

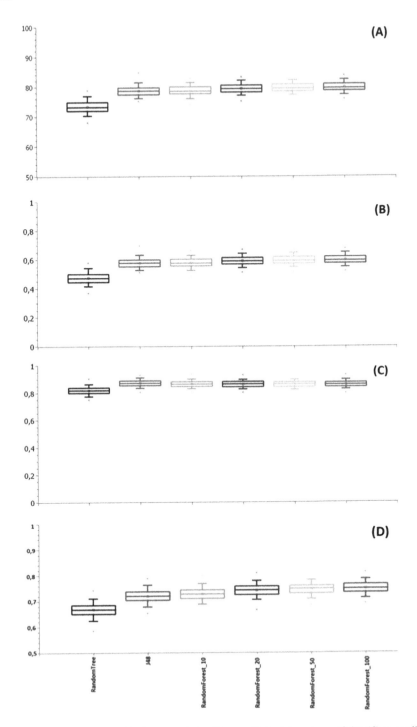

Fig. 7 Results of accuracy (**a**), kappa statistics (**b**), sensitivity (**c**), and specificity (**d**) to predict semi-intensive care unit hospitalization using the original database, without pre-processing (Step 5)

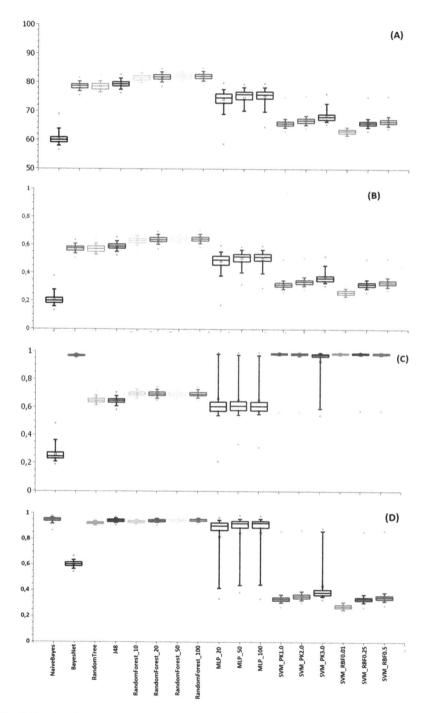

Fig. 8 Results of accuracy (**a**), kappa statistics (**b**), sensitivity (**c**) and specificity (**d**) to predict intensive care unit hospitalization using the class-balanced database (Step 3)

with RBF kernel and gamma of 0.01 (63%), followed by SVM with linear kernel (64%).

We highlight that the kappa statistics presented a behavior very similar to the accuracy. The Random Forest model stood out from the other methods, with kappa reaching 0.55. The SVM models with RBF kernel with 0.01 gamma and SVM with linear kernel performed worse again, with kappa of 0.25 and 0.29, respectively.

Unlike what has been analyzed and reported so far, for sensitivity (Fig. 9c) in Step 4, we see that the models with the best sensitivity were Naive Bayes (0.98), followed by SVM with linear kernel (0.97) and the SVM with RBF kernel and gamma of 0.01 (0.89), while Random Tree and J48 showed sensitivity of 0.60 and 0.61, respectively. We also noticed that the same models that performed better in sensitivity achieved worse values for specificity (Fig. 9d), indicating a discrepancy between true-positive and negative predictions.

For comparative purposes, as shown in Fig. 10, only the Random Forest, J48, and Random Tree models were executed on the database without pre-processing (Step 5), since these algorithms showed better overall performances in the previous steps. As a result, the accuracies of J48 (94%) and Random Forest models (from 90% to 93%) were outstanding. Random Tree, on the other hand, achieved 87% of accuracy, decreasing in relation to the other models.

Still at this step, during the analysis of the kappa statistics, the Random Forest models of 50 and 100 trees continued to stand out, with average values of 0.38, followed by J48 with kappa of 0.37. The Random Tree classifier did not show good performance, with an average kappa equal to 0.27.

Figure 10 also presents the results related to the sensitivity in Step 5. In this metric, the findings demonstrate that J48 stands out with an average value of 0.99, closely followed by Random Forest, with 0.98. Random Tree obtained 0.95 for the same indicator. As for specificity, the discrepancy with the pattern presented so far is visible. This time, the J48 model performed worse among the models, with a true-negative rate of 0.22. The Random Forest model with 100 trees, on the other hand, reached the highest average specificity value (0.26).

After a comparative and multimodal analysis in each of the stages, to predict the three types of hospitalization proposed, the best models were selected. In order to further show these results, the indicators for the best models in each scenario were compiled in Table 5, together with the averages and standard deviations of accuracy (ACC), kappa statistics (KPP), sensitivity (SEN), specificity (SPE), area under the ROC curve (AUC), and the training time (TT).

Finally, it is of great value to mention that at all steps for the three different types of care unit (regular, semi-intensive, and intensive), the model that stood out positively regarding the evaluation metrics was the Random Forest, mainly with the configuration of 50 and 100 trees. On the other hand, the models that had less satisfactory results, in all three conditions, were Naive Bayes, SVMs, and Random Tree (Stage 5). However, in order to obtain a more accurate and general analysis, the discussion of these results will be further explored in the following section.

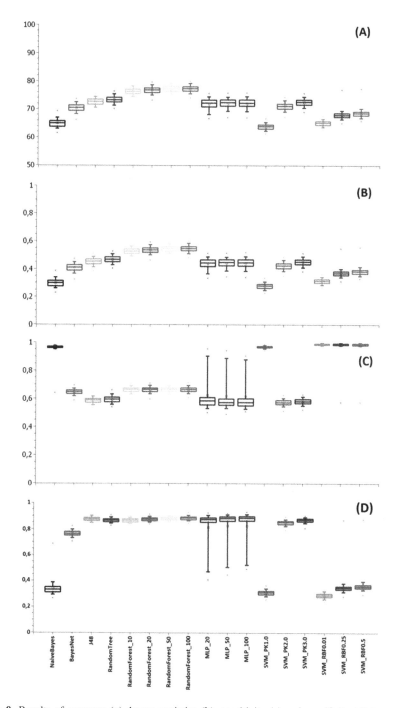

Fig. 9 Results of accuracy (**a**), kappa statistics (**b**), sensitivity (**c**) and specificity (**d**) to predict intensive care unit hospitalization using the database with selected features (Step 4)

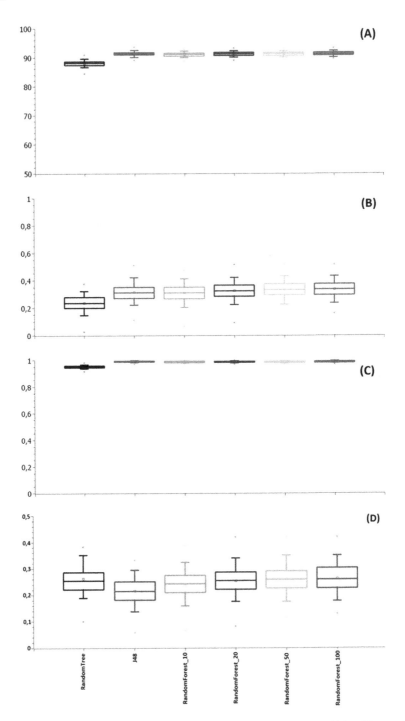

Fig. 10 Results of accuracy (**a**), kappa statistics (**b**), sensitivity (**c**), and specificity (**d**) to predict intensive care unit hospitalization using the original database, without pre-processing (Step 5)

6 Discussion

In general, the Random Forest model with its four configurations (10, 20, 50, and 100 trees), had better performance. In all the generated databases (original, balanced, and with attributes selection), its performance was satisfactory, not only considering the average values of its metrics, but also the standard deviations and the constancy of these models. About the training time, it was identified that very high degree parameters (such as SVM and MLP) showed an exponential growth for an insignificant performance gain. Although the re-training is done on average once a week, the metrics were not positive enough to encourage the use of these models.

In the context of regular ward care, the Random Forest model was the most successful. Both with the configuration of 100 trees, as well as those of 50, 20, and 10 trees. Another classifier that also showed positive results was the J48. It is worth mentioning that in this context, as shown in the results section, the SVMs and MLPs have not shown satisfactory results.

The prediction of hospitalization in semi-intensive care units shares the same findings as the regular ward. In both, Random Forest with its four configurations performed better. On the other hand, SVMs and Bayes Net declined in the results of the evaluation indicators. It is important to note that most of the classifiers were more promising with the balanced database.

Considering the intensive care unit hospitalization, the Random Forest models with 100 and 50 trees showed better accuracy and sensitivity with the original database. Kappa index, specificity, and area under the ROC curve were best evaluated on the balanced database. Still for this type of care, the Naive Bayes model was the worst classifier, taking into account that its metrics are the lowest.

One curiosity found, which is worth reporting, refers to the calculated sensitivity for recommendation in care in semi-intensive and intensive care units. In this indicator, SVM models were the best evaluated, SVM with RBF kernel and 0.01 gamma, followed by SVM with linear kernel. Despite their good performance for the identification of true positives, these classifiers showed discrepancies in terms of accuracy, kappa index, specificity, and area of the ROC curve.

Finally, in view of the evidence, the results weighed against the original bases. This leads to belief that the balance of bases, more than the reduction in the number of attributes, result in better classifications of this problem. This leaves the possibility that, perhaps, the MLPAutocode is not the most appropriate technique in this context of prediction. In testing this hypothesis, future works may use other techniques and, consequently, different configurations to carry out a comparative analysis of the results.

7 Conclusion

Considering this atypical pandemic moment caused by the new coronavirus, the use of predictive models has been helping health professionals in order to control the spread of the virus and the burden on the health system. In this context, our work contributes to this theme by proposing a comparative analysis of seven classic classifiers capable of making predictions regarding care and assessments of the severity of patients with and without COVID-19, seen at the Unified Health System (SUS) units of the municipality of Paudalho, Brazil, based on the evaluation of hematological data (blood tests).

The results obtained through the evaluative metrics show that among the seven classifiers used, both for predicting regular appointments and for attending a semi-intensive care unit, the Random Forest algorithm showed better performance with all configurations, in relation to the other models. In both cases, the SVM also stood out, but negatively, thus being the least suitable model to be used in this scenario.

Besides the findings, we highlight that analyzing the results from different quantitative and qualitative perspectives is important both for a better understanding of the problem, as well as for choosing the best solution in the researched context. For this reason, several aspects, ranging from the robustness of the model to the time of its execution, must be taken into account.

It is worth noting that the predictive models should assist the health professional in making decisions. The model is only for support and streamline the process. In the case of this work, the prediction can help by assisting in the screening of regular patients, so that those with moderate and severe cases receive care as soon as possible.

Finally, as perspectives for future work, it is intended to apply to the three databases (referring to regular, intensive, and semi-intensive care) other pre-processing techniques, as well as other methods for reducing and selecting the number of attributes with different configurations, in order to verify comparatively the classifiers' performance in relation to the original knowledge base and the base with the pre-processed data.

References

1. Zhu, N., Zhang, D., Wang, W., Li, X., Yang, B., Song, J., et al. (2019). A novel coronavirus from patients with pneumonia in China. *The New England Journal of Medicine, 382*, 727–733.
2. Croda, J. H. R., & Garcia, L. P. (2020). Resposta imediata da Vigilância em Saúde à epidemia da COVID-19. *Epidemiologia e Serviços de Saúde, 29*(1), Brasília.
3. Iser, B. P. M., Sliva, I., Raymundo, V. T., Poleto, M. B., Scuelter-Trevisol, F., & Bobinski, F. (2020). Definição de caso suspeito da COVID-19: uma revisão narrativa dos sinais e sintomas mais frequentes entre os casos confirmados. *Epidemiologia e Serviços de Saúde, 29*(3), e2020233.
4. World Health Organization (WHO). (2019). *Coronavirus disease (COVID-19) pandemic*. Geneva: World Health Organization. Available from: https://www.who.int/emergencies/diseases/novel-coronavirus-2019.

5. Teodoro, L. A., & Kappel, M. A. A. (2020). Aplicação de Técnicas de Aprendizado de Máquina para Predição de Risco de Evasão Escolar em Instituições Públicas de Ensino Superior no Brasil. *Revista Brasileira de Informática na Educação, 28*, 838–863.
6. Nemati, M., Ansary, J., & Nemati, N. (2020). Machine-learning approaches in COVID-19 survival analysis and discharge-time likelihood prediction using clinical data. *ScienceDirect, Patterns, 1*(5), 100074.
7. Kumar, M., Patel, A. K., Shah, A. V., Raval, J., Rajpara, N., Joshi, M., & Joshi, C. G. (2020). First proof of the capability of wastewater surveillance for COVID-19 in India through detection of genetic material of SARS-CoV-2. *Science of the Total Environment, 746*, 141326.
8. Woelfel, R., Corman, V. M., Guggemos, W., Seilmaier, M., Zange, S., Mueller, M. A., Niemeyer, D., Kelly, T. C. J., Vollmar, P., Rothe, C., Hoelscher, M., Bleicker, T., Brunink, S., Schneider, J., Ehmann, R., Zwirglmaier, K., Drosten, C., & Wendtner, C. (2020). Clinical presentation and virological assessment of hospitalized cases of coronavirus disease 2019 in a travel-associated transmission cluster. *medRxiv*. Available from: https://doi.org/10.1101/2020.03.05.20030502.
9. Tolia, V. M., Chan, T. C., & Castillo, E. M. (2020). Preliminary results of initial testing for coronavirus (COVID-19) in the emergency department. *Western Journal of Emergency Medicine, 21*(3), 503–506. Available from: https://doi.org/10.5811/westjem.2020.3.47348.
10. Hadaya, J., Schumm, M., & Livingston, E. H. (2020). Testing individuals for coronavirus disease 2019 (COVID-19). *JAMA, 323*(19), 1981. Available from: https://doi.org/10.1001/jama.2020.5388.
11. Batista, A. F. M.; Miraglia, J. L.; Donato, T. H. R., & Chiavegatto Filho, A. D. P. (2020). COVID-19 diagnosis prediction in emergency care patients: a machine learning approach. *medRxiv*.
12. Torcate, A. S.; Barbosa, J. C. F., & de Oliveira Rodrigues, C. M. (2020). Utilizando o Learning Analytics com o K-Means para Análise de Dificuldades de Aprendizagem na Educação Básica. In *Anais do XXVI Workshop de Informática na Escola* (pp. 31–40). SBC, November.
13. de Barbosa, V. A. F., Gomes, J. C., Santana, M. A., Albuquerque, J. E. A., Souza, R. G., Souza, R. E., & Santos, W. P. (2020). Heg.IA: Um sistema inteligente para apoiar o diagnóstico de Covid-19 com base em exames de sangue. *medRxiv preprint*. https://doi.org/10.1101/2020.05.14.20102533. this version posted May 18, 2020.
14. Jordan, M. I., & Mitchell, T. M. (2015). Aprendizagem de máquina: tendências, perspectivas e perspectivas. *Science, 349*, 255–260. Available from: https://doi.org/10.1126/science.aaa8415.
15. Gunčar, G., Kukar, M., Notar, M., Brvar, M., Černelč, P., Notar, M., & Notar, M. (2018). An application of machine learning to haematological diagnosis. *Scientific Reports, 8*, 1–12.
16. Gomes, J. C., Barbosa, V. A. D. F., Santana, M. A., Bandeira, J., Valença, M. J. S., Souza, R. E., Ismael, A. M., & Santos, W. P. (2020). IKONOS: uma ferramenta inteligente para apoiar o diagnóstico de COVID-19 por análise de textura de imagens de raios-X [publicado online antes da impressão, em 3 de setembro de 2020]. *Pesquisa em Engenharia Biomédica*; 1–14. https://doi.org/10.1007/s42600-020-00091-7.
17. Tanner, L., Schreiber, M., Low, J. G., Ong, A., Tolfvenstam, T., Lai, Y. L., Ng, L. C., Leo, Y. S., Thi Puong, L., Vasudevan, S. G., Simmons, C. P., Hibberd, M. L., & Ooi, E. E. (2008). Decision tree algorithms predict the diagnosis and outcome of dengue fever in the early phase of illness. *PLoS Neglected Tropical Diseases, 2*(3), e196. https://doi.org/10.1371/journal.pntd.0000196. PMID: 18335069; PMCID: PMC2263124.
18. Luo, Y., Szolovits, P., Dighe, A. S., & Baron, J. M. (2016). Using machine learning to predict laboratory test results. *American Journal of Clinical Pathology, 145*, 778–788.
19. Cordeiro, F. R., Santos, W. P. S., & Silva-Filho, A. G. (2017). Analysis of supervised and semi-supervised growcut applied to segmentation of masses in 635 mammography images. *Computer Methods in Biomechanics and Biomedical Engineering: Imaging & Visualization, 5*, 297–315.
20. Silva, W. W. A., Santana, M. A., Silva Filho, A. G., Lima, S. M. L., & Santos, W. P. (2020). Morphological extreme learning machines applied to the detection and classification of mammary lesions. In T. K. Gandhi, S. Bhattacharyya, S. De, D. Konar, & S. Dey (Eds.), *Advanced machine vision paradigms for medical image analysis*. London: Elsevier.

21. Ji, Y., Ma, Z., Peppelenbosch, M. P., & Pan, Q. (2020). Potential association between COVID-19 mortality and health-care resource availability. *The Lancet Global Health, 8*(4), e480.
22. Yan, L., Zhang, H. T., Goncalves, J. et al. (2020, May 14). An interpretable mortality prediction model for COVID-19 patients. *Nature Machine Intelligence, ed 2*, 283–288. Available from: https://doi.org/10.1038/s42256-020-0180-7.
23. Ardabili, S. F., Mosavi, A., Ghamisi, P., Ferdinand, F., Varkonyi-Koczy, A. R., Reuter, U., Rabczuk, T., & Atkinson, P. M. (2020). Covid-19 outbreak prediction with machine learning. *Algorithms, MedRxiv, 13*(10), 249.
24. Cohen, J. P., Morrison, P., & Dao, L. (2020). COVID-19 image data collection. arXiv:2003.11597. Available from: https://arxiv.org/abs/2003.11597.
25. Di Radiologia Medica and Intervencionista. (2020). *Covid-19 Database*. Available from: https://www.sirm.org/category/senza-categoria/covid-19/.
26. Witten, I. H., & Frank, E. (2020). Data mining: Practical machine learning tools and techniques with Java implementations. *ACM SIGMOD Record, 31*(1), 76–77.
27. Chawla, N. V., Bowyer, K. W., Hall, L. O., & Kegelmeyer, W. P. (2020). SMOTE: Synthetic minority over-sampling technique. *Journal of Artificial Intelligence Research, 16*, 321–357.
28. Hernández, E., Sanchez-Anguix, V., Julian, V., Palanca, J., & Duque, N. (2016). Rainfall prediction: A deep learning approach. In *International Conference on Hybrid Artificial Intelligence Systems* (pp. 151–162). Springer, Cham; April.

Bioinformatics in Diagnosis of COVID-19

Sanjana Sharma, Saanya Aroura, Archana Gupta, and Anjali Priyadarshini

1 Introduction

Coronavirus disease 2019 (COVID-19) pandemic has an exponential growth rate and a partially understood dissemination procedure requiring a lot of interventions to control the menace. Commonly COVID-19 infection is observed with very few or almost no symptom, but can lead to a quick transmission followed by fatal pneumonia in about 2–8% of those infected individuals [1–3] thus listing it among the deadly infections.

The undefined mortality rate, prevalence, and transmission dynamics are still vague due to various factors such as peak virulence at or just preceding manifestations at onset accompanied by ill-defined multi-organ pathophysiology with presiding characteristics and fatality in the lungs [4].

The very much enhanced rate of spread has exhausted healthcare system globally due to scarcity in PPE (personal protective equipment) and certified providers [5]. Access to point-of-care testing methodologies such as reverse transcription polymerase chain reaction (RT-PCR) has improved the rate of diagnosis. As rapid RT-PCR testing becomes more accessible, the results have been marred by inclusiveness of high false negative rates and low sensitivity [6, 7].

Immunoenzymatic assays or agglutination tests are also available for tracing viral antigens, and nucleic acid amplification tests are also being performed for tracing viral genetic material [8] thus improving the diagnostics.

The diagnostic methods mentioned above detect the virus and thus can be said to be primary detection methods. A secondary way to detect viral infections is the recognition of a particular immune system response. The humoral response (based

S. Sharma · S. Aroura · A. Gupta · A. Priyadarshini (✉)
SRM University Delhi-NCR, Rajiv Gandhi Education City, Sonepat, Haryana, India
e-mail: archana.v@srmuniversity.ac.in

© The Author(s), under exclusive license to Springer Nature Switzerland AG 2022
S. K. Pani et al. (eds.), *Assessing COVID-19 and Other Pandemics and Epidemics using Computational Modelling and Data Analysis*,
https://doi.org/10.1007/978-3-030-79753-9_12

on antibody production) is the one of the simplest techniques to detect infectious conditions. There are various techniques available in market for recognizing antibodies that are directed against different sections of viral bodies [9, 10] which are produced by immune cell upon viral encounter.

Early detection and late detection of COVID-19: A nasopharyngeal (NP) swab and/or an oropharyngeal (OP) swab are usually advised by healthcare professionals for screening or detection of early infection. This can be further improved by the application of Machine Learning and Artificial Intelligence to tackle novel coronavirus outbreak.

2 Detection and Annotation

2.1 UniProt COVID-19 Protein Portal

UniProt (UniProt Consortium), or Universal Protein Resource, is a basic but indispensable tool of bioinformatics in the form of a comprehensive, freely accessible, high-quality central repository of manually annotated or computationally analyzed non-redundant protein records. This has been brought out by the UniPort Consortium which is further a collaboration of European Bioinformatics Institute (EBI), Protein Information Resource (PIR), and the Swiss Institute of Bioinformatics (SIB). This platform enables us to study protein sequences and their functional information. Being the final product of The Central Dogma of Life, the study of proteins empowers us to infer evolutionary and phylogenetic information. Figure 1 presents us with a list of materials provided by the UniProt COVID-19 protein portal to tackle this pandemic.

Understanding the urgency of current situation, the COVID-19 portal of UniProt has provided early pre-release access to the following:

- First and foremost, annotated protein sequences of novel coronavirus or the SARS CoV-2. Figure 2 is an example which shows the result of searching the protein sequence sp.|P0DTC2|SPIKE_SARS2.
- The closest SARS proteins from SARS 2003. Finding similarities with the precursor helps in speeding up the process of finding diagnostic tests and drug-delivery targets.
- The human proteins which are relevant to the science of viral septicity are the generally enzymes and receptors. On the basis of similarities between SARS CoV and SARS Cov-2, it is said that ACE-2 has homology to ACE. This is a type I transmembrane metallocarboxypeptidase and the cellular entry receptor for COVID-19. (https://www.rndsystems.com/resources/articles/ace-2-sars-receptor-identified).
- Visualization of sequences for each protein via ProtVista [9, 12]. It is a powerful tool which can be incorporated easily with web applications.
- The portal also delivers links to sequence analysis tools and pertinent resources.

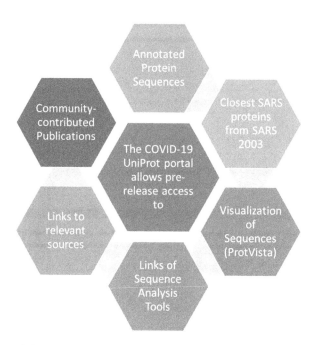

Fig. 1 Hexagonal chart showing the early pre-release accessible data provided by UniProt in the present state of pandemic on its COVID-19 portal [8]

- UniProt allows retrieval of not only collated but also community-contributed publications which are relevant to COVID-19 pandemic. Using "Add a Publication" option within any of the entries, UniProt enables community crowdsourcing of publications. Th mechanism of ORCID in UniProt COVID-19 portal not only validates user credentials but also recognizes contribution for these publications. There has been submission of 16 publications that have significantly contributed in understanding of the COVID-19 virology. https://community.uniprot.org/bbsub/bbsubinfo.html?covid=true

Thus, COVID-19 UniProt portal progresses the research on the novel coronavirus by imparting latest knowledge of Proteomics, in connection to both, the novel coronavirus and the receptors of the human host.

The portal can be accessed through https://covid-19.uniprot.org/uniprotkb?query=*. The webinars that are hosted on this platform can be accessed via https://www.youtube.com/watch?v=EY69TjnVhRs and Community Bibliography Submissions can be reached out through https://community.uniprot.org/bbsub/bbsubinfo.html?covid=true.

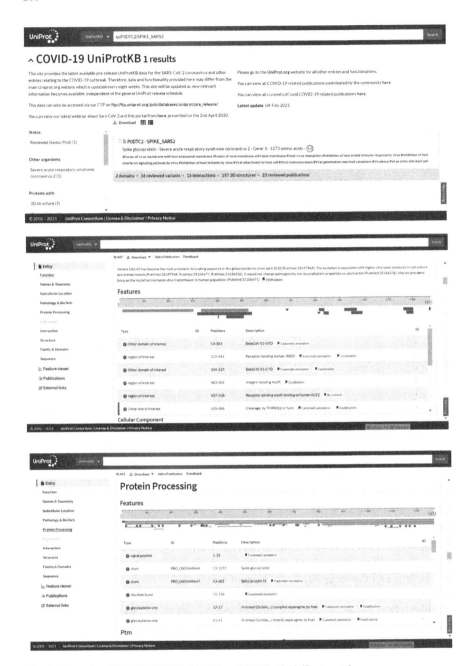

Fig. 2 Data of sp.|P0DTC2|SPIKE_SARS2 on COVID-19 UniProt portal

2.2 Rfam COVID-19 Resources

The Rfam database (Fig. 4) [13] is an accumulated cluster of RNA sequence families inclusive of ncRNA genes along with cis regulatory elements. These RNA families are depicted by MSA (by tools such as ClustalX and Omega) and by CMfinder (co-variance model). Similar to UniProt COVID-19 portal, Rfam also introduced a special release 14.2 in dedication to the coronavirus RNA families. This version has been made in collaboration with Marz Group and EVBC. It includes 4 revised and 10 new families that portray the entire 5'and 3' UTRs from alpha-, beta-, gamma-, and delta coronaviruses which are then used to annotate novel coronavirus, SARS-CoV-2, and many other RNA families with coronavirus genomes. A special edition of models representing Sarbecovirus is also arranged comprising of SARS-CoV-1 and novel coronavirus SARS CoV-2 fasta sequences. These RNA families are standardized on a set of high-quality whole genome sequencing, produced with LocARNA after being reviewed by experts. Furthermore, the updated compilation of non-UTR novel coronavirus structured RNAs namely the frameshift stimulating element, for example, s2m RNA, as well as the 3' UTRs has also been incorporated. (https://xfam.wordpress.com/2020/04/27/rfam-coronavirus-release/).

Infernal Software [14] can be used in accordance with the new RNA families to explain structured RNAs in these coronavirus sequences and anticipate their secondary structures as shown in Fig. 3. Additionally, this platform of coronavirus-specific RNA families permits users to perceive RNA elements by scanning genomic sequences. Besides, they help us to study and analyze clans, genomes, motifs, sequences, and families with or without their 3D-structures (Fig. 4).

The portal of Rfam COVID-19 Resources has open access to the public now under the Creative Commons Zero (CCO) license through http://rfam.xfam.org/covid-19.

2.3 Viral Annotation DefineR: SARS-CoV-2 Genome Interpretation and Validation

Viral Annotation DefineR (VADR: Fig. 5) authenticates and interpretates viral sequences in agreement at GenBank. This interpretation system is dependent on the inspection of the given nucleotide sequence using representative models built from categorized Reference sequences (RefSeq). VADR can be easily accessible in public domain, accessed by github (https://github.com/nawrockie/vadr), including specific edicts for use on SARS-CoV-2 nucleotide sequences (https://github.com/nawrockie/vadr/wiki/Coronavirus-annotation).

In this (Fig. 6), HM models (*Hidden Markov*) are utilized to group input sequences by considering the Reference Sequences that they are most known to, and feature annotation from the Reference Sequences is overlapped based on a nucleotide alignment of the full input sequence to a covariance model (CM). Such

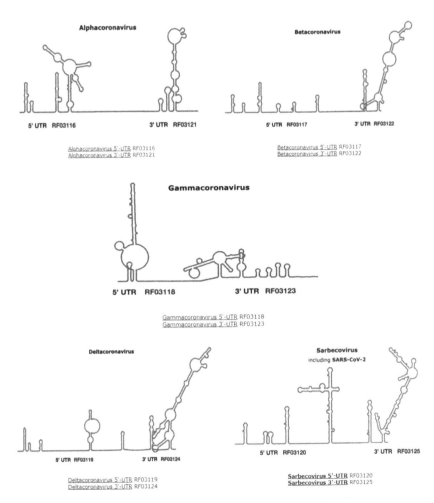

Fig. 3 The 5′- and 3′-UTRs of the coronavirus genomes as a representation of the latest advanced families (http://rfam.xfam.org/covid-19)

covariance representative models are based on NCBI RefSeq entries [30], one such model for SARS-CoV-2 (NC_045512.2) can be easily accessed for inspecting coronavirus sequences. Predicted peptides are encoded by these coronavirus sequences and are validated with nucleotide-to-peptide alignments using BLAST tool (blastx) [31]. The tool recognizes and produces information with about 43 kinds of different issues [31] with sequences at maximum level, for example, early stop codons in CDS, and have been utilized by GenBank for identifying and interpretating SARS-CoV-2 sequence acceptance *since* March 2020 [11].

VADR has been unified into GenBank's submission pipeline, which further permits for viral submissions after qualifying all the experiments to be accepted and

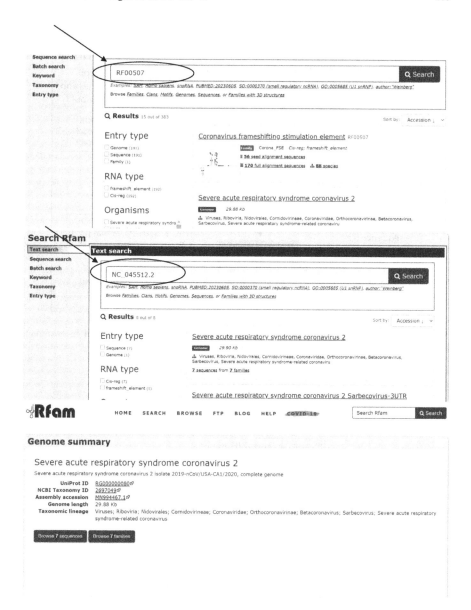

Fig. 4 The homepage and search procedure of Rfam version 14.2 [11, http://rfam.xfam.org/family/RF00507]

interpretated simultaneously, without the requirement for any human interference [28].

VADR (v1.1.3; Feb 2021) involves heuristics for speeding up interpretation and for including with extensions of obscure N nucleotides that were precisely appended

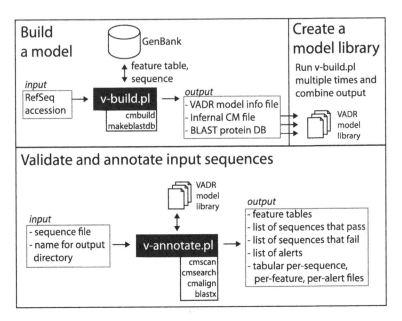

Fig. 5 VADR schematic illustrations of its workflow defining the functioning of its two main python scripts: v-build.pl and v-annotate.pl [28]

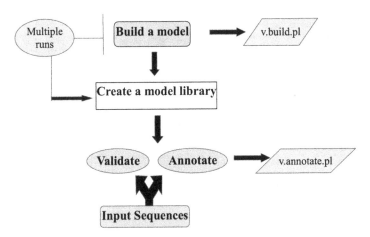

Fig. 6 Outline of VADAR Workflow

for SARS-CoV-2 examination. It has helped foster SARS-CoV-2 research by systematizing the interpretation of SARS-CoV-2 sequences submitted in GenBank and various databases, by empowering scientists to completely interpret as well as identify their sequences for errors because of faulty assembly or any other issue [31].

3 Tracking, Epidemiology, and Evolution

3.1 Covidex

Since viruses are the connecting link between the living and the non-living, there is no standardized approach to classify the genetic diversity of a virus (phylogenetic clades, may also be termed as "subtypes," "genotypes," etc.). Phylogenetic assignment plays a crucial role for various virology aspects such as epidemiology, evolution, and pathogenesis thus, making it necessary to develop a system of nomenclature to define the immensely expanding number of phylogenetic lineages which marks the heterogeneity in the population of SARS-CoV-2 virus [11]. One such nomenclature has already been established [15].

Viral clades characterize aggregates among the isolates from the overall population of a specific species. The process of subtypification is another important phenomenon in virology. Covidex (Fig. 7) is one such open-source subtyping tool in the form of a shiny app [16] that enables quick and precise classification of viral genomes in collections which have been defined beforehand. With a capacity of compartmentalizing 16,000 genome sequences in less than a minute, it is one of the mechanisms being used for tracking and studying epidemiology and evolution of novel coronavirus. (https://sourceforge.net/projects/covidex/).

Mostly, the method of subtype classification requires aligning the given input of data set against an arrangement of pre-established subtype reference sequences. But these are extremely expensive, especially, when the sequences are lengthy like SARS-CoV-2 which is about 30 kb per genome. So, here comes in the crucial role of Machine Learning aids for virus subtyping [17].

With respect to SARS-CoV-2 novel coronavirus, the default model has been formed on the basis of Nextstrain [18] and GISAID data [19]. In addition, the models uploaded by the user can be utilized as well. Covidex has been built on a fast execution of arbitrary forest classifier, trained over a k-mer datasets [20, 21]. Through this training of stratification algorithms over the k-mer database, Covidex gradually decreases computational and temporal prerequisites which classify about hundreds of SARS-CoV-2 novel coronavirus genomes in just a few seconds. In the current milieu of global pandemic, where the number of available SARS-CoV-2 genome is multiplying and spreading tremendously, Covidex has emerged as a significant tool in reducing time in data analysis for SARS 2020 research.

Covidex is freely and easily accessible under GPLv3 through SourceForge via https://sourceforge.net/projects/covidex/. This web application is also available with the link https://cacciabue.shinyapps.io/shiny2/.

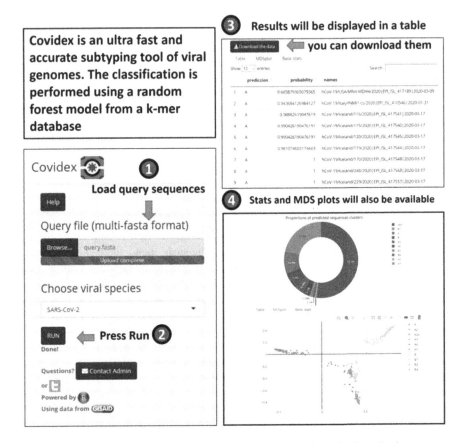

Fig. 7 The outline of Covidex. On the left, we can see that the user is required to upload a sequence and choose a representative model to be tested for a sp. classifier. The user can select these models from the provided default list or upload by himself/herself. The figure shows output of this program in the form of table, plots, and graphs (pie chart) on the right side

3.2 Covid Simulation Tool (CovidSIM): Epidemiological Models of Viral Spread

CovidSIM (Fig. 8) is a sophisticated extension of a classical SEIR or SEIRD (individuals in a community are grouped into four groups: Susceptible (S), Exposed (E), Infected (I), Recovered (R), and Deceased (D)) model. This was development by Martin Eichner (Epimos GmbH) and Markus Schwehm (ExploSYS GmbH) supported by University of Tübingen and IMAAC NEXT Association. The latest Version 2.0, the further development of CovidSIM, is sponsored by the German Federal Ministry of Education and Research. This simulation tool CovidSIM (Fig. 9) is an openly accessible web interface to interactively conduct simulations of this model. This simulation tool is used to assess the effects of various interventions,

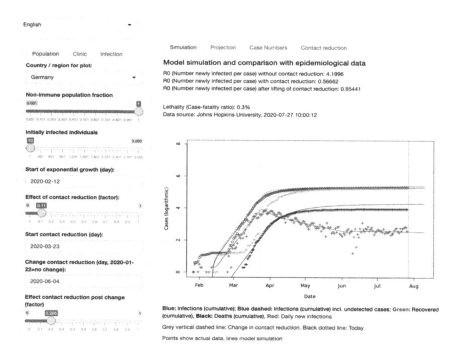

Fig. 8 COVID-19 Simulator: Web interface of the CovidSIM simulator (http://www.kaderali.org: 3838/covidsim/)

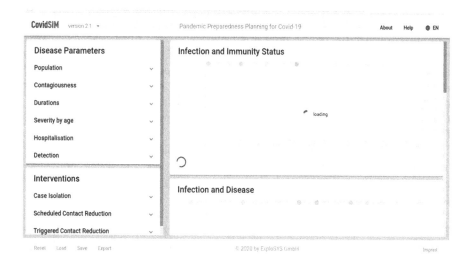

Fig. 9 Web Interface of the CovidSIM version 2.1 showing its various disease parameters and interventions and pictorial representation in order to determine the infection and immunity status (http://covidsim.eu/)

assuming parameters that supposedly reflect a situation of particular location (say, India) [32]. The main purpose of this simulation tool is to provide a realistic and an easy-to-use simulation tool with the capacity to support decision-making in public and global health, epidemiology, and economy.

It can also be used to put into action, the pandemic community's infection dynamics using parameters (Fig. 10) such as the incubation time or the mean disease duration. These (SEIR or SEIRD) models have been used to speculate the COVID-19 dynamics, for example, in Spain and Italy, and to scrutinize the effect of control strategies [33]. Extended versions of the SEIR model were even developed to guide political decision-making such as in the case of Germany; this model is being implemented in CovidSIM, including hospitalized or intensive care patients thereby implementing effects of contact reduction. It can be overlaid with the data from different German federal states and other countries [34]. This model has an easy-to-use web interface (Figs. 9 and 10), allowing the user to change model parameters and get an instinctive feeling for the model dynamics – allowing it to estimate infection parameters and to analyze effects of contact reduction measures and guide political decision-making [11].

4 Drug Design

4.1 *VirHostNet SARS-CoV-2 Release*

It is of paramount importance to develop therapy as well as preventive strategies against COVID-19 to limit the pandemic peril. The understanding of underlying molecular mechanism of the disease's pathogenesis is principal in identifying potential drug candidates for clinical trials. Screening of known drug compounds and PPIs also referred to as drug repurposing is usually more economical and comparatively quicker than designing drugs from scratch [22, 23]. Being a member of RNA virus family that has been studied in depth, this is especially true for SARS-CoV-2 as we can infer a lot of information from beta coronaviruses and SARS CoV-1.

VirHostNet is one such database containing information about virus-host PPIs. Viral-host protein-protein interactions (PPIs) play a pivotal role during viral infection and hold optimistic therapeutic aspects thus making this database indispensable for such therapeutic interventions.

Complete description and understanding of interactions between SARS-CoV-2 and host cellular proteins at molecular level is needed to understand root mechanisms of viral replication and pathogenesis of COVID-19 outbreak. In March 2020, VirHostNet released its 38.642 version which includes virus-host and virus-virus manually bio-curated protein-protein interactions, including more than 4.000 new interactions since its last January 2019 release (http://virhostnet.prabi.fr/). Annotation of this this information has been done manually from literature reviews.

Fig. 10 The various parameters involved in CovidSIM 2.0 version has been mentioned here (http://covidsim.eu/)

Population parameters and initial infections

$N_{a,r}$	Population size by age a and risk status r
a_{max}	Number of age groups
r_{max}	Number of risk groups
$E_{a,r}^{(init)}$	Number of initial infections

Contagiousness parameters

R_0	All-year average of the basic reproduction number
A	Amplitude of the seasonal fluctuation of the basic reproduction number
$t_{max_{R_0}}$	Day when the seasonal fluctuation of the basic reproduction number reaches its maximum
$\beta_{a_1,a_2}^{(Gen)}$	Contact rate between age a_1 and a_2 in the general population (in the absence of contact reduction)
$\beta_{a_1,a_2}^{(School)}$	Contact rate between age a_1 and a_2 in school or kindergarten (in the absence of contact reduction)
$\beta_{a_1,a_2}^{(Work)}$	Contact rate between age a_1 and a_2 at work (in the absence of contact reduction)
$\beta_{a_1,a_2}^{(Home)}$	Contact rate between age a_1 and a_2 at home
$\beta_{a_1,a_2}(t)$	Combined contact rate (all contacts combined; may be modified by contact reduction)
$\lambda_a(t)$	Total force of infection for susceptible individuals of age a at time t
$\lambda^{(Ext)}$	Force of infection which originates from outside of the population (e.g. via travellers)
c_P	Contagiousness in the prodromal period (relative to the contagiousness in the early infectious period)
c_L	Contagiousness in the late infectious period (relative to the contagiousness in the early infectious period)

SEIR transition rates

D_E	Average duration of the latent period
n_E	Number of stages for the latent period
z	Stage transition rate for the latent period ($z = n_E / D_E$)
D_P	Average duration of the prodromal period
n_P	Number of stages for the prodromal period
φ	Stage transition rate for the prodromal period ($\varphi = n_P / D_P$)
D_I	Average duration of the early infectious period
n_I	Number of stages for the early infectious period
γ	Stage transition rate for the early infectious period ($\gamma = n_I / D_I$)
D_L	Average duration of the late infectious period
n_L	Number of stages for the late infectious period
δ	Stage transition rate for the late infectious period ($\delta = n_L / D_L$)

Disease parameters

$f_{a,r}^{(Sick)}$	Fraction of individuals in the (early and late) infectious period who have symptoms (i.e. who are sick)
$f_{a,r}^{(Consult)}$	Fraction of sick cases of age a and risk status r who seek medical help
$f_{a,r}^{(Hosp)}$	Fraction of sick cases of age a and risk status r who are hospitalized
$f_{a,r}^{(ICU)}$	Fraction of hospitalized cases of age a and risk status r who need intensive care
$f_{a,r}^{(Dead)}$	Fraction of sick cases of age a and risk status r who die from the disease
$D_{a,r}^{(Hosp)}$	Average duration of hospitalization for cases of age a and risk status r
$D_{a,r}^{(ICU)}$	Average duration of stay at the ICU for cases of age a and risk status r

Isolation settings

Q_{max}	Maximum capacity of the isolation units
$t_1^{(Iso)}$	Day when the case isolation measures start in the population
$t_2^{(Iso)}$	Day when the case isolation measures end in the population
$p^{(Home)}$	Prevented fraction of "home contacts" for cases who are isolated at home

Contact reduction settings

$p^{(Gen)}$	Prevented fraction of general contacts (when this intervention is activated)
$t_1^{(Gen)}$	Day when general contact reduction measures start
$t_2^{(Gen)}$	Day when general contact reduction measures end
$p^{(Gen,Trig)}$	Prevented fraction of general contacts (when triggered contact reduction is activated)
$T^{(Gen,Sick)}$	Threshold number of sick cases which triggers general contact reduction
$T^{(Gen,Hosp)}$	Threshold number of hospitalized cases which triggers general contact reduction
$T^{(Gen,ICU)}$	Threshold number of ICU cases which triggers general contact reduction
$t_1^{(Gen,Trig)}$	Start day when triggered general social distancing measures are enabled
$t_2^{(Gen,Trig)}$	End day when triggered general social distancing measures are enabled

The review search made use of ORFeomes from multiple coronaviruses strains such as MERS-CoV, SARS-CoV-1, and SARS-CoV-2. Apart from literature search, some experimental data obtained through affinity-purification mass spectrometry by the Korgan laboratory has found to be integrated instantaneously [24] which is of great use.

The lean web application which is an open data resource enables fast and reproducible in silico prediction of interaction of SARS-CoV-2 with human interactome thus ameliorating the cost of expensive experimentation. This application has led to the interactome speculation for SARS-CoV-2 which was connected to an anti-apoptotic switch regulated by Bcl-2 family members. This antiapoptotic switch could be an effective therapeutic target which needs to be validated. The network reconstruction had also marked the pro-survival protein Bcl-xL and the autophagy effector Beclin 1 as exposed nodes in the host cellular defense system against SARS-CoV-2. Both proteins harbor Bcl-2 homology 3 (BH3)-like motif, which is entailed in homotypic and heterotypic interactions with the other domains [11] which could be harnessed to design the therapeutics.

VirHostNet SARS CoV-2 platform release will be helpful for advanced examination and analysis of underlying molecular mechanisms of viral replication and COVID-19 pathogenesis. These examinations and analysis may give a system virology framework which can prioritize drug candidates repurposing and repositioning. VirHost Net web application is freely available (open access) via http://virhostnet. prabi.fr/.

4.2   P-HIPSTer

Viral-host protein-protein interactions play a key role by assimilating host cellular processes which hold propitious prospects related to therapeutics. With respect to this, P-HIPSTer portal aids in designing drugs on the basis of the prior stored information of other coronaviruses related to their underlying molecular pathways, potential drug targets, and infection treatments as explained in Fig. 11.

P-HIPSTer involves around 282,000 predicted viral-human PPIs on 1000 viruses with an experimental validation rate of about 76% [25]. Its anticipatory algorithm is an alteration of PrePPI [26, 27] and amalgamate sequence and structural informative statistics and other data to work out viral human PPIs mediated by peptide-domain or domain-domain contacts (Fig. 12). Additionally, P-HIPSTer builds all-atom interaction models for high-confidence PPI predictions involving folded domains and integrating sequence- and structure-based functional annotations for viral proteins at different levels, comprising of host biological pathways based on the predicted PPI models [24, 28, 29]. Thereby, P-HIPSTer accounts for a complimentary resource to high-throughput experimental validation approaches [38]. P-HIPSTer contains predictions for 15 coronaviruses with varying pathogenic likelihood (alpha- and beta coronaviruses) and intimates 4587 viral-host PPIs which involve 397 human peptides. This distinctive collection of anticipated viral-human PPIs has empowered

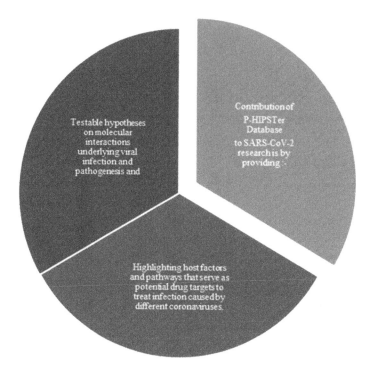

Fig. 11 Pie chart depicting the two major contributions of P-HIPSTer database to novel coronavirus research [8]

the discovery of PPIs models, usually employed within the *Coronaviridae* family and PPIs in correspondence with their pathogenicity.

The database provides an open access through the link http://www.phipster.org/.

5   Drug Design for COVID

It is of paramount significance to develop therapy and vaccination strategies against COVID-19 to control the pandemic. To achieve this understanding, the molecular and biochemical pathways underlying the disease's pathogenesis play a vital role to identify potential drug candidates for clinical trials. Viral-host protein-protein interactions have played a crucial role during viral infection and had proved its potential in therapeutic roles as discussed earlier with respect to P-HIPSTer.

Identifying new therapeutic use(s) for old/existing/available drugs is a process called as drug repurposing (also known as drug repositioning) has proved to be an effective strategy in discovering or developing drug molecules with new pharmacological/therapeutic indications.

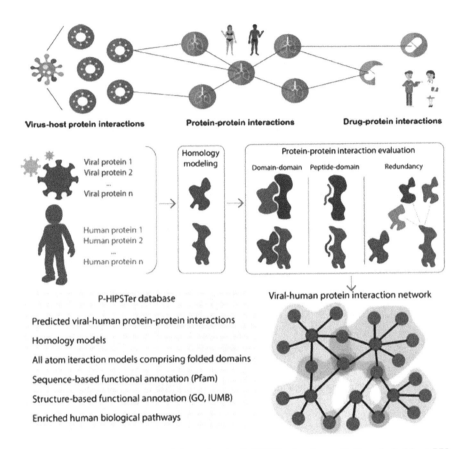

Fig. 12 Sequence and structural information by P-HIPSTer for the prediction of viral-host PPI models and evaluation of the likelihood ratio for the potential interaction between a viral protein (red) and a human protein (blue) merging three proofs: (i) Domain-domain, (ii) Peptide-domain, (iii) Redundancy [11]

Repurposing of the existing drugs is usually cheaper and time-efficient than designing drugs from scratch [22, 35]. This is especially true in case of SARS-CoV-2, where the drugs used for the same viral RNA genus can be tested in silico to test the efficacy. Thereby, we may infer information and potential drug targets from other beta coronaviruses as SARS-CoV-2 belongs to this group.

The following databases contain information about the various host and virus protein-protein interactions.

5.1   *CORDITE: CORona Drug InTERactions Database*

CORDITE is the largest curated database for drug interactions for SARS-CoV-2. It assembles information on potential drugs, drug targets, and their interactions for SARS-CoV-2 from published articles and preprints [36]. It provides an easy-to-use web interface and application programming interface that allows a wide use for other software and applications, such as for meta-analysis or tracking any clinical studies. Therefore, CORDITE offers an ample opportunity to facilitate research in the domains of virology and drug design (Fig. 13).

It automatically incorporates publications that facilitate information on computational, in vitro, or case studies on potential drugs for COVID-19 from the sites listed:

- PubMed (https://www.ncbi.nlm.nih.gov/pubmed/).
- bioRxiv (https://www.biorxiv.org/).
- chemRxiv (https://www.chemrxiv.org/).
- medRxiv (https://www.medrxiv.org/).

The current summary for interactions, drugs, etc., is reported in the Table 1 (https://cordite.mathematik.uni-marburg.de/#/, as of February 24, 2021):

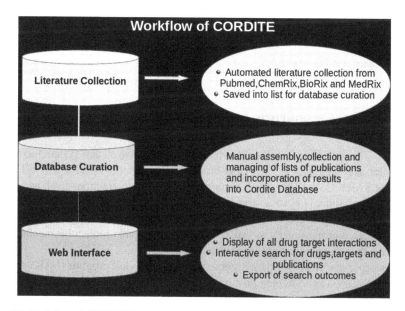

Fig. 13 Workflow of CORDITE database in a stepwise manner [32] (Martin et al.)

Table 1 [https://cordite.
mathematik.uni-marburg.
de/#/, as of February 24,
2021] Summary for
interaction of drugs and
relevant publication and
clinical trials

Type	Numbers
Interactions	1272
Targets	29
Drugs	891
Publications	307
Clinical trials	247

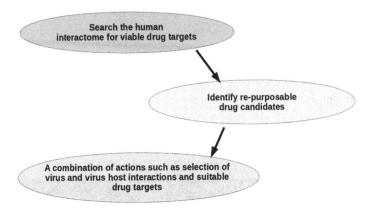

Fig. 14 Functions of CoVex

5.2 CoVex: Coronavirus Explorer

CoVex [37] is an online web interface that involves a network and system medicine.
This interface integrates in vitro virus-human protein interactions for SARS-CoV-2
[21] and SARS-CoV-1 [39, 40], human protein-protein interactions [39], and drug-
protein interactions [11] to a large-scale interaction system. It is used to identify
potential drug targets and drug repurposing candidates by allowing the researchers
and healthcare professionals to visually explore druggable molecular mechanisms
that drive the interactions between a virus and a host.

It incorporates protein-protein interactions between viral proteins (SARS-CoV-2
and SARS-CoV-1) and human host proteins. There are three different datasets that
are integrated (Fig. 14):

- SARS-CoV-2 [24].
- SARS-CoV-1 [38] (VirHostNet).
- Integrated database (protein-protein interactions, drug-target interactions, tissue-
  specific interactions).

General workflow of CoVex involves the following:

1. Selection of proteins: human/host, viral, and customized proteins.
2. Identification of drug targets: CoVex integrates six different algorithms in order to find drug targets starting from the selected seeds and the protein-protein interaction network (mentioned in Fig. 15).
3. Identification of drugs: The finding of drugs involves network analysis with drug interactions for approved and non-approved drugs.

Then, the last step in the workflow is to inspect and download our results. One can click on results section in the Tasks panel to explore the results. It displays the results consisting of the following:

1. Table view of drugs and proteins (Fig. 16a).
2. Network visualization (Fig. 16b).

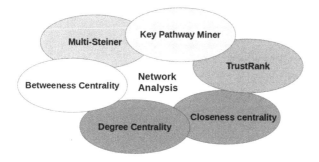

Fig. 15 Six algorithms that are integrated in CoVex and helps in identification of drug targets [33, 37]

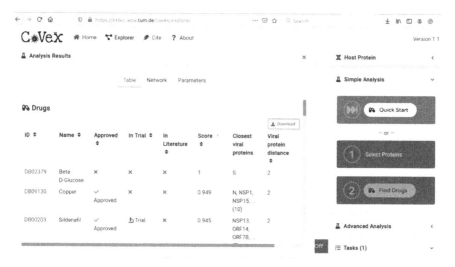

Fig. 16 (continued)

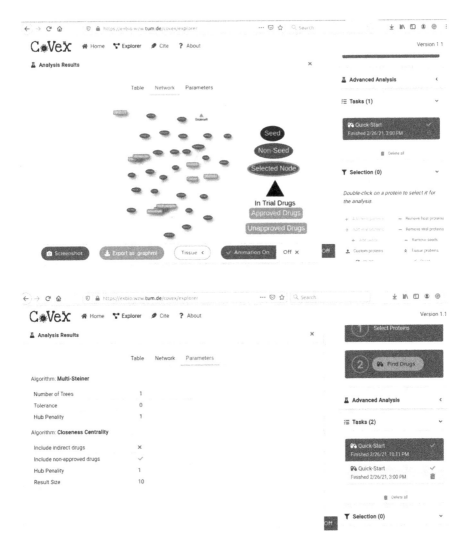

Fig. 16 The result of an example analysis in order to find drug for the example host-virus interaction, (**a**) shows the table having drugs listed for the example analysis, (**b**) determines the network involved, while (**c**) mentions the various algorithms and their expected results possible for the example network analysis

3. Description of the parameters used for the analysis (Fig. 16c).

One can navigate among the results using the navigation bar (Table, Network, Parameters) across the top of the results window (https://exbio.wzw.tum.de/covex/explorer).

6 Conclusion

Since the outbreak of the novel SARS-CoV-2, scientists and medical industries around the globe ubiquitously urged to fight against the pandemic, searching alternative method of rapid screening and prediction process, contact tracing, forecasting, and development of vaccine or drugs with the more accurate and reliable operation. Machine Learning and Artificial Intelligence are such promising methods employed by various healthcare providers. The ongoing development in the field of Artificial Intelligence and Machine Learning has greatly improved treatment, medication, screening, prediction, forecasting, contact tracing, and drug/vaccine development process for COVID-19. This has also reduced the human interference in the clinical practice. However, most of the models are not deployed enough to showcase their real-world operation, yet these are up to the mark to tackle the SARS-CoV-2 pandemic and there is a scope for more research in this field for its improvement and applications. The medical fraternity, scientific researchers, biomedical engineers, and computational biologists need to work in tandem to have a fruitful insight into the diagnostics, therapeutics, and prevention of COVID-19.

References

1. Hoehl, S., et al. (2020). Evidence of SARS-CoV-2 infection in returning travelers from Wuhan, China. *The New England Journal of Medicine*. https://doi.org/10.1056/NEJMc2001899.
2. Lai, C. C., et al. (2020). Asymptomatic carrier state, acute respiratory disease, and pneumonia due to severe acute respiratory syndrome coronavirus 2 (SARS-CoV-2): Facts and myths. *Journal of Microbiology, Immunology, and Infection*. https://doi.org/10.1016/j.jmii.2020.02.012.
3. Sohrabi, C., et al. (2020). World Health Organization declares global emergency: A review of the 2019 novel coronavirus (COVID-19). *International Journal of Surgery, 76*, 71–76.
4. He, X., et al. (2020). Temporal dynamics in viral shedding and transmissibility of COVID-19. *Nature Medicine*. https://doi.org/10.1038/s41591-020-0869-5.
5. Ranney, M. L., Griffeth, V., & Jha, A. K. (2020). Critical supply shortages—The need for ventilators and personal protective equipment during the Covid-19 pandemic. *The New England Journal of Medicine*. https://doi.org/10.1056/NEJMp2006141.
6. Yang, Y. et al. (2020). Evaluating the accuracy of different respiratory specimens in the laboratory diagnosis and monitoring the viral shedding of 2019-nCoV infections. *medRxiv*. Preprint at https://www.medrxiv.org/content/10.1101/2020.02.11.20021493v2.
7. Fang, Y., et al. (2020). Sensitivity of chest CT for COVID-19: Comparison to RT-PCR. *Radiology*. https://doi.org/10.1148/radiol.2020200432.
8. Murray, P. R. (2015). The clinician and the microbiology laboratory. In J. E. Bennett, R. Dolin, & M. J. Blaser (Eds.), *Mandell, Douglas and Bennett's principles and practice of infectious diseases* (pp. 191–223). Philadelphia: Elsevier. [Google Scholar].
9. Pretorius M., Venter M. (2017) Diagnosis of Viral Infections. In: Green R. (eds) Viral Infections in Children, Volume I. (pp.151–182) Springer,[Google Scholar].
10. Rouse, B. T., & Muller, S. N. (2019). Host defense to viruses. In R. R. Rich, T. A. Fleisher, W. T. Shearer, H. W. Schroeder, A. J. Frew, & C. M. Weyand (Eds.), *Clinical immunology: Principles and practice* (5th ed., pp. 365–74e). Elsevier Limited. [Google Scholar].
11. Hufsky, F., Lamkiewicz, K., Almeida, A., et al. (2020). Computational strategies to combat COVID-19: useful tools to accelerate SARS-CoV-2 and coronavirus research. *Briefings in Bioinformatics, 22*(2), 642–663. bbaa232.

12. Watkins, X., Garcia, L. J., Pundir, S., et al. (2017). ProtVista: Visualization of protein sequence annotations. *Bioinformatics, 33*(13), 2040–2041.
13. Kalvari, I., Argasinska, J., Quinones-Olvera, N., et al. (2018). Rfam 13.0: Shifting to a genome-centric resource for non-coding RNA families. *Nucleic Acids Research, 46*(D1), D335–D342.
14. Nawrocki, E. P., & Eddy, S. R. (2013). Infernal 1.1: 100-fold faster RNA homology searches. *Bioinformatics, 29*(22), 2933–2935.
15. Rambaut, A., Holmes, E. C., O'Toole, Á., et al. (2020). A dynamic nomenclature proposal for SARS-CoV-2 lineages to assist genomic epidemiology. *Nature Microbiology, 5*(11), 1403–1407.
16. Chang, W., Cheng, J., Allaire, J., Xie, Y., & Mcpherson, J.. (2020). Shiny: Web application framework for r. r package version 1.4.0.2.
17. Solis-Reyes, S., Avino, M., Poon, A., et al. (2018). An open-source k-mer based machine learning tool for fast and accurate subtyping of HIV-1 genomes. *PLoS One, 13*(11), e0206409.
18. Hadfield, J., Megill, C., Bell, S. M., et al. (2018). Nextstrain: Realtime tracking of pathogen evolution. *Bioinformatics, 34*(23), 4121–4123.
19. Elbe, S., & Buckland-Merrett, G. (2017). Data, disease and diplomacy: GISAID's innovative contribution to global health. *Global Challenges, 1*(1), 33–46.
20. Breiman, L. (2001). Random forests. *Machine Learning, 45*(1), 5–32.
21. Wright, M. N., & Ziegler, A. (2017). Ranger: A fast implementation of random forests for high dimensional data in C++ and R. *Journal of Statistical Software, 77*(1), 1–17.
22. Schneider, G., & Fechner, U. (2005). Computer-based *de novo* design of drug-like molecules. *Nature Reviews Drug Discovery, 4*(8), 649–663.
23. Kapetanovic, I. (2008). Computer-aided drug discovery and development (CADDD): In silico-chemico-biological approach. *Chemico-Biological Interactions, 171*(2), 165–176.
24. Gordon, D. E., Jang, G. M., Bouhaddou, M., et al. (2020). ASARS-CoV-2 protein interaction map reveals targets for drug repurposing. *Nature, 583*(7816), 459–468.
25. Lasso, G., Mayer, S. V., Winkelmann, E. R., et al. (2019). A structureinformed atlas of human-virus interactions. *Cell, 178*(6), 1526–41.e16.
26. Garzón, J. I., Deng, L., Murray, D., et al. (2016). A computational interactome and functional annotation for the human proteome. *eLife, 5*, e18715.
27. Zhang, Q. C., Petrey, D., Deng, L., et al. (2012). Structure-based prediction of protein–protein interactions on a genome-wide scale. *Nature, 490*(7421), 556–560.
28. Bairoch, A. (2000). The ENZYME database in 2000. *Nucleic Acids Research, 28*(1), 304–305.
29. Finn, R. D., Coggill, P., Eberhardt, R. Y., et al. (2015). The pfam protein families database: Towards a more sustainable future. *Nucleic Acids Research, 44*(D1), D279–D285.
30. O'Leary, N. A., Wright, M. W., Brister, J. R., et al. (2016). Reference sequence (refseq) database at ncbi: Current status, taxo-nomic expansion, and functional annotation. *Nucleic Acids Research, 44*, D733–D745.
31. Schäffer, A. A., Hatcher, E. L., Yankie, L., et al. (2020). VADR: Validation and annotation of virus sequence submissions to GenBank. *BMC Bioinformatics, 21*, 211.
32. Kermack, W. O., & McKendrick, A. G. (1927). A contribution to the mathematical theory of epidemics. *Proc R Soc A, 115*(772), 700–721.
33. Lopez, L. R., & Rodo, X. (2020). A modified SEIR model to predict the COVID-19 outbreak in spain and italy: simulating control scenarios and multi-scale epidemics. *MedRxiv*.
34. Khailaie, S., Mitra, T., Bandyopadhyay, A, et al. (2020). Estimate of the development of the epidemic reproduction number Rt from coronavirus SARS-CoV-2 case data and implications for political measures based on prognostics. *MedRxiv*.
35. Martin, R., Löchel, H. F., Welzel, M., et al. (2020). CORDITE: The curated CORona drug InTERactions database for SARS-CoV-2. *IScience, 23*(7), 101297.
36. Sadegh, S., Matschinske, J., Blumenthal, D. B., et al. (2020). Exploring the SARS-CoV-2 virus-host-drug interactome for drug repurposing. *Nature Communications, 11*(1), 3518.
37. Guirimand, T., Delmotte, S., & Navratil, V. (2014). VirHostNet 2.0: Surfing on the web of virus/host molecular interactions data. *Nucleic Acids Research, 43*(D1), D583–D587.

38. Pfefferle, S., Schöpf, J., Kögl, M., et al. (2011). The SARS-coronavirus-host interactome: Identification of cyclophilins as target for pan-coronavirus inhibitors. *PLoS Pathogens, 7*(10), e1002331.
39. Kühnert, D., Stadler, T., Vaughan, T. G., et al. (2016). Phylodynamics with migration: A computational framework to quantify population structure from genomic data. *Molecular Biology and Evolution, 33*(8), 2102–2116.
40. Alcaraz, N., et al. (2016). Robust de novo pathway enrichment with KeyPathwayMiner 5. *F1000Research, 5*, 1531.

COVID-19 Detection Using Discrete Particle Swarm Optimization Clustering with Image Processing

Bhimavarapu Usharani

1 Introduction

COVID-19 is an infectious disease caused by the Severe Acute Respiratory Syndrome Corona Virus2 (SARS-COV2), and this disease was first identified in Wuhan Province in China in December 2019 [1]. In 2019, the World Health Organization (WHO) declared that an unknown disease has been detected in the city of the Wuhan and named that virus as the novel corona virus (ncov-2019), and its main symptoms is the severe pneumonia [2]. Globally around 50 million people are affected by corona virus, among them 1.5 million people unfortunately lost their lives as of December 2020 [3]. The symptoms of the COVID-19 are given in Table 1.

Air pollution also one of the factors of COVID-19 infection spread. According to the US Environmental Science and Forestry, an increase in hazardous air pollutants is correlated with around 20% in humanity among COVID-19 patients. Figure 1 shows the COVID-19 mortality association with the mass pollution and the hazardous air pollution.

Andhra Pradesh stood at the fourth place in COVID-19 cases in India as of April 07, 2021. The mass population is one of the major factors of COVID-19 spread in Andhra Pradesh. The mortality and morbidity in Andhra Pradesh and its capital district Krishna is shown in Fig. 2.

This chapter discusses the automatic detection model of mortality and morbidity in a patient with COVID-19. The confirmed, active, recovered, and deceased cases in India are shown in Fig. 3.

B. Usharani (✉)
Department of Computer Science and Engineering, Koneru Lakshmaiah Education Foundation, Vaddeswaram, AP, India

© The Author(s), under exclusive license to Springer Nature Switzerland AG 2022
S. K. Pani et al. (eds.), *Assessing COVID-19 and Other Pandemics and Epidemics using Computational Modelling and Data Analysis*,
https://doi.org/10.1007/978-3-030-79753-9_13

Table 1 Symptoms of COVID-19

COVID-19 category	Symptoms
Asymptomatic or pre-symptomatic	No symptoms
Mild	Fever, fatigue, cough
Moderate	Nasal congestion, malaise, muscle aches, diarrhoea, nausea, vomiting, lack of appetite
Severe	High fever, difficulty breathing, chest pain, trouble in walking
Critical	Weak pulse, respiratory failure, organ failure

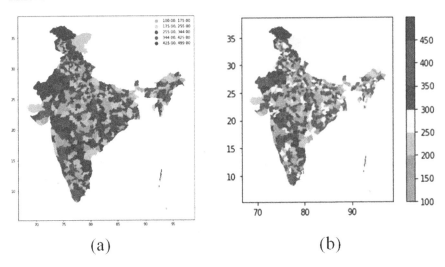

Fig. 1 The mortality and morbidity of COVID-19 in India district wise. (**a**) Morbidity of COVID-19 in India. (**b**) Mortality of COVID-19 in India

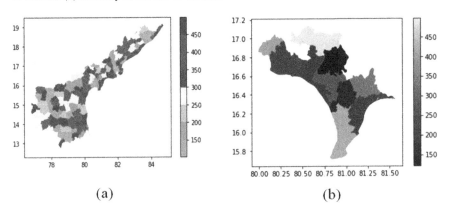

Fig. 2 COVID-19 mortality and morbidity in Andhra Pradesh and Krishna district in Andhra Pradesh. (**a**) COVID-19 cases in Andhra Pradesh. (**b**) COVID-19 Cases in Krishna district in Andhra Pradesh COVID-19 Cases in Krishna district in Andhra Pradesh

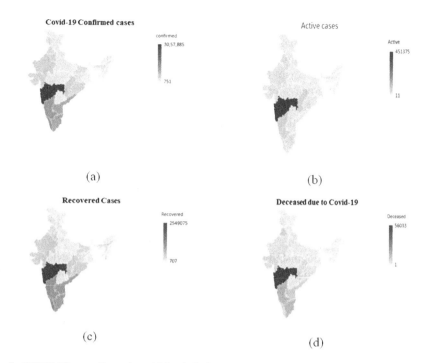

Fig. 3 COVID-19 mortality and morbidity in India. (**a**) Confirmed cases. (**b**) Active classes. (**c**) Recovered cases. (**d**) Deceased cases

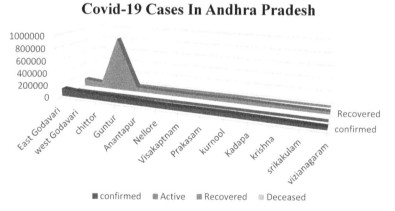

Fig. 4 COVID-19 morbidity cases in Andhra Pradesh, district wise

According to the Andhra Pradesh government reports as of April 07, 2021, the mortality rate in the Krishna district is high, and it stood as the second place in Andhra Pradesh. COVID-19 cases report of Andhra Pradesh is shown in Fig. 4.

The contribution is summarized as follows:

- To diagnose the premature detection of COVID-19 utilizing the particle swarm optimization clustering for segmentation of computed tomography image which improves the accuracy of the detection of COVID-19.
- Performed pre-processing to improve the contrast of the image by applying the Contrast Limited Adaptive Histogram Equalization algorithm.
- Blended discrete particle swarm optimization clustering algorithm is applied in the segmentation phase.
- To distinguish the candidate morbidity as COVID-19 or non-COVID-19 seven different features are extracted.
- Extensive experiments were conducted using computed tomography image dataset from Kaggle.

This chapter aims at the computed tomography image to identify and root out COVID-19 infection. This chapter proposes blended discrete particle swarm optimization to distinguish the infection rate by reflecting the pixel evidence of the infected lung. The ratio between the pixels of the infected lung and the normal lung segments is measured to identify the seriousness of COVID-19 pneumonia.

Medical imaging techniques plays a vital role for diagnosing the life-threating diseases. Computed tomography (CT) plays an important role in detecting the life-threating diseases like COVID-19, and these techniques are sensitive as well as a fast procedure in diagnosing the disease [4, 5]. It is observed that COVID-19 shows its impact on the lungs after 2 days and its severity after 10 days [6, 7].

This chapter is organized as follows. Section 2 gives an overview of the particle swarm optimization clustering, fitness measures, and particle swarm optimization terminology. Section 3 summarizes the literature survey of the contributions made by several researchers in the premature detection of pneumonia and COVID-19 detection. The proposed technology is discussed in Sect. 4. Finally, this chapter concludes with the conclusion section.

2 Preliminaries

2.1 Overview of Particle Swarm Optimization

Particle swarm optimization (PSO) [8] is used for continuous optimization. Each particle $pi, 1 \leq i \leq Nd$ is associated with a position X_i, d and a velocity V_i, d, $1 \leq d \leq D$ in the dth dimension of the search space, the dimension D is same for all particles. First, all the particles must be initialized with a random position and velocity to move in the search space. At each iteration t, every particle finds its own best solution ($Pbest_i$) and the global best solution (Gbest). The formulas for the particles updated position, updated velocity is given as:

Velocity Update:

Velocity Update Formula

$$v_{i,k}(t+1) = W v_{i,k}(t) + C_1 r_{1,k}(t)(y_{i,k}(t) - x_{i,k}(t)) + C_2 r_{2,k}(t)(y(t) - x_{i,k}(t))$$

In the above equation, W is called the inertia weight, $C1$ and $C2$ are the global constants.

Position Update:

Position Update Formula

$$x_i(t+1) = x_i(t) + v_i(t+1)$$

The particle updates Pbest$_i$ as well as Gbest as follows:

Pbest Formula

$$Pbest_i = \begin{cases} P_i, if(Fitness(P_i) < Fitness(Pbest_i)) \\ Pbest_i, otherwise \end{cases}$$

Gbest Formula

$$Gbest = \begin{cases} Pbest_i, if(Fitness(Pbest_i) < Fitness(Gbest)) \\ Gbest, otherwise \end{cases}$$

Particle swarm optimization algorithm [8]:

1. Evaluate the fitness measure for each particle.
2. Evaluate the velocity and position updates.
3. Find the Pbest and Gbest. Particle swarm optimization algorithm continues its iteration until it reaches a specified number of iterations or when the velocity value becomes zero.

2.2 Particle Swarm Optimization Clustering

Clustering is a technology that groups similar data points into the same group from the given data set [9]. Some applications of the clustering algorithms are image segmentation [10], pattern analysis [11], and image compression [12]. Clustering based on swarm intelligence, that is, particle swarm optimization can be used for clustering analysis [13]. Particle swarm optimization–based algorithms consider datapoints as particles. The initial clusters of particles are collected from other clustering algorithms. The cluster of particles is updated continuously based on the centre of the clusters, velocity, and position of the cluster until the cluster centre converges. The main disadvantages of the hierarchical clustering are as follows:

1. Time complexity.
2. Does work good for large datasets, challenging to determine the number of clusters with dendrograms.

The main disadvantages of the density-based clustering are as follows:

1. DBSCAN does not work well for high-dimensional data.
2. DBSCAN fails for varying density clusters.
3. Does not work well for large datasets.
4. User must give in advance radius and minimum number of points in the cluster's neighbourhood.

Particle swarm optimization clustering algorithm [14].

1. Evaluate the fitness measure for each particle in each cluster. $\forall x_i \varepsilon C_j$ (Cj is a cluster j).
2. Evaluate the velocity and position updates for each particle in a cluster $\forall C_j$.
3. Replace Pbest with individual i^{th} cluster centre and Gbest with neighbour cluster centres.

Particle swarm optimization clustering algorithm iterates until it reaches maximum iteration, or it converges. The formulas for velocity updates, position updates, Pbest, Gbest are already discussed in the above section.

The fitness measure used for particle swarm optimization clustering is.

Fitness Measure

$$J_e = \frac{\sum_{j=1}^{N_c} \left[\frac{\sum_{\forall z_p \varepsilon C_j} d(z_p, m_j)}{|C_j|} \right]}{N_c}$$

where

$$d(z_p, m_j) = \sqrt{\sum_{k=1}^{N_b} (z_{pk} - m_{jk})^2}$$

The cluster centroids used for the particle swarm optimization image clustering algorithm [14] are as follows:

(a)

$$v_i(t+1) = w v_i(t) + c1 r1(t) (y_i(t) - x_i(t)) + c2 r2(t) (\hat{y}(t) - x_i(t))$$

(b)

$$xi\,(t+1) = xi(t) + vi\,(t+1)$$

3 Reviews of Literature

This chapter focuses on the particle swarm optimization–based image clustering. Each particle is a set of n coordinates, where each one corresponds to the dimensional position of a cluster centroid. Each particle represents a position the Nd, then adjust its position based on the particle local best position, particle global best position.

COVID-19 Detection

Wang et al. [15] examined the lung computed tomography of patients and collected 366 computed tomography slices and confirmed that the infection will be severe within the duration of 6 to 11 days from the infection. The authors [16] implemented the image feature as well as the laboratory-supported analytic techniques to distinguish COVID-19 disease using the computed tomography of 81 patients and established that the evaluation of imaging feature along with the clinical and laboratory parameters will support in premature recognition of pneumonia triggered by COVID-19. The authors [17] examined 81 patients, implemented the computed tomography image feature as well as the laboratory-assisted techniques, and they classified the input features as COVID-19 and non-COVID-19 with sensitivity of 98%. The authors [18] examined around 250 patients and considered the computed tomography features and classified input features as COVID-19 and non-COVID-19 with sensitivity of 93%.

The authors [19] examined 36 patients with computed tomography and identified premature COVID-19 infection with accuracy of 91%. Liu et al. [20] employed a technique to identify the virus using the computed tomography dataset and classified COVID-19 infection as the mild, moderate, severe, and critical. The authors [21] examined 62 patients and executed the investigation with computed tomography images and confirmed that the proposed evaluation shows better accuracy on identifying COVID-19 in the initial phase. The authors [22] examined the chest X-ray and the computed tomography of COVID-19 patients and confirmed that the image-based assessment is better to categorize COVID-19 with improved accuracy.

Anomaly detection also helps to detect COVID-19. Convolution neural networks established generative types such as Generative Adversarial Networks [23] and Variational Auto Encoders [24] are used for unsupervised anomaly detection. The expansions such as the context encoder [25], Constrained Variational Auto Encoders [25], Adversarial Auto Encoder [26], and Bayesian Variational Auto Encoders [27] improve the accuracy of the projections. The authors [28] proposed a CAAD (Confidence Aware Anomaly Detection) model and a confidence prediction network for viral pneumonia screening and achieved accuracy of 83.61%. The author [29] employed a discipline approach because of the array of COVID-19, pneumonia, or breast X-ray. They concluded that their approach shows the accuracy of 0.9725.

In [30], the authors implemented the Resnet architecture to differentiate the covid and non-covid for MRI scan images. In Resnet architecture the authors used 121 hidden layers. In [31] to notice pneumonia and the technique is after prolonged after observing 14 different ailments in chest X-ray images. In [32], employed a pre-trained InceptionV3 because the abstraction of an image embeddings. The acknowledged configuration was able to distinct unique respiratory illnesses expertly or carried out an exceptionally excessive accuracy regarding. A Deep learning algorithm used to be recommended to classify pneumonia in accordance with observe COVID 19 disorder by means of working usage of both mutual and awesome functions on COVID-19 yet pneumonia [33].

In [34] the authors proposed a new pre-processing technique on the xray images to identify the covid-19 using the convolution neural network. In [35], the authors implemented the Chexnet version, to spot anomalies within chest X-rays after aligning into usual, pneumonia, Covid-19. An SE-ResNext101 encoder besides SSD RetinaNet is implemented in [36]. The authors in [38] used convolution neural network architectures, inclusive of InceptionV3, InceptionResnetV2, Xception, by considering the different hyper parameters to identify the ailments in the chest x-ray images. This model was beyond adjusted regarding a database including heart X-rays around 27,000 special patients. Every examined x-ray image is labelled from the related radiological reviews as No Lung Opacity (or) Not Normal, Normal, Lung Opacity, in view that lung imprecision is a considerable symptom of pneumonia [37]

4 Proposed Methodology

The main aim of this research intends to identify pneumonia in the early stage with high accuracy. This research focuses on the detection of COVID-19. Detecting the pneumonia with the particle swarm optimization discrete clustering segmentation increases the accuracy in the pneumonia detection. The proposed methodology is given in Fig. 5.

4.1 Pre-processing

The pre-processing stage helps to enhance the quality of the computed tomography image. The spatial domain techniques are the techniques that operate on pixels. In a pixel-based approach, the input information is the pixel itself. The image produced after applying the pixel-based approach is of high contrast, and these enhancement techniques depend on the Gray levels. In this chapter, the pre-processing step uses histogram processing, as this technique works on pixels. The stages in pre-processing used are resizing, conversion of Red Blue Green (R GB) image to

Fig. 5 Flowchart of the
proposed methodology

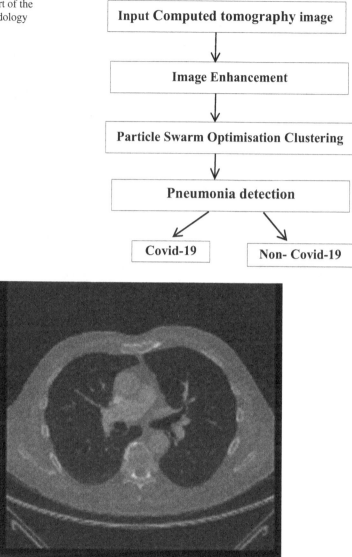

Fig. 6 Pre-processed image

the Gray image, applying the Contrast Limited Adaptive Histogram Equalization
(CLAHE) [39], and techniques working. The pre-processing phase of the computed
tomography image is shown in Fig. 6.

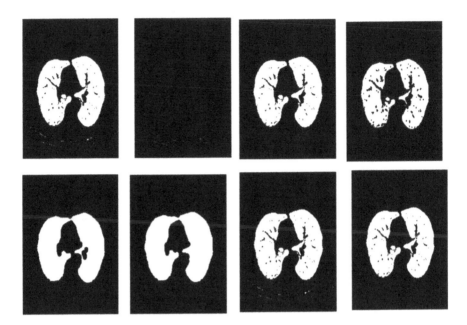

Fig. 7 Particle swarm optimization clustering

4.2   Segmentation

The main objective of the segmentation is to rearrange or change the computed tomography image appearance to make it more significant, simple to differentiate, and extract the features of the computed tomography scan image. In this work, blended particle swarm optimization clustering is used to segment the pre-processed image. The segmentation phase helps for the recognition of pneumonia from the computed tomography scan image. After applying the blended discrete PSO clustering, the results are shown in Fig. 7.

4.2.1   Particle Swarm Optimization Clustering

Blended discrete particle swarm optimization clustering is used in this chapter to effectively segment the microaneurysms.

Algorithm 1: Blended Discrete Particle Swarm Optimization (BPSO)
   Result: Gbest of the whole swarm.
   Initialization: Generate Swarm size particles

  (i) Initialize all particles in the swarm.
 (ii) Estimate the fitness rate for all particle.
(iii) For every single particle.

1. Calculate the velocity in the cluster block.
2. Apply the neighbour search scheme on each particle.
3. Evaluate the fitness measure.
4. Update Pbest, Gbest.

(iv) Return Gbest.

4.2.2 Fitness Measures

The fitness measure is the primary accuracy of the segmented computed tomography scan image. The fitness measure measures the quality of the clusters. Each particle x_n in a cluster is represented as $x_n = (x_{n,1}, ...x_{n,k})$. A swarm represents several data clustering solutions. The quality of the particle in the image is evaluated using the fitness function. The fitness measures that are used in the above discussed elevated particle swarm optimization clustering is discussed in Table 2.

4.3 Feature Extraction

In the feature extraction stage of the proposed system, the lesion is extracted from the segmented image. A set of features, namely, area, perimeter, circularity, and diameter are used to extract lesions. Irrelevant and redundant features should be eliminated from the segmented image to improve accuracy. The features used in this chapter are listed in Table 3. The result of feature extraction is shown in Fig. 8.

5 Experimental Result

The performance of pneumonia detection technique on publicly available datasets was evaluated on Kaggle. The execution of the algorithms was tested using 75-25 train test split methodology. Blended discrete particle swarm optimization algorithm in python was implemented and used the IDE Anaconda. The metric levels are given in Table 4.

Performance measures used in the proposed technique are discussed in Table 5. We compared the accuracy of the present method with other research works in Table

Table 2 Fitness measures used for particle swarm optimization clustering in proposed section

Measure	Formulae
Entropy	$\sum_{i=0}^{\infty}\sum_{j=0}^{\infty}\frac{p}{\log p}$
Kurtosis	$\frac{1/n\sum(y-\bar{y})^2}{s^4}$
Skewness	$\frac{1/n\sum(y-\bar{y})^2}{s^3}$

Table 3 List of features

S. No	Feature description	Formula
f1	Entropy	$\sum_{i=0}^{\infty}\sum_{j=0}^{\infty}\frac{p}{\log p}$
f2	Max. Probability	$Max\{p_{ij}(x,y)\}$
f3	Energy	$\sum\sum P_{IJ}^2$
f4	Mean	$\sum_{i=0}^{\infty}\sum_{j=0}^{\infty}\frac{p}{NXM}$
f5	Variance	$\sum_{i=0}^{\infty}\sum_{j=0}^{\infty}\left(P(I\text{-}J)\text{-}\mu\right)^2$
f6	Dissimilarity	$\sum_{i=0}^{\infty}\sum_{j=0}^{\infty}P(I\text{-}J)$
f7	Contrast	$\sum_{i=0}^{\infty}\sum_{j=0}^{\infty}P(I\text{-}J)^2$

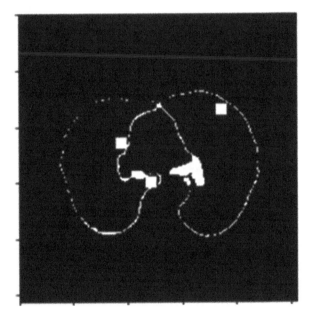

Fig. 8 Result of feature extraction

Table 4 Metric levels

Metric	Description
True positive	Correctly classified COVID-19 pixels
True negative	Correctly classified non-COVID-19 pixels
False positive	Misclassified non-COVID-19 treated as COVID-19
False negative	Misclassified COVID-19 treated as non-COVID-19

6. The statistic properties used to measure the similarity and the distribution of the data are kurtosis and skewness. Kurtosis confirms the symmetry of the data distribution. To identify the symmetry of the data distribution in the cluster, we are using the measures kurtosis and skewness to assess the prediction about the

Table 5 Performance metrics

Measure	Formulae
Sensitivity	$\frac{TP}{(TP+FN)}$
Specificity	$\frac{TN}{(TN+FP)}$
False positive rate	$\frac{FP}{(FP+TN)}$
False negative rate	$\frac{FN}{(TP+FN)}$
Accuracy (Acc)	$\frac{(TP+FN)}{(TP+FP+TN+FN)}$

Table 6 Comparison of accuracy, AUC, and F1-score with existing technique

Study	Classifier	Accuracy	AUC	F1-score
[38]	SVM, RF	–	87%	72%
[40]	XGBOOST	–	66%	–
[41]	SVM	80%	–	–
Present work	BDPSO	93.7%	92.4%	96.5%

distribution of the data in the clusters. The outcome on the test data is depicted in Fig. 9.

The present work is compared with the existing approaches and is represented in Fig. 10. The kurtosis and skewness measures are used as the accuracy measures, and these statistical measures for the present work are compared with the existing PSO algorithm and are given these values in Table 7.

The results of experiments on datasets that are publicly available and validate the suggested method surpass well-renowned techniques in provisions of high-order statistics measures. To compare clusters, we use high-order statistics such as kurtosis and skewness. These two high-order statistical measures give a more accurate description of the shape of the cluster. The main challenge in image clustering is how to manipulate clusters based on the previous complex representations and to identify the new cluster. Existing cluster algorithms used the mean, median, variance, and covariance to represent the cluster. In this work, we use kurtosis and skewness for the representation of the clusters. The main advantage of using these two high-order statistics is to identify low complexity cluster. The skewness represents the asymmetry of the cluster, whereas the kurtosis represents the concentration of the cluster. In case, if the two different clusters possess similar statistical properties, then the clusters are merged. The challenging task is the detection of new clusters and merging old and new clusters. Aggregation of clusters is done by examining the statistics of the entropy, kurtosis, and skewness of the clusters. COVID-19 detection is identified and shown in Fig. 11.

6 Conclusion

This study used the blended discrete particle swarm optimization clustering algorithms to segment the pre-processed image and extracted the features to detect the pneumonia in the segmented computed tomography scan image. Automatic

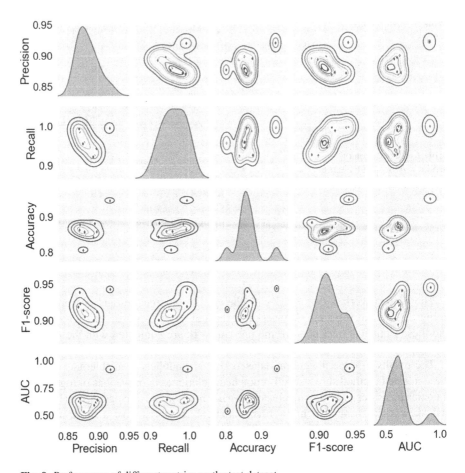

Fig. 9 Performance of different metrics on the test dataset

detection of pneumonia in the computed tomography scan image reduces the risk of life. In this chapter, a technique is proposed and demonstrated through experimental outcomes, enhancing the pneumonia detection compared to the existing methods. The future work on the proposed technique can extend its scope to the diagnosis of pneumonia for the premature recognition of COVID-19. In this chapter, there is an automated estimate of COVID-19 employing a blended particle swarm optimization learning technique. For this, we applied blended particle swarm optimization learning pretrained modes to obtain a higher accuracy.

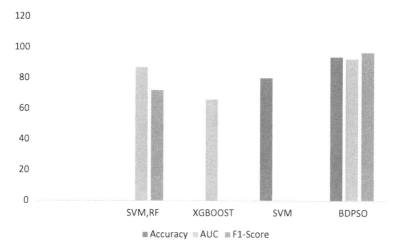

Fig. 10 Comparison of different techniques

Table 7 Similarity measures

Image Id	Particle swarm optimization				Blended discrete particle swarm optimization			
	Entropy	kurtosis	skewness	Runtime	Entropy	kurtosis	skewness	Runtime
Image001	3.57	−0.75	−1.11	16.45	4.28	5.56	2.44	15.64
Image002	4.33	−1.8	0.39	19.51	3.25	2.41	2.54	14.45
Image003	4.10	−1.94	−9.25	19.47	3.59	3.15	2.13	16.49
Image004	4.91	−0.87	1.06	20.64	3.21	0.12	0.99	16.58
Image005	4.20	−1.82	0.41	18.28	1.13	0.65	1.09	15.84
Image006	1.24	−1.46	−0.73	17.88	1.07	0.99	0.13	16.19
Image007	4.10	−1.85	−0.38	18.19	2.54	0.99	0.98	14.01
Image008	3.57	−1.89	−0.34	18.30	3.31	1.15	1.37	13.11
Image009	4.07	−1.78	0.47	18.85	3.86	−1.85	0.35	14.38
Image010	4.78	−0.45	1.25	18.14	2.45	−1.71	0.44	16.31
Image011	4.23	−1.79	−0.46	18.25	3.56	−1.79	0.48	15.47
Image012	3.25	0.94	−1.72	18.41	3.68	−1.87	0.45	18.94
Image013	4.52	6.13	5.28	18.08	2.78	−1.91	0.41	16.62
Image014	4.01	−1.99	−0.07	18.25	3.59	−2.11	0.65	18.16
Image015	4.23	−1.94	0.24	18.24	3.45	0.99	0.98	17.85
Image016	4.23	3.07	2.25	18.01	3.47	−1.63	0.15	16.54
Image017	4.18	−0.89	1.05	18.53	3.84	−1.97	0.07	15.69
Image018	3.15	0.73	−1.65	18.03	3.45	−1.99	0.28	16.61
Image019	4.16	−1.99	−0.10	18.08	3.47	2.12	2.01	15.62
Image020	4.28	−1.99	0.01	18.33	3.18	−1.69	0.53	15.69

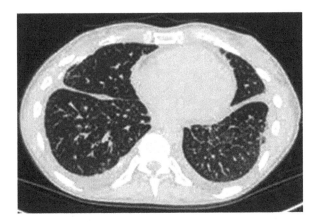

Fig. 11 COVID-19 detection

References

1. Song, F., Shi, N., Shan, F., Zhang, Z., Shen, J., Lu, H., et al. (2020). Emerging coronavirus 2019-nCoV pneumonia. *Radiology, 295*, 210–217.
2. Singhal, T. (2020). A review of coronavirus Disease-2019 (COVID-19). *Indian Journal of Pediatrics, 87*(4), 281–286.
3. WHO Coronavirus Disease (COVID-19) Dashboard. Available online: https://covid19.who.int/. Online accessed on 5 December 2020.
4. Zu, Z. Y., Jiang, M. D., Xu, P. P., Chen, W., Ni, Q. Q., Lu, G. M., & Zhang, L. J. (2020). Coronavirus disease 2019 (COVID-19): A perspective from China. *Radiology, 296*(2), 15–25.
5. Lee, E. Y., Ng, M. Y., & Khong, P. L. (2020). COVID-19 pneumonia: What has CT taught us? *The Lancet Infectious Diseases, 20*(4), 384–385.
6. Bernheim, A., & Mei, X. (2020). Chest CT findings in coronavirus disease-19 (COVID19): Relationship to duration of infection. *Radiology, 1*(1) 1–19.
7. Pan, F., & Ye, T. (2020). Time course of lung changes on chest CT during recovery from 2019 novel coronavirus (COVID-19) pneumonia. *Radiology 3*(295), 715–721.
8. Kennedy, J., & Eberhart, R. (1995). Particle swarm optimization, ieee international of first conference on neural networks.
9. Anderberg, M. R. (1973). The broad view of cluster analysis. *Cluster Analysis for Applications, 1*(1), 1–9.
10. Ray, S., & Turi, R. H. (1999). Determination of number of clusters in kmeans clustering and application in colour image segmentation. In *Proceedings of the 4th international conference on advances in pattern recognition and digital techniques*, pp. 137–143. Calcutta, India.
11. Li, X., & Fang, Z. (1989). Parallel clustering algorithms. *Parallel Computing, 11*(3), 275–290.
12. Lee, R. C. T. (1981). Clustering analysis and its applications. In *Advances in information systems science* (pp. 169–292). Springer.
13. Kennedy, J (2000) Stereotyping: Improving particle swarm performance with cluster analysis. In *Proceedings of the 2000 Congress on Evolutionary Computation. CEC00 (Cat. No. 00TH8512)* (Vol. 2, pp. 1507–1512). IEEE.
14. Omran, M. G., Engelbrecht, A. P., & Salman, A. (2004). Image classification using particle swarm optimization. In *Recent advances in simulated evolution and learning* (pp. 347–365). World Scientific.

15. Wang, Y., et al. (2020). Temporal changes of CT findings in 90 patients with COVID-19 pneumonia: A longitudinal study. *Thoracic Imaging, 2*(296), 55–64.
16. Shi, H., et al. (2020). Radiological findings from 81 patients with COVID-19 pneumonia in Wuhan, China: A descriptive study. *The Lancet Infectious Diseases, 20*(4), 425–434.
17. Fang, Y., et al. (2020). Sensitivity of chest CT for COVID-19: Comparison to RT-PCR. *Radiology.*
18. Bai, H. X., et al. (2020). Performance of radiologists in differentiating COVID-19 from viral pneumonia on chest CT. *Radiology.*
19. Chua, F., et al. (2020). The role of CT in case ascertainment and management of COVID-19 pneumonia in the UK: Insights from high-incidence regions. *Lancet Resp Med.*
20. Liu, K.-C., et al. (2020). CT manifestations of coronavirus disease-2019: A retrospective analysis of 73 cases by disease severity. *European Journal of Radiology, 126*, 108941.
21. Zhou, Z., Guo, D., Li, C., et al. (2020). Coronavirus disease 2019: Initial chest CT findings. *European Radiology, 1*(1), 1–19.
22. Yoon, S. H., et al. (2020). Chest radiographic and CT findings of the 2019 novel coronavirus disease (COVID-19): Analysis of nine patients treated in Korea. *Korean Journal of Radiology, 21*(4), 494–500. https://doi.org/10.3348/kjr.2020.0132
23. Goodfellow, I. J. (2014). On distinguishability criteria for estimating generative models. arXiv, *1*(1), 1–6.
24. Kingma, D. P., & Welling, M. (2013). Auto-encoding variational bayes. arXiv.
25. Zimmerer, D., Kohl, S. A. A., Petersen, J., Isensee, F., & Maier-Hein, K. H. (2018). Context-encoding variational autoencoder for unsupervised anomaly detection. arXiv.
26. Chen, X., & Konukoglu, E.. (2018). Unsupervised detection of lesions in brain MRI using constrained adversarial auto-encoders. arXiv.
27. Pawlowski, N., Lee, M. C. H., Rajchl, M., McDonagh, S., Ferrante, E., Kamnitsas, K., Cooke, S., Stevenson, S., Khetani, A., Newman, T., et al. (2018) Unsupervised lesion detection in brain CT using Bayesian convolutional autoencoders. arXiv.
28. Zhang, J., Xie, Y., Liao, Z., Pang, G., Verjans, J., Li, W., Sun, Z., He, J., & Li, C. S. Y. (2020). Viral pneumonia screening on chest x-ray images using confidence-aware anomaly detection. arXiv.
29. de Moura, J., Novo, J., & Ortega, M. (2020). Fully automatic deep convolutional approaches for the analysis of COVID-19 using chest X-ray images. medRxiv.
30. Pathak, Y., Shukla, P. K., Tiwari, A., Stalin, S., & Singh, S. (2020). Deep transfer learning based classification model for COVID-19 disease. *Irbm, 1*, 1–6.
31. Rajpurkar, P., Irvin, J., Zhu, K., Yang, B., Mehta, H., Duan, T., Ding, D., Bagul, A., Langlotz, C., Shpanskaya, K., Lungren, M. P., & Ng, A. Y. (2017). CheXNet: RAdiologist-level pneumonia detection on chest X-rays with deep learning. *ArXiv*, 3–9.
32. Verma, D., Bose, C., Tufchi, N., Pant, K., Tripathi, V., & Thapliyal, A. (2020). An efficient framework for identification of tuberculosis and pneumonia in chest X-ray images using neural network. *Procedia Computer Science, 171*(1), 217–224.
33. Zhang, Y., Niu, S., Qiu, Z., Wei, Y., Zhao, P., Yao, J., Huang, J., Wu, Q., & Tan, M. (2020). COVID-DA: Deep domain adaptation from typical pneumonia to COVID-19, XX, 2020, pp. 1–8.
34. Butt, C., Gill, J., Chun, D., & Babu, B. A. (2020). Deep learning system to screen coronavirus disease 2019 pneumonia. *Applied Intelligence, 2019*, 1–29.
35. Mangal, A., Kalia, S., Rajgopal, H., Rangarajan, K., Namboodiri, V., Banerjee, S., & Arora, C. (2020). CovidAID: COVID-19 detection using chest X-ray. *ArXiv.*
36. Gabruseva, T., Poplavskiy, D., & Kalinin, A. A. (2020). Deep learning for automatic pneumonia detection, 2019.
37. Mohammed, I., Singh, N., Area, B., & National, L.B. (n.d.). Computer-assisted detection and diagnosis of pediatric pneumonia in chest X-ray images, *1*(1), 1–9.
38. de Moraes Batista, A. F., Miraglia, J. L., Donato, T. H. R., & Filho, A. D. P. C. (2020). Covid-19 diagnosis prediction in emergency care patients: a machine learning approach. *medRxiv, 1*(1), 1–13.

39. Jenifer, S., Parasuraman, S., & Kadirvelu, A. (2016). Contrast enhancement and brightness preserving of digital mammograms using fuzzy clipped contrastlimited adaptive histogram equalization algorithm. *Applied Soft Computing, 42*, 167–177.
40. Schwab, P., Schütte, A. D. M., Dietz, B., & Bauer, S. (2020). predcovid-19: A systematic study of clinical predictive models for coronavirus disease 2019. *arXiv*.
41. Jiang, X., Coffee, M., Bari, A., Wang, J., Jiang, X., Huang, J., Shi, J., Dai, J., Cai, J., Zhang, T., et al. (2020). Towards an artificial intelligence framework for data-driven prediction of coronavirus clinical severity. *CMC: Computers, Materials & Continua, 63*, 537–551.

LSTM-CNN Deep Learning–Based Hybrid System for Real-Time COVID-19 Data Analysis and Prediction Using Twitter Data

Sitanath Biswas and Sujata Dash

1 Introduction

The world would remember the year 2020 as a catastrophic year for humanity on this planet earth. Pneumonia of unknown etiology (novel coronavirus) identified in the city of Wuhan, China, in December 2019 [1] with its first mortality reported on January 10, 2020, has become a pandemic [2] and quickly gulping the entire world under its net. It is named as COVID-19 (Coronavirus disease 2019) by the World Health Organization (WHO) [3]. As per John Hopkins University, 4,563,458 confirmed cases of COVID-19 [4] reported as of May 16, 2020. India contributes 1.9% with 86,508 cases and has fatality rate of 3.2% with 0.2 deaths per 100 k population [5]. All countries are trying to save their people lives by implementing measures such as travel restrictions, quarantines, event postponements and cancellations, social distancing, testing, and hard and soft lockdowns [6]. More than the lives this virus has taken, the economic and social impact is far more disastrous and especially for developing and underdeveloped countries. It is terrifying to imagine the disaster this COVID-19 may cause in India where 18% of the world's population resides [7] with a population density of 32,303 people per square kilometer in cities such as Mumbai [8]. So, this novel coronavirus may spread at a very high pace in India's large population. The Government of India is proposing multiple lockdowns to prevent the spread of this virus. Initially, in lockdown 1.0 (March 25, 2020, to April 14, 2020), the entire nation was under complete lockdown except for essential services and lockdown 2.0 (April 15, 2020, to May 3, 2020) was implemented with relaxation in areas where the virus was contained and lockdown 3.0 (May 4, 2020, to May 17, 2020) with more relaxations in areas where there were fewer number of

S. Biswas · S. Dash (✉)
Department of Computer Science, Maharaja Sriram Chandra Bhanja Deo, University (Erstwhile North Orissa University), Wellington dos Santos Takatpur, Baripada, Odisha, India

© The Author(s), under exclusive license to Springer Nature Switzerland AG 2022
S. K. Pani et al. (eds.), *Assessing COVID-19 and Other Pandemics and Epidemics using Computational Modelling and Data Analysis*,
https://doi.org/10.1007/978-3-030-79753-9_14

coronavirus cases. Due to these lockdowns, there has been a decrease in the number of cases from 11.8% to 6.3% on a daily basis [9]. But the government cannot shut the entire nation forever as the economy may fall drastically. Hence, a practical solution could be to quarantine the very critical zones, so that the people affected by this virus shall remain in that zone only.

As per the World Health Organization, no vaccine and anti-viral treatments are yet available for this virus [10], and medical organizations are trying hard to find out the vaccine for this novel coronavirus. However, even after fast-tracking the usual vaccine period of 5–10 years, the vaccine may take at least 18–24 months before it is available and may take further more time to produce it enough for the majority of the world [11]. Also, we do not know how long a vaccine would stay effective as the virus mutates. Every effort is made to slow down the spread of the coronavirus and prepare medical response systems to tackle the increase in patient loads and to protect the front-line medical staff with adequate supplies of personal protective gears such as personal protective equipment (PPE), masks, and other essentials. So, if we know beforehand the number of novel coronavirus cases for the next few days, we can plan our inventory accordingly. There is less number of papers on the prediction of novel coronavirus cases in the literature, and few of them are reviewed below in the literature review section.

Although various governments, private agencies, and NGOs are working relentlessly to control the pandemic, a single-window system for real-time data tracking, extraction, analysis, and prediction is the need of the hour. Therefore, there is a need for an effective and accurate system and method for predicting the spread of an infectious disease. Thus, in view of the above, there is a long-felt need in the industry to address the aforementioned deficiencies and inadequacies.

The organization of this chapter is as follows: The Sect. 2 describes the literature review in detail, Sect. 3 explains the various materials and methods that have been implemented for the proposed work, Sect. 4 essentially deals with the detailed architecture and its working principles, Sect. 5 is about the experimental result analysis, Sect. 6 highlights the discussion of the work, Sect. 7 explains the conclusion, limitations, and future work of this chapter followed by an exhaustive bibliography.

Objectives and Contributions

In this chapter, we are proposing a novel LSTM-CNN deep learning–based hybrid single-window system for real-time COVID-19 data analysis and prediction using Twitter data. The system is capable of generating accurate data related to COVID-19 such as the number of active patients in a particular location and how many have been recovered. These reports may be helpful for decision-making by the central government/local government and its officials to take next course of actions. The following are the goals of this chapter:

I. To prepare the dataset, IEEE data port is used to obtain the Twitter data.
II. To develop a hybrid deep machine learning model by combining long short-term memory (LSTM) model and convolution neural network (CNN) model to analyze the Twitter data for useful prediction, that is, the affected regions of

India, infected people, recovered patients, and medical facilities available for COVID-19.

III. The findings of the problem can prepare a detailed report on the current status of infection, recovery, quarantine, affected regions of COVID-19 in India, which may be useful for all government officials, private companies, other stakeholders, decision-makers, and individuals. The analysis is done by using long short-term memory (LSTM) model, convolution neural network (CNN) model, and Twitter data.

IV. A complete experimental report of the performance of the proposed system is depicted in form of accuracy, specificity, precision, recall, and f1-score.

2 Literature Review

The decision-makers require the accurate forecasting of the spread and recovery rate of COVID-19 for better understanding and formulation of approach for decreasing its spread stated by Velásquez RMA, Yousaf M, and Ribeiro MHDM. COVID-19 has become the most serious problem today as it is having an adverse effect on health and aftermath discussed by M. Ergen B. Cömert Z. The impact of COVID-19 ranges from the child to the elderly person, normal to chronic diseases. Hence it has become a global multidisciplinary issue involving people from all walks of life. The biggest challenges to all are augmenting the process of understanding and formulating an approach of slowing down its spread and reducing the impact. This has led to new approaches of modeling estimation and forecasting for better understanding and managing it. There are instances of using various mathematical models for estimating and forecasting based on the infected cases [5, 6]. Based on quarantined and recovery states, a generalized SEIR model has been developed for analyzing and predicting COVID-19 was discussed by Peng L., Yang W., Zhang D. As per Roosa K et al., both SEIR and SIR models are applied to model the predictions and representing the confirmed cases data information. SEIR and SIR models are used for prediction and representing the cases which are confirmed. It has been shown that the SIR model outperforms the SEIR model in terms of Akaike Information Criteria (AIC). Furthermore, an extended SIR version has been used by using another parameter, number of reported and unreported cases discussed by Liu Z et al. and Biswas, K. SIR model is furthermore stretched on Euclidean network to improve the quality of confirmed cases and exemplify role of spatial factor in widespread transmission of COVID-19 reported by Biswas K., Khaleque A. Roosa K, Lee Y had implemented three phenomenological models, namely, generalized logistic growth model (GLM), Richards growth model, and sub-epidemic wave model are suggested for short-term prediction of confirmed cases of COVID-19. The apprehension of the sub-exponential growth is handled using the GLM model, the deviation between symmetric logistics curve using the Richard model, and finally the complex trajectories are dealt with sub-epidemic curve. Jia L., Li K., Jiang Y had implemented Logistic, Bertalanffy, and Gompertz models that have been used adequately to investigate epidemic likelihoods. The logistic model exhibited improved prediction performance than the two other considered models. However, the main drawback is that they have been applied on few outbreak stages with enough data. To lessen this inadequacy, described by Wu K., Darcet D, generalized versions are suggested by comprising few extra parameters on the previous models. Sha He LR et al. and Roosa K et al. discussed a discrete-time stochastic model is established to pronounce the dynamic of the epidemic spread. This model verified the ability to capture the epidemiological status. Additional studies used time-series methods, such as Auto-Regressive Integrated Moving Average (ARIMA) to project the number of confirmed cases by Dehesh T, Mardani-Fard H. Gupta R et al. reintroduced a traditional ARIMA modeling and Exponential Smoothing method was used to examine and estimate the trends of the COVID-19 outbreak in India. Chintalapudi N. has implemented the ARIMA model is used to examine the registered and recovered COVID-19 cases in Italy after the completion of 60 days lockdown. Many earlier studies based on traditional time-series forecasting models have been explored to estimate future COVID cases in China and a few other countries was described by Kucharski AJ, Wu JT, and Zhuang Z. Accurate estimation of the number of COVID-19 cases is becoming the mainstay to enable the use of the available resources in hospitals and improve management strategies to optimally manage infected patients. Recently, machine learning and deep learning have been looked as a capable field of research in a wide range of applications, both in academia and industry as described by Tuncer AD. Rustam F et al. have discussed four supervised machine learning algorithms, namely, linear regression, LASSO regression, Exponential Smoothing (ES), and Support Vector Machine (SVM) have been used to forecast the future of COVID-19. The ES has outdone other models in forecasting the number of newly contaminated cases, recoveries, and the deaths. The main reason is that ES can handle time-series data by including information from past data in the prediction process. The study by Tuli S., Wu K et al. and Sha he et al. exhibited that machine learning and cloud computing provided potential solutions in improving proactively the prediction of the growth of the epidemic proactively. Chimmula VKR et al. have demonstrated how the countries such as Canada, Italy, and the USA, a deep learning approach based on long short-term memory (LSTM) is examined in the predicting of COVID-19 transmission. With the help of time-dependent datasets, the LSTM achieved respectable predicting performance due to its ability in handling time-dependent datasets. Rekha Hanumanthu S has described a stacked auto-encoder model that is introduced to fit the dynamical propagation of the epidemic and real-time forecasting of confirmed cases in China. Pal R. proposed a shallow long short-term memory in which the risk category, trend, and weather data are used as input for the prediction. In [12], a greater detail about intelligent computing–based research for COVID-19 has been discussed. A comparative study between the five most advanced data-driven forecasting methods in forecasting COVID 19 cases has been discussed in some articles authored by Hochreiter S., Cruz G., Ashour AS., and Harrou F. Few important deep learning models , namely, simple Recurrent Neural Network (RNN), long short-term memory (LSTM), Bidirectional LSTM (Bi-LSTM) were described by Schuster M. Variational Auto

Encoder (VAE) was introduced by Graves A. and Gated recurrent units (GRUs) was implemented compared by Liu Z., to predict the time series of the number of newly affected COVID 19 cases and recovered cases. These models have various striking features, such as handling temporal dependencies in time-series data, distribution-free learning models, and their elasticity in modeling nonlinear features. The VAE model has not been examined for any pandemic, before for COVID-19 prediction, but later it is implemented by Sharfuddin, Zhang B, and Wang Sand learning models have been evaluated on the publicly available COVID-19 patient statistics dataset provided by Johns Hopkins recorded from the starting of COVID-19 till June 17, 2020.

Based on these literature reviews, we have handpicked seven state-of-the-art works which have been used to model our system and evaluate the performance. We have tried to overcome all the challenges, we found in these papers.

Ricardo Manuel and Arias Velásquez have implemented reduced-space Gaussian process regression in their research work, where they have achieved 98.10% accuracy. Although their findings are promising, the main limitation of their research work was using multiple datasets. The relative absolute error was also high, that is, 99.19%. In the second paper, M. Yousaf and S. Zahir has implemented Auto-Regressive Integrated Moving Average Model (ARIMA), where their system is achieved 75.50% accuracy. The main limitation of their research was that their analysis is based on the assumptions and if the assumptions are not true, it may lead to an inaccurate forecast. Furthermore, to forecast using time-series model, which actually requires huge historical data, is not available in their work, therefore there is a potential chance that may lead to inaccurate prediction. The current pattern may not continue in the future. In another paper authored by Mesut Toğaçar who has implemented deep learning models, using fuzzy color and stacking approaches, achieved 99.27% accuracy. The main limitation of their research was that they have used very limited dataset, majorly collected from COVID cases in hospitals. As deep learning requires huge training data, there is a possibility of having reduced accuracy while handing large test data. Kucharski AJ et al. have implemented stochastic transmission dynamic model, achieved 77.05% accuracy. Although the result is promising, the main limitation of their research was that they had considered limited data, limited parameters, limited to only one city. Prediction for a small city with limited data may not be accurate and appropriate for few more cities if implemented. The paper authored by Ismail Kirbas implemented ARIMA, NARNN, and LSTM approaches and achieved 82.03% accuracy. The main limitation of their research was that they had limited data. But in their research, it was proved that the LSTM approach has much higher success compared to ARIMA and NARNN. In the paper authored by Vinay Kumar, Reddy Chimmula, they have implemented LSTM and achieved 92.67% accuracy. The main limitation of their research was that they have used static dataset by John Hopkins and the RMSE error was much higher 34.85%. As stated in the paper, the model could not handle the rapid growth and dynamic behavior of COVID-19 in Canada with the small datset. In the paper authored by BoZhang, Hanwen Zhang, they implemented Bi-LSTM neural networks. The main

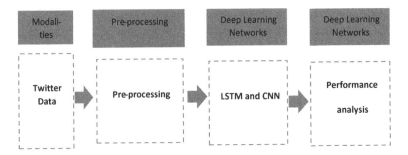

Fig. 1 The overall architecture of the proposed system

limitation of their research was that they used the model for different domain but results were very promising.

3   Materials and Methods

The proposed system for the detection of COVID-19 consists of different phases which are described in Fig. 1. The different phases include the first phase: pre-processing of data, data shuffling, and data normalization. In the second phase, data partitioning for training and test was done. In the third phase, we have prepared the LSTM-CNN model for training. After each epoch, the model needs to be tuned based on accuracy and other parameters. Finally, the proposed model was evaluated based on the precision, recall, and F-Score.

We have used CORONAVIRUS (COVID-19) TWEETS DATASET developed and maintained by IEEE Data port. We have used the dataset both for training and validating purpose.

4   Detailed Architecture of the Proposed Model

From the literature review, it is clearly evident that machine learning algorithms are very much helpful for the analysis and prediction of COVID-19. It was observed that few deep learning algorithms are outperforming particularly LSTM and CNN. LSTM and CNN are also helpful in another related domain [13–16] as well. But none of a single author has tested the combination of CNN and LSTM. During our initial setup, it was observed that implementing CNN and LSTM together may produce better result. We have tried with both the model CNN+ LSTM and LSTM+ CNN.

4.1   Pre-processing

As the collected tweets were crawled from social media, the data are expected to be noisy and should be cleaned up before performing any of the NLP tasks to get better results. We have removed URLs, mentions, media, duplicate Tweets, punctuations, numbers, and special characters. Data shuffling and data normalization were also done to smoothen the data for better accuracy of the proposed system.

4.2   Deep Neural Network

Artificial intelligence essentially has two types of machine learning: traditional machine learning (ML) and deep neural network learning (DL). Deep learning now a days become very popular among researchers and is being extensively implemented to analyze, predict, and diagnose COVID-19 accurately by using various public data sets. Deep learning architectures have recurrent neural networks (RNNs), convolutional neural networks (CNNs), autoencoders (AEs), generative adversarial networks (GANs), and deep belief networks (DBNs). Apart from these, DL also has hybrid models such as CNN-AE, CNN-RNN, and LSTM-CNN to name a few.

4.3   LSTM Models

LSTM is a classy gated memory unit designed to alleviate the waning gradient problems restraining the proficiency of a simple RNN [17] (Fig. 2). More precisely, the gradient becomes too small or large, which results in a waning gradient problem during the case of the important time step, this can be observed during the training, where the optimizer back propagates and makes the procedure run, while the weights almost do not change at all. LSTM consists of three gates regulating the information flow termed input, forge, and output gates. These gates are formed basically with logistic functions of weighted sums. The cell state is succeeded via the input gate and the forget gate. The output gate or the hidden state generates the output representing the memory directed for use thereby allowing the network memorizing for a long time which is not seen in the conventional single RNNS. Thereby the most looked-for traits of LSTM are their extended capacity to apprehend long-term dependencies and immense capability of handling time-series data. Given the input time-series X_t, and the number of hidden units as h, the gates have the following equations:

$$\text{Input Gate}: I_t = \sigma\left(X_t W_{xi} + H_{t-1} W_{hi} + bi\right),$$

$$\text{Forget Gate}: F_t = \sigma\left(X_t W_{xf} + H_{t-1} W_{hf} + bf\right),$$

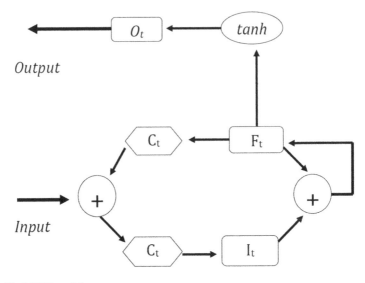

Fig. 2 The LSTM module

$$\text{Output Gate}: O_t = \sigma \left(X_t W_{xo} + H_{t-1} W_{ho} + \text{bo} \right),$$

$$\text{Intermediate Cell State}: C_t = \tanh \left(X_t W_{xc} + H_{t-1} W_{hc} + \text{bc} \right)$$

$$\text{Cell State (next memory input) } C_t = F_t \cdot C_{t-1} \cdot C_t$$

$$\text{New State}: H_t = O_t \cdot \tanh \left(C_t \right), \text{ where,}$$

- W_{xi}, W_{xf}, W_{xo} and W_{hc}, W_{hf}, which refer respectively to the weight parameters and bi, bf, bo denote bias parameters.
- W_{xc}, W_{hc} denote weight parameters, bc is bias parameter, o refer to the element-wise multiplication. The estimation of C_t depends on the output information's from memory cells (C_{t-1}) and the current time step C_t.

4.4 Convolutional Neural Networks (CNNs)

The working of the basic CNNs (as outlined in Fig. 3) is based on serving the multidimensional data (e.g., images and word embedding) to a convolutional layer comprising of multiple filters that helps in learning different features. From the figure itself, it is visible that these filters are sequentially applied to diverse sections of the input. The output is typically pooled or sub-sampled to tiny dimensions which are later served into a connected layer. The idea behind using CNNs on text

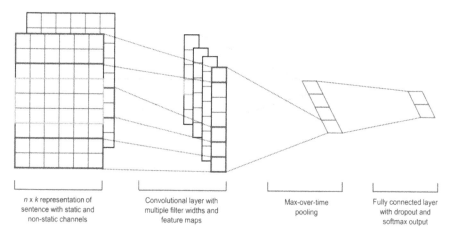

n x k representation of sentence with static and non-static channels

Convolutional layer with multiple filter widths and feature maps

Max-over-time pooling

Fully connected layer with dropout and softmax output

Fig. 3 The CNN model for classifying sentences present in Twitter

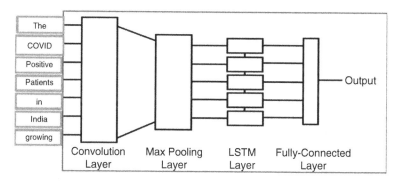

Fig. 4 The CNN-LSTM model

depends on structured and organized nature of the text thus helping in discovering and learning patterns that would else be gone missing in a feed-forward network. For example, the use of "down" in the context of "down to earth" is actually of positive sentiment as compared to the other phrases such as "feeling down." Moreover, it will be able to extract these features irrespective their position of occurrence in the sentence.

4.5 LSTM-CNN Model

In this research, a combined model was developed by integrating LSTM and CNN network to automatically analyze and predict the COVID-19 situation in India. Our LSTM-CNN model (Fig. 4) combination comprises of an initial convolution layer, Max Pooling layer, LSTM layer, and fully connected Layer. Every convolution

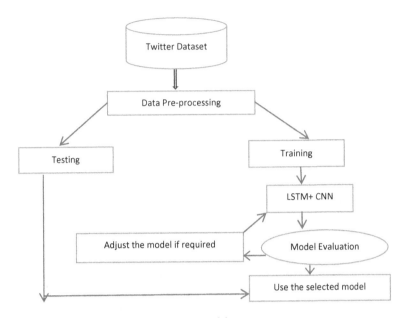

Fig. 5 Framework of our proposed forecasting model

block integrates 2–3 two-dimensional CNN and one pooling layer. The extraction of convolution kernel is done by multiplying the matrix (superposition) in every convolution operation. For feature extraction from the Twitter data, the max-pooled filter of the feature map is used (Figs. 5 and 6).

4.6 Twitter Dataset

We have used CORONAVIRUS (COVID-19) TWEETS DATASET developed and maintained by IEEE Data port. We have used the dataset both for training and validating purpose. We have used the most recent dataset "corona_tweets_257.csv: 3,144,302 tweets (November 30, 2020, 10:10 AM–December 01, 2020, 10:10 AM). The data are present in the following link:

https://ieee-dataport.org/open-access/coronavirus-covid-19-tweets-dataset

The dataset was initially in CSV format with two columns for two parameters: Twitter IDs and sentiment scores related to COVID-19 pandemic. The existing model keep track of the real-time Twitter feed related to coronavirus tweets applying more than 90 specialized keywords and hash tags that are generally used to reference the global pandemic. The dataset was completely redesigned on March 20, 2020, as per the standard content redistribution policy set by Twitter. The tweets are geo-tagged. Few keywords are mentioned in Table 1:

Fig. 6 Working flowchart of CNN-LSTM model

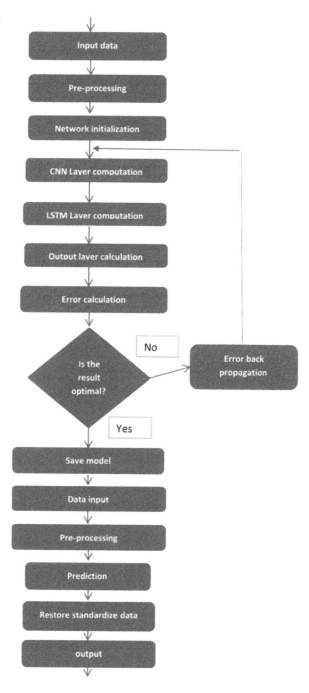

Table 1 Few trending hashtag/keywords frequently used on Twitter

Twitter hashtag/keyword (most trending)		
"corona"	"#coronavirus"	"covid19"
"#handsanitizer"	"working from home"	"flattening the curve"
"#n95"	"covid"	"2019ncov"
"covid-19"	"#covid"	"#covid19"
"#socialdistancing"	"pandemic"	"#covid-19"
"#corona"	"#ppe"	"flatten the curve"
"#sarscov2"	"#workfromhome"	"sarscov2"
"coronavirus"	"#covidiots"	"n95"
"sarscov 2"	"ncov"	"sars cov2"
"#flatteningthecurve"	"#workingfromhome"	"covid_19"

As per the content redistribution policy of Twitter, there is a restriction of sharing tweets other than tweet user IDs or IDs. As per Twitter, scientists and researchers may every time pull fresh data, because user may update/delete a tweet time and again. We have used Python/Panda code to hydrate the Twitter IDs. A picture of the data is given below (Fig. 7):

5 Experimental Result Analysis

5.1 Experiment

Table 3 refers the average accuracy of individual network recorded after five successful experiments. From our dataset, we have extracted around 12,000 tweets as the training set. We have labeled 3000 tweets for testing purpose. We have carefully maintained exactly equal numbers of +ve and −ve tweets. The parameters described in Table 1 were used for the purpose of the training.

5.2 Parameter Selection

The parameters described in Table 1 were chosen very carefully to obtain state-of-the-art results. After testing (Manual), we have improvised our parameters as when required. The different parameters are dimension of embedding, epoch, batch size, filters, kernel size, pool size, dropout rate, and word embeddings. The detailed analysis of the improvisation of parameters is described in Table 2.

Fig. 7 Twitter IDs and sentiment scores related to COVID-19 pandemic

Table 2 Selected parameter for the proposed CNN-LSTM and LSTM-CNN architecture

Dimension of embedding	34
Epoch	12
Size (batch)	126
Filters	34
Size of the kernel	3
Size of the pool	2
Dropout rate	0.55
Word embeddings	Not previously trained

5.3 Selection of Different ML Models

After preparing the training data, we have done an extensive testing on different model like only CNN, Only LSTM, CNN-LSTM, LSTM-CNN, Feature Vector–based feed-forward network, and Word Embedding–based feed-forward–based network in order to evaluate the performance of each of the network. Then we

Table 3 Accuracy (average) obtained through different NN

Neural network model (NN)	Avg. accuracy
LSTM-CNN	75.72%
CNN-LSTM	67.79%
LSTM	70.53%
CNN	67.57%
Feature Vector–based feed-forward network	67.89%
Word Embedding–based feed-forward network	61.39%

Table 4 Different range of epoch gives different results (average accuracy)

Epochs	LSTM-CNN	CNN-LSTM	LSTM	CNN
4	61.50%	58.50%	63.45%	53.75%
12	75.72%	67.79%	70.53%	67.57%
22	71.25%	66.50%	53.90%	71.45%

compared their accuracy and performance thoroughly and found out that LSTM-CNN-based deep neural network is outperforming than other networks. The result of the findings is recorded in Table 3.

5.4 Result Analysis

As compared to normal only CNN, Only LSTM, CNN-LSTM, LSTM-CNN, Feature Vector–based feed-forward network, and Word Embedding–based feed-forward–based network model, the LSTM-CNN is giving around 3% higher accuracy but when we compared our model with LSTM, there is drop of 3.2% in our model. Our actual proposed model is crossing all the percentage of accuracy by securing highest accuracy of 75.72% which is even much higher than CNN, LSTM, and CNN [17–19]. The higher accuracy was only obtained due to combining both the network LSTM and CNN. CNN has an advantage of identifying local patters, and LSTM has an advantage of finding out the order of the text. The convolution layer actually underplayed in the CNN-LSTM model, but our proposed model looks very promising due to the encoding functionality of LSTM so that for each input token, there will be at least one output token. The benefit of this is that the token holds the information of all previous tokens thereby potential increase in the accuracy. We have observed that for our proposed model, the number of epoch is 12 [14–16]. As various networks have various learning rates, thereby avoiding the problem of over fitting and under fitting, we have tested our model with different epoch and the same is described in Table 4. Table 5 describes the month-wise actual COVID-19 cases and the prediction of our system (Fig. 8).

Table 5 Month-wise actual COVID-19 cases and the prediction of our system with errors

Date	Actual COVID positive	Predicted cases	Error (%)
05.04.2020	3374	3310	1.896858
07.05.2020	52951	51912	1.962191
08.06.2020	246628	244118	1.017727
02.07.2020	625.544	621.122	0.706905
03.08.2020	1750723	1741723	0.514073
04.09.2020	3853406	3831305	0.573545
08.10.2020	6623815	6622517	0.0204
07.11.2020	8364086	8355011	0.0021
07.12.2020	9677203	9672001	0.0006

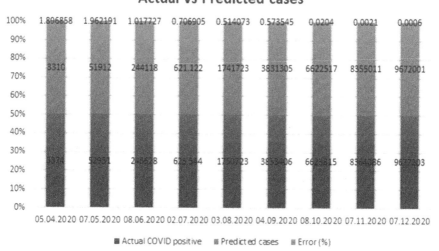

Fig. 8 Month-wise actual COVID-19 cases and the prediction of our system

5.5 Evaluation Metrics

To assess the performance of our system, we had chosen four evaluation parameters: accuracy score, precision score, recall score, and F_1 score. The accuracy score is the fraction of correct predictions. The maximum accuracy score can be 1 and the minimum accuracy score can be 0. It is given as

$$\text{Accuracy} = \text{TP} + \text{TN}/\text{TP} + \text{TN} + \text{FP} + \text{FN} \qquad (1)$$

Four basic notations are explained as follows.

- *True Positives (TP):* The number of correct positive predictions of a class.
- *True Negatives (TN):* The number of correct negative predictions of a class.

- *False Positives (FP):* The number of incorrect positive predictions of a class.
- *False Negatives (FN):* The number of incorrect negative predictions of a class.

Precision indicates the exactness of the classifiers. It lies in [0, 1] and calculated as

$$Precision = TP/ (TP + FP) \tag{2}$$

Recall indicates about the completeness of a classifier. It lies in [0, 1] and calculated as

$$Recall = TP/ (TP + FN) \tag{3}$$

F_1 score is a harmonic mean of precision and recall scores. It lies in [0, 1] and calculated as

$$F_1 = 2^*(precision^*recall) / (precision + recall) \tag{4}$$

6 Discussion of the Work

In this paper, we are proposing a LSTM-CNN deep learning–based hybrid single-window system for real-time COVID-19 data analysis and prediction using Twitter data. The system may be helpful for administration/government agencies/private healthcare providers. Our proposed model LSTM-CNN has achieved the highest accuracy of 75.72% which is much higher comparing with other available models. The system is capable of generating accurate data related to COVID-19 such as the number of active patients in a particular location and how many have been recovered.

To model our LSTM-CNN model, we have considered Twitter data. The existing model keeps track of the real-time Twitter feed related to coronavirus tweets applying more than 90 specialized keywords and hash tags that are generally used to reference the global pandemic. From our dataset, we have extracted around 12,000 tweets as the training set. We have labeled 3000 tweets for testing purpose. We have carefully maintained exactly equal numbers of +ve and –ve tweets. The parameters described in Table 1 were chosen very carefully to obtain the state-of-the-art results. After testing (manual), we have improvised our parameters as when required. We have balanced each of the following parameter to obtain best of the results: epoch, size (batch), filters, size of the kernel, size of the pool, dropout rate, and word embeddings. We have observed that for our proposed model, the number of epochs is 12. As because various networks have various learning rates, thereby avoiding the problem of over fitting and under fitting we have tested our model with different epochs and the same is described in the Table 3. Table 5 describes the month-wise actual COVID-19 cases and the prediction of our system which is much better

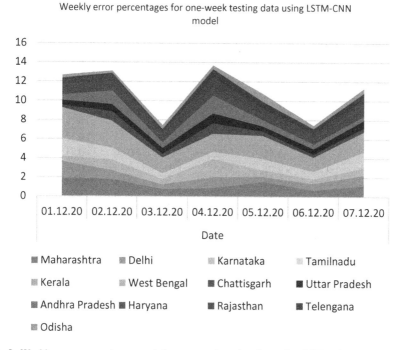

Fig. 9 Weekly error percentages graph for one-week testing data using LSTM-CNN model

than the other model. As per Table 5 and Fig. 9, the month-wise actual COVID-19 cases and the prediction of our system is nearly achieving 96%. As per Table 6, the weekly error percentages for one-week testing data using LSTM-CNN model are also achieving state of the art.

As compared to the normal CNN network model, the CNN-LSTM is giving around 3% higher accuracy but when we compared our model with LSTM, there is a drop of 3.2% in our model. Our actual proposed model is crossing all the percentage of accuracy by securing highest accuracy of 75.72% which is even much higher than CNN, LSTM, and CNN [17–19]. The higher accuracy was only obtained due to combining both the network LSTM and CNN.

7 Conclusion, Limitations, and Future Work

In this chapter, we have demonstrated LSTM-CNN deep learning–based hybrid single-window system for real-time COVID-19 data analysis and prediction using Twitter data. The very first objective of our work was to prepare the machine learning module to analyze the data received from Twitter and give useful prediction on COVID-19. We have trusted Twitter as because it gives precise and accurate

Table 6 Weekly error percentages for one-week testing data using LSTM-CNN model

Place name	Date						
	01.12.20	02.12.20	03.12.20	04.12.20	05.12.20	06.12.20	07.12.20
Maharashtra	1.91	1.82	0.77	0.89	1.55	0.67	1.16
Delhi	1.75	0.89	0.51	1.15	0.55	0.69	1.11
Karnataka	0.55	1.13	0.58	1.89	0.7	0.51	0.77
Tamilnadu	1.81	1.23	0.57	0.77	1.17	0.79	1.65
Kerala	1.82	0.92	0.79	0.82	1.55	0.77	1.11
West Bengal	1.45	1.89	0.87	1.05	0.86	0.71	0.98
Chhattisgarh	0.25	1.01	0.38	1.09	0.43	0.31	0.43
Uttar Pradesh	0.51	0.73	0.61	1.03	0.53	0.56	0.77
Andhra Pradesh	0.55	1.43	0.58	1.89	0.67	0.51	0.45
Haryana	1.15	0.99	0.71	0.85	0.55	0.69	1.31
Rajasthan	0.35	0.71	0.39	0.99	0.63	0.42	0.49
Telangana	0.25	0.13	0.28	0.89	0.73	0.47	0.61
Odisha	0.31	0.23	0.41	0.38	0.82	0.33	0.47

information. Our proposed model is crossing all the percentage of accuracy by securing highest accuracy of 75.72% which is even much higher than individual CNN, LSTM, and CNN-LSTM. The higher accuracy was only obtained due to combining both the network LSTM and CNN.

Although the proposed model as of now giving state-of-the-art results, implementing bidirectional LSTM for achieving more accurate result can be a good choice in future. Apart from that we can exploit some hand-crafted rules and more feature vectors into the system for getting more accuracy. In the next chapter, we will demonstrate the functionality and detailed implementation of the handheld digital device. We will demonstrate the integration and aggregation of the data received form the user of the digital device and the data received from Twitter. Based on these data integration, a set customized report may be prepared for the different stake holders/ beneficiaries such as government officials/ decision-makers/doctors/hospital administration/health workers, and common people who will be able to access the customized report through different platforms such as web dashboard and mobile application.

References

1. Velásquez, R. M. A., & Lara, J. V. M. (2020). Forecast and evaluation of COVID-19 spreading in USA with reduced-space Gaussian process regression. *Chaos Solitons Fractals, 136*, 109924.
2. Yousaf, M., Zahir, S., Riaz, M., Hussain, S. M., & Shah, K. (2020). Statistical analysis of forecasting COVID-19 for upcoming month in Pakistan. *Chaos Solutions Fractals, 138*, 109926.

3. Ribeiro, M. H. D. M., da Silva, R. G., Mariani, V. C., & dos Santos Coelho, L. (2020). Short-term forecasting COVID-19 cumulative confirmed cases: Perspectives for Brazil. *Chaos Solitons Fractals, 135*, 109853.

4. Toğaçar, M., Ergen, B., & Cömert, Z. (2020). COVID-19 detection using deep learning models to exploit social mimic optimization and structured chest x-ray images using fuzzy colour and stacking approaches. *Computers in Biology and Medicine, 121*, 103805.

5. Benvenuto, D., Giovanetti, M., Vassallo, L., Angeletti, S., & Ciccozzi, M. (2020). Application of ARIMA model for the COVID-2019 epidemic dataset. *Data in Brief, 29*, 105340.

6. Ceylan, Z. (2020). Estimation of COVID-19 prevalence in Italy, Spain, and France. *Science of the Total Environment, 729*, 138817.

7. Jia, L., Li, K., Jiang, Y., Guo, X., & Zhao, T. (2020). Prediction and analysis of corona virus disease 2019. *arXiv:05447*.

8. Peng, L., Yang, W., Zhang, D., Zhuge, C., & Hong, L. (2020). Epidemic analysis of COVID-19 in China by dynamical modelling. *arXiv:200206563*.

9. Roosa, K., Lee, Y., Luo, R., Kirpich, A., Rothenberg, R., & Hyman, J. (2020). Real-time forecasts of the COVID-19 epidemic in China from February 5th to February 24th, 2020. *Infectious Disease Modelling, 5*, 256–263.

10. Liu, Z., Magal, P., Seydi, O., & Webb, G. (2020). Understanding unreported cases in the COVID-19 epidemic outbreak in Wuhan, China, and the importance of major public health interventions. *Biology, 9*, 50.

11. Liu, Z., Magal, P., Seydi, O., & Webb, G. (2020). Predicting the cumulative number of cases for the COVID-19 epidemic in China from early data. *medRxiv*. https://doi.org/10.1101/2020.03.11.20034314

12. Rekha Hanumanthu, S. (2020). Role of intelligent computing in COVID-19 prognosis: A state-of-the-art review. *Chaos, Solitons and Fractals, 138*, 109947.

13. Sharfuddin, A. A., Tihami, M. N., & Islam, M. S. (2018). A deep recurrent neural network with BiLSTM model for sentiment classification. In *2018 International conference on Bangla speech and language processing (ICBSLP)* (pp. 1–4). IEEE.

14. Graves, A., Jaitly, N., & Mohamed, A. R. (2013). Hybrid speech recognition with deep bidirectional LSTM. In *2013 IEEE workshop on automatic speech recognition and understanding* (pp. 273–285). IEEE.

15. Zhang, B., Zhang, H., Zhao, G., & Lian, J. (2020). Constructing a PM 2.5 concentration-prediction model by combining auto-encoder with Bi-LSTM neural networks. *Environmental Modelling & Software, 124*, 104600.

16. Wang, S., Wang, X., Wang, S., & Wang, D. (2019). Bi-directional long short-term memory method based on attention mechanism and rolling update for short-term load forecasting. *Journal of Electrical Power & Energy Systems, 109*, 470–479.

17. Hochreiter, S., & Schmidhuber, J. (1997). Long short-term memory. *Neural Computation, 9*(8), 1735–1780.

18. Ashour, A. S., El-Attar, A., Dey, N., El-Kader, H. A., & El-Naby, M. M. A. (2020). Long short-term memory-based patient-dependent model for FOG detection in Parkinson's disease. *Pattern Recognition Letters, 131*, 23–29.

19. Harrou, F., Kadri, F., & Sun, Y. (2020). Forecasting of photovoltaic solar power production using LSTM approach. In *Advanced statistical modelling, forecasting, and fault detection in renewable energy systems*. IntechOpen.

An Intelligent Tool to Support Diagnosis of Covid-19 by Texture Analysis of Computerized Tomography X-ray Images and Machine Learning

Maíra Araújo de Santana, Juliana Carneiro Gomes,
Valter Augusto de Freitas Barbosa, Clarisse Lins de Lima,
Jonathan Bandeira, Mêuser Jorge Silva Valença, Ricardo
Emmanuel de Souza, Aras Ismael Masood, and Wellington P. dos Santos

1 Introduction

In December 2019, in the city of Wuhan, capital of the Province of Central China, a new specimen of coronavirus crossed the barriers between species and hit humans for the first time. A member of the Coronaviridae family and also associated with Severe Acute Respiratory Syndrome (SARS), similarly to its predecessor, SARS-CoV, the virus was named SARS-CoV-2 [46, 51]. The new coronavirus is responsible for 2019 coronavirus disease, or Covid-19, a blood disorder that strongly affects the respiratory system, causing, in mild and moderate cases, fever, dry cough, decreased or loss of sense of smell and taste.

In the most severe cases, the disease leads to decreased oxygen saturation in the blood and destruction of the surfactant inside the alveoli, which can lead to collapse, causing a respiratory deficiency that can worsen until death. SARS-CoV-2

M. A. de Santana · J. C. Gomes · C. L. de Lima · J. Bandeira · M. J. S. Valença
Graduate Program in Computer Engineering, Polytechnique School of the University of Pernambuco, Recife, Brazil

V. A. de Freitas Barbosa
Academic Unit of Serra Talhada, Rural Federal University of Pernambuco, Serra Talhada, Brazil

R. E. de Souza
Department of Biomedical Engineering, Federal University of Pernambuco, Recife, Brazil

A. I. Masood
Information Technology Department, Technical College of Informatics, Sulaimani Polytechnic University, Sulaymaniyah, Iraq

W. P. dos Santos (✉)
Department of Biomedical Engineering, Federal University of Pernambuco, Recife, Pernambuco, Brazil
e-mail: wellington.santos@ufpe.br

also affects blood clotting, being responsible for small clots, thrombopenias, and thrombosis. There are cases where patients recovered from Covid-19 have had late thrombosis and strokes [1, 16, 37, 48, 49].

Until the end of April 2020, the SARS-CoV-2 had spread from 213 countries, infecting almost three million people worldwide and causing more than 200 thousand deaths [19]. Until February 2021, the world has counted more than 100 million Covid-19 positive cases, about 61 million recovered cases, and more than 2 million deaths. The USA has registered more than 27 million positive cases and about 485 thousand deaths. Brazil has registered about 10 million infected cases, about 9 million recovered cases, and about 240 thousand deaths, in a daily sample average superior to a thousand deaths. The USA and Brazil are the new world epicenters of Covid-19. Therefore, precise diagnosis methods are urgently needed, in order to plan and implement health policies and measures to control the spread of Covid-19 and reinforce the public health systems.

Considering the high virus spread rate, rapid and precise diagnostic tests are increasingly needed, in order to provide the patients the necessary medical attention and isolation [7]. The ground-truth test for Covid-19 diagnosis is the quantitative polymerase chain reaction (qPCR). However, although it is precise, this exam needs several hours to confirm positivity [25].

Rapid tests based on antibodies like IgM/IgG are nonspecific for Covid-19, frequently presenting low sensitivity and specificity [13, 25, 26, 52, 62]. This is due to the fact that IgM/IgG detects the serological evidence of recent infection, not the virus presence. Thus it is not possible to ensure that the positive response is due to antibodies developed from the contact with SARS-CoV-2 or other coronaviruses and flu viruses [44]. Consequently, the use of IgM/IgG rapid test kits as definitive diagnosis of COVID-19 in currently symptomatic patients is not recommended [13]. IgM/IgG tests made in samples collected in the first week of illness have only 18.8% of sensitivity and 77.8% of specificity [47]. Döhla et al. [25] found a low rate of sensitivity for IgM/IgG rapid tests as well. They compared the results of IgM/IgG with reverse transcription polymerase chain reaction (RT-PCR) in 59 patients and concluded that the rapid test obtained 36.4% of sensitivity and 88.9% of specificity.

However, IgG/IgM tests can reach high sensitivities and specificities when the viral charge is high. But it is possible only when the disease is in its advanced stage [30, 34]. Liu et al. [47] confirmed this fact: tests performed during the second week of the disease have 100% of sensitivity and 50% of specificity.

Computerized Tomography (CT) X-ray scans combined with RT-PCR have a great clinical value, since in CT images it is possible to analyze the Covid-19 effects as bilateral pulmonary parenchymal ground-glass and consolidative pulmonary opacities in a precise way [42]. As a disadvantage, CT is an expensive exam, requiring a dedicated room with difficult isolation, becoming a risk factor for contamination.

Recent works have been showing that chest X-ray image can also be used to detect the Covid-19 with high accuracy [2, 3, 50, 60]. Chest X-ray Covid-19 image diagnosis can be a feasible alternative, since it is the standard for diagnosing pneumonia. Furthermore, X-ray radiography availability is much larger than CT's, due to its relatively low price and the power to furnish useful images, resulting a good cost-effectiveness ratio. However, CT images present higher resolution and interesting perspectives that can be useful to see lung alterations made by Covid-19 and their differences in contrast to other pneumonias. Additionally, the medical community has built a robust knowledge on CT interpretation for diagnosis support of lung cancer and respiratory diseases [9, 31, 36, 64].

In clinical practice, whitish and lungs' opacity are the main findings associated with pneumonias. These image artifacts are commonly related to the production of mucus. Covid-19 can lead to thickening of the blood and thrombosis. Alveoli gas exchanges are impaired. Patients experience breathing difficulties, as well as blood saturation is diminished. Lung surfactant is damaged, leading alveoli to collapse, compromising respiratory capacities. The tendency is both lungs be affected in the same manner. Since differential image diagnosis is based on lungs' opacity, textures assume an important role to aid the diagnosis of both common bacterial and viral pneumonias and Covid-19. The combination of texture and complex abstract shapes moments like Haralick and Zernike moments, respectively, can lead to interesting and effective image diagnosis support methodologies based on machine learning, e.g. breast cancer support diagnosis based on mammographies, mammary thermographies, and pattern recognition [22, 23, 45, 57].

Computational intelligence, specially Machine Learning techniques, have been indicated to be used in several clinical tasks involving biomedical image classification [4, 6, 17, 18, 21–24, 45, 53–55, 57, 59, 61]. Such techniques could provide a secure and semi-automatic way to diagnosis Covid-19 in CT X-ray images.

In this study, we propose an automatic system for Covid-19 diagnosis using machine learning techniques and CT X-ray images. In our experiments we used Multilayer Perceptron [33, 58], Support Vector Machine [11, 20], Random Trees, Random Forest [12, 28], Bayesian networks, and Naive Bayes [15, 33]. As descriptors we used Haralick [32] and Zernike [35] moments.

This work is organized as follows. Section 2 reviews related works in the diagnosis of Covid-19 using CT X-ray image. Section 3 presents dataset information and reviews the theoretical concepts necessary to understand the work. Section 4 shows the experimental results and the resulting desktop application we developed. Section 5 analyzes the experimental results. In Sect. 6 we present our general conclusions and draw some future work possibilities.

2   Related Works

Several researchers have investigated tomography images for Covid-19 indicators during this pandemic [14, 27, 39, 43]. Bernheim et al. [8] analyzed CT images of 121 patients infected with Covid-19 in China. To be included in the research, patients needed to be over 18 years old and with Covid-19 diagnosis confirmed by the RT-PCR method. The authors' hypothesis is that certain findings on CT images are more common at each stage of the disease or infection. Thus, they grouped the images according to the period of the first symptoms and the exam, dividing them between early (0–2 days), intermediate (3–5 days), and late (6–12 days). All images were then analyzed independently by two trained radiologists. The authors observed that 56% of patients in the initial phase had normal images. However, as the time period grew, abnormalities became more frequent. The most commonly found abnormalities were consolidation, bilateral and peripheral disease, linear opacities, and the reverse halo sign. Total lung involvement was observed in 28% of patients in the early stage, 76% in the intermediate stage, and 88% in patients in the late stage. Finally, the study highlighted that chest CT images are essential in the diagnosis of suspected cases of Covid-19, especially in contexts with a limited number of RT-PCR testing kits. In contrast, the authors pointed out that CT images have limited sensitivity and negative results in the first days of symptoms, and should not be used alone for diagnosis.

Thus, it is possible to perceive the relevance of chest CT images in the context of Covid-19, as well as their limitations in detecting the disease. The studies also suggest that well-trained radiologists are needed to find abnormalities in the images. Therefore, considering the context of the pandemic, the overcrowding of hospitals and the overload of medical teams, support systems for diagnosis can play an important role. Bai et al. [5], for instance, evaluated the performance of seven radiologists from the USA and China in distinguishing Covid-19 in 205 CT images. Radiologists showed a performance in the range of 72–94% of sensitivity, and specificity between 24% and 100%. The authors state that despite the relative high sensitivity, when compared to RT-PCR, the images may not reveal patterns that allow differentiation in all cases, as with other viral pneumonias. In cases of influenza, for example, ground-glass opacity and consolidation have been found, also present in cases of Covid-19. Therefore, the study points to the use of artificial intelligence techniques as a way to improve the performance of radiologists.

Gozes et al. [29] proposed an AI solution for Covid-19 screening using CT images. They believed that combining CT power to deep learning techniques could not only improve Covid-19 detection, but also provide information regarding disease quantification and evolution. The study was conducted using data from 157 patients from China and USA. Their approach achieved AUC of 0.996, sensitivity of 98.2%, and specificity of 92.2%. They also found that heatmap images may provide better visualization of pulmonary opacity. Finally, the study proposes a "Corona Score," which measures the disease progression over time. However, such as mentioned by the authors themselves, the database is small and very restricted. Therefore, studies with greater population variability are needed, since the disease has manifested itself in different ways worldwide.

The study from Wang et al. [63] found a deep AI architecture able to automatically extract graphical features and provide Covid-19 diagnosis with accuracy of 82.9%, specificity of 80.5%, and sensitivity of 84%. They used a total of 453 CT images divided into both Covid-19 and typical viral pneumonia. This amount of images was further splitted to build training, validation, and testing sets, resulting in a relatively small database to be used in a deep learning approach. Nevertheless, their study found promising results that point to the relevance of using AI techniques to detect Covid-19 from CT chest scans.

A larger database was assessed in Li et al. [40] approach. Their study used 4352 CT images from 3322 patients to train a deep learning model to detect Covid-19 and differentiate it from community-acquired pneumonia (CAP). By using a convolutional network to extract visual features, they were able to reach sensitivity of 90% and specificity of 96% in differentiating Covid-19 from CAP. The relevance of their study could be increased by assessing model ability to differentiate Covid-19 from other diseases that lead to lung lesion with similar imaging characteristics (e.g. viral pneumonias).

In other approach, Li et al. [41] used CT chest scans to acquire information regarding the temporal progression of Covid-19. Images from three Chinese hospitals were sorted into four phases: early phase, progressive phase, severe phase, and dissipative phase. From the study, they found that an AI model could successfully segment the lesion area and calculate its volume. This system may play an important role in clinical analysis, supporting and easing physicians' decisions regarding disease severity. However, the paper did not mention the amount of images used to achieve the results, nor detailed the AI techniques they applied to find and quantify lungs lesions.

3 Methods

3.1 Proposed Method

In this context, this work proposes the development of an intelligent system called IKONOS-CT. This system aims to support and optimize the diagnosis of Covid-19 through chest CT images. We also aim to produce a tool for easy maintenance and scalability, using algorithms of low computational complexity. Thus, we seek to provide one more diagnostic method to combat the current pandemic, in order to complement this process and minimize costs.

The basic functioning of this system is this: The medical team of the health institution must request chest CT examinations from patients with symptoms characteristic of Covid-19. After receiving the digital images, the healthcare professional can login and then upload the image into IKONOS-CT. The images will then be analyzed by an intelligent system. It will be able to carry out binary classification, differentiating Covid-19 patients from Non-Covid-19 ones. For this, machine learning techniques will be used, aiming at good results. The methods tested in this work will be: Haralick and Zernike for feature extraction and multiple classical classifiers will be tested and compared. The classifiers to be tested are: Multilayer Perceptron, Support Vector Machines, Decision trees such as Random Tree and Random Forest,

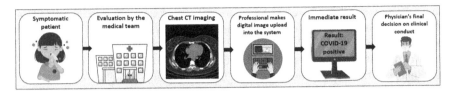

Fig. 1 Diagram of the proposed method. Chest CT images of symptomatic patients can be loaded into the IKONOS-CT system. It consists of an intelligent system capable of extracting features and classifying the image among two cases: Covid-19 positive and negative. The immediate result can be viewed on any computer. Performance metrics will also be available, helping the professional in decision making

and Bayesian Networks. The chosen classifier for the system development will be the one with the best performance according to the following evaluation metrics: Accuracy, Kappa Index, Specificity, Sensibility, Precision, Recall, and Confusion matrix. Finally, the system will provide a diagnosis, which can be viewed on the computer screen. The diagnosis will be available with performance information, so that the health professional can make the final decision of the ideal clinical conduct. This proposal is summarized in Fig. 1. In this present work, we will focus on computational experiments on the feature extraction and classification. The development of the final intelligent system will be covered in future works.

3.2 Dataset

For the development of this project, we used CT images from two different databases. The first one is available on Github repository. It contains 349 CT images from 216 Covid-19 positive patients, and 463 non-Covid-19 CT scans [65]. This database was created using images published in 760 different articles. The second dataset is from Peshmarga Hospital, Erbil, located in Iran. It contains 160 Covid-19 CT images. The joining of the bases resulted in 972 CT images. Figure 2 presents sample images of Covid-19 and Non-Covid-19 patients (Table 1).

There are some limitations to the databases that we used for this project. First, Covid-19 images have no information about the severity of the patients. Thus, we believe that it is possible that the images are of more severe cases. We also emphasize that we do not have patient demographic information, such as sex, age, and presence of comorbidities.

3.3 Feature Extraction: Haralick and Zernike

The descriptor of Haralick extracts feature related to the textures of the images. Texture is an intrinsic property of surfaces. It contains important information about

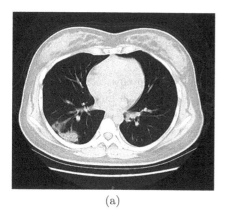

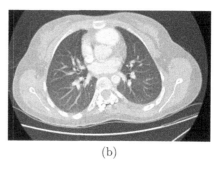

(b)

(a)

Fig. 2 Sample images of the datasets used in this work: Covid-19 positive (**a**), Covid-19 negative (**b**)

Table 1 Number of CT images of each class: Covid-19 and non-Covid-19

Class	Total
Covid-19	509
Non-Covid-19	463

their structural composition. From Haralick's moments, it is possible to differentiate textures that do not follow a certain pattern of repetition throughout the image. This method calculates statistical information associated with the co-occurrence matrices from the gray scale image. These matrices show the occurrence of certain pixel intensities. Each $p(i, j)$ of the matrix consists of the probability of going from one pixel of intensity i to another pixel of intensity j, according to a certain distance and an angle of the neighborhood [32].

In this way, each matrix considers the relationship between a reference pixel and its neighborhood. Thus representing the spatial distribution and dependence of gray levels in regions of the image. In this study, we considered two versions of the image to perform the feature extraction. The first was the gray scale image and the second was a pre-processed image, using Kohonen maps as filter. This process resulted in 104 features per image.

The Zernike descriptor is another widely used tool for feature extraction. It extracts information related to shape or geometry from an image. Zernike moments are invariant to rotation, not redundant, and robust to noise [35]. To calculate these moments, we consider the center of the image as the center of a unit disk. The moments are calculated from the projections of the intensity function of an image on the orthogonal base functions. So we calculate each of the 64 moments from the Zernike family of polynomials, $V_{n,m}$, described by Eqs. 1 and 2.

$$V_{n,m}(\rho, \theta) = R_{n,m}(\rho)^{-jm\theta} \tag{1}$$

$$R_{n,m} = \sum_{s=0}^{\frac{n-|m|}{2}} (-1)^s \frac{(n-s)!}{s!(\frac{n+|m|}{2}-s)!(\frac{n-|m|}{2}-s)!} \rho^{n-2s}. \tag{2}$$

The 64 descriptors are equally divided into two groups, according to the polynomial order (n). To calculate the features, n and m from Eqs. 1 and 2 assume the values in Table 2. So, at the end of the process, we have 32 moments of low order and 32 of high order. Shape-related features of an image are also relevant in the context of identifying pathologies, since these conditions usually result in changes in geometric patterns (Fig. 3).

3.4   Classification

In this section we briefly discussed the machine learning methods used to classify CT images.

3.4.1   Multilayer Perceptron

The psychologist Rosenblatt [58] was one of the pioneers in the concept of artificial neural networks. In 1958 he proposed the perceptron model for supervised learning. Perceptron is the simplest form of neural network used for binary classifications of

Table 2 Moments of Zernike according to group and parameters n and m

Group	n	m	Number of moments
1	3	1,3	32
	4	0,2,4	
	5	1,3,5	
	6	0,2,4,6	
	7	1,3,5,7	
	8	0,2,4,6,8	
	9	1,3,5,7,9	
	10	0,2,4,6,8,10	
2	10	2,6,10	32
	11	3,7,11	
	12	0,4,8,12	
	13	1,5,9,13	
	14	2,6,10,14	
	15	3,7,11,15	
	16	0,4,8,12,16	
	17	1,5,9,13,17	

linearly separable patterns. It consists of a single neuron with adjustable synaptic weights and a bias [33].

Multilayer Perceptron (MLP) is a generalization of the single layer perceptron. It consists of a feed-forward network with an input layer, hidden layers, and one output layer. The addition of hidden layers allows to the network the ability to classify more complex problems than single layer perceptron such as image classification [6, 38, 56].

The main algorithm for training MLPs is the error backpropagation algorithm. Based on a gradient, backpropagation proceeds in two phases: propagation and backpropagation. In the propagation phase, an output is obtained for a given input pattern. In backpropagation phase, an error is calculated using the desired output and the output obtained in the previous phase. Then the error is used to update the connection weights. Thus backpropagation aims to iteratively minimize the error between the network output obtained and the desired output [33].

3.4.2   Support Vector Machine

Created by Vapnik [11, 20] the support vector machine (SVM) performs a nonlinear mapping on the dataset in a space of high dimension called feature space. So a linear decision surface, called hyperplane, is constructed in order to separate distinct classes [20].

Thus, the training process of a support vector machine aims to find the hyperplane equation which maximizes the distance between it and the nearest data point. That hyperplane is called optimal hyperplane [33].

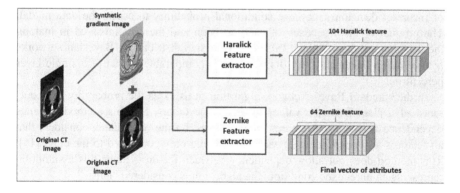

Fig. 3 Block diagram of the proposed solution: the 168-dimension feature vectors are composed of the fusion of texture (52 Haralick moments) and shape (32 Zernike moments) features extracted from the original (84 features) and the Sobel's gradient version (84 features) of the specified CT slice

3.4.3 Decision Trees

Decision trees are a type of supervised machine learning model. They are widely used to solve both classification and regression problems. In general, trees have nodes, which are structures that store information. In a tree there are basically four types of nodes: root, leaf, parent, and child. The root node is the starting point and has the highest hierarchical level. One node may connect to another, establishing a parent–child relationship, in which a parent node generates a child node. Leaf nodes, in turn, are terminal nodes, so they have no children, and represent a decision. In this way, using such trees, the algorithm makes a decision after following a path that starts from the root node and reaches a leaf node. There are several types of decision trees. They usually differ from the way the method goes through the tree structure. Among these types, the methods Random Tree and Random Forest are two of the main ones.

Random Tree algorithm uses a tree built by a stochastic process. This method considers only a few features in each node of the tree, which are randomly selected [28]. The Random Forest algorithm, in turn, consists in a collection of trees. These trees hierarchically divide the data, so that, each tree votes for a class of the problem. At the end, the algorithm chooses the most voted class as the prediction of the classifier [12].

3.4.4 Bayesian Network and Naive Bayes

Bayes Net and Naive Bayes are classifiers based on Bayes' Decision Theory. Bayesian classifiers, also called the test procedure by the Bayes hypothesis, seek to find a minimum mean risk. By considering a set of correct decisions and a set of incorrect decisions, they use conditional probability to create the data model. The product of the frequency of each decision and the cost involved in making the decision are the weights [33]. For a Gaussian distribution, Bayesian networks behave like a linear classifier. Its behavior is comparable to that of a single layer perceptron.

In the standard Bayes Network algorithm, it assesses the probability of occurrence of a class from the values given by the others. So, this method assumes dependence between the features. Naive Bayes, on the other hand, considers that all features are independent of each other, being only connected to the class [15]. This method does not allow dependency between features. Since this assumption represents an unrealistic condition, the algorithm is considered "naive."

3.4.5 Parameters Settings of the Classifiers

All experiments were performed using the Weka software. For each configuration described in the table below 25 simulations were performed, using tenfold cross-

Table 3 Classifiers parameters: SVMs with linear, 2-degree and 3-degree polynomials, and RBF kernels; MLPs with 50 and 100 neurons in the hidden layer; random forests with 10–100 trees; random tree; and standard Bayesian networks and Naive Bayes classifiers

Classifier	Parameters
SVM	Kernel Linear
	Kernel Polynomial E2
	Kernel Polynomial E3
	Kernel RBF
MLP	50 neurons in the hidden layer
	100 neurons in the hidden layer
Random Forest	Number of trees: 10, 20, 30, 40, 50, 60, 70, 80, 90, 100
	Batch size: 100
Bayesian networks	Batch size: 100
Naive Bayes	Batch size: 100
Random tree	Batch size: 100
	Seed = 1

validation. The parameters used in each machine learning method are shown in Table 3.

3.5 Metrics

In order to analyze the classification performance, we used six metrics: Accuracy, Recall, Sensitivity, Precision, Specificity, and Kappa Index. Accuracy assesses the proportion of images correctly classified on all results. It can be calculated according to Eq. 3. In a machine learning context, the term Recall is common. However, in the medical world, the use of the sensitivity metric is more frequent.

Mathematically, both terms are the same. They are the rate of true positives, and indicate the system's ability to correctly detect people who are sick (with Covid-19, for example). They can be calculated according to Eq. 4. Precision, on the other hand, is the fraction of the positive predictions that are actually positive. The precision can be calculated according to Eq. 5.

Specificity is the metric that evaluates a model's ability to predict true negatives of each available category or the rate of true negatives. This means that specificity indicates the classifier's ability to correctly exclude healthy or disease-free people. It can be calculated as Eq. 6. Finally, the Kappa index is a very good measure that can handle very well both multi-class and imbalanced class problems, as the one proposed here. It can be calculated according to Eq. 7. These four metrics allow to discriminate between the target condition and health, in addition to quantifying the diagnostic exactitude [10]. These metrics are described as following:

$$Accuracy = \frac{TP + TN}{TP + TN + FP + FN}, \tag{3}$$

$$Sensitivity = Recall = \frac{TP}{TP + FN}, \tag{4}$$

$$Precision = \frac{TP}{TP + FP}, \tag{5}$$

$$Specificity = \frac{TN}{TN + FP}, \tag{6}$$

where TP is the true positives, TN is the true negatives, FP is the false positives, and FN the false negatives.

The κ coefficient (kappa) is defined as follows:

$$\kappa = \frac{\rho_o - \rho_e}{1 - \rho_e}, \tag{7}$$

where ρ_o is observed agreement, or accuracy, and p_e is the expected agreement, defined as follows:

$$\rho_e = \frac{(TP + FP)(TP + FN) + (FN + TN)(FP + TN)}{(TP + FP + FN + TN)^2}. \tag{8}$$

Finally, we also used the area under the ROC curve (AUC). It is a measure of a classifier's discriminating ability and is calculated from Eq. 9. Given two classes—a sick individual and a non-sick individual—chosen at random, the area below the ROC curve indicates a probability of the latter being correctly classified. If the classifier cannot discriminate between these two separately, an area under a curve is equal to 0.5. When this value is the next 1, it indicates that the classifier is able to discriminate these two cases.

$$AUC = \int TPd(FP). \tag{9}$$

4 Results

From feature extraction process, we created two databases. The first one uses Haralick moments, and the second combines Haralick and Zernike features. Both databases were trained using several classic machine learning methods.

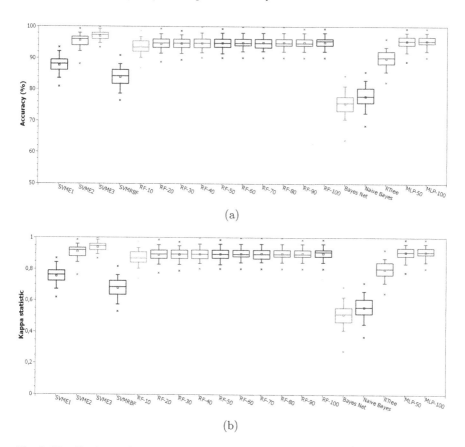

Fig. 4 Classification performance for detection of SARS-CoV-2 using Haralick as feature extractor. In (**a**) are the results of accuracy while (**b**) shows the kappa results

4.1 Classifiers Experiments Results

4.1.1 Results Using Haralick for Feature Extraction

Figures 4 and 5 show boxplots referring to classification performance of all classifiers based on accuracy, kappa statistic, sensitivity, and specificity. Considering this database using Haralick moments, the SVM with polynomial kernel of exponent 3 showed the best overall performance. In contrast, networks based on Bayes' theory showed the worst results. Table 4 illustrates the comparison of these two cases.

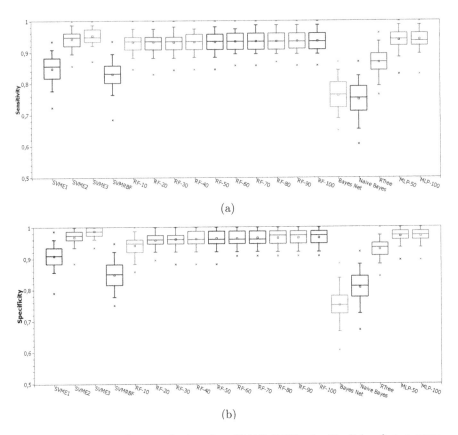

Fig. 5 Classification performance for detection of SARS-CoV-2 using Haralick as feature extractor. In (a) and (b) are the results of sensitivity and specificity, respectively

4.1.2 Results Using Haralick and Zernike for Feature Extraction

Figures 6 and 7, on the other hand, show boxplots with the results obtained with the second database, using moments from Haralick and Zernike. As in the previous case, the best model was achieved using a Support Vector Machine. However, in this case, the best performance was using degree 2 for the polynomial kernel. Furthermore, the Naive Bayes algorithm presented the worst performance this time. Table 5 summarizes these results.

5   Discussion

From the results we noticed that Haralick feature extractor was overall more appropriate to represent the CT images used in this study than the combination

Table 4 Classification performance for the best and the worst models using Haralick for feature extraction

Model	Accuracy (%)		Kappa statistic		Recall/Sensitivity		Precision		Specificity		Area under ROC	
	Average	StdDev	Average	StdDev	Average	StdDev	Average	StdDev	Average	StdDev	Average	StdDev
SVM (E3)	96.9936	1.3751	0.9399	0.0275	0.9524	0.0242	0.9872	0.0138	0.9874	0.0137	0.9699	0.0137
Bayes Net	75.5414	3.4129	0.5108	0.0682	0.7612	0.0480	0.7539	0.0393	0.7497	0.0504	0.8421	0.0312

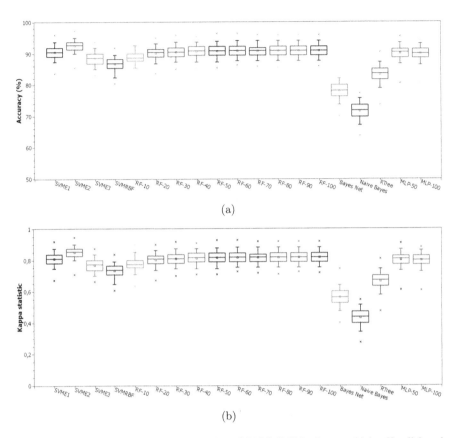

Fig. 6 Classification performance for detection of SARS-CoV-2 when combining Haralick and Zernike for feature extraction. In (**a**) are the results of accuracy, while in (**b**) we have the kappa results for each model

of these features and the ones extracted by Zernike moments. As can be seen in Figures 4, 5, 6, and 7, the combination of Haralick and Zernike features led to a decrease on classification performance for all methods, considering the values of accuracy and kappa statistic. Nevertheless, by assessing sensitivity and specificity results, we noticed a greater data dispersion when using Haralick features alone. Furthermore, for specificity, the overall performance of the models was higher after combining Haralick and Zernike features. By assessing sensitivity results, in both conditions we still found better results when using only features from Haralick. Considering this, we believe that since Covid-19 identification from pulmonary image is mostly based on the detection of ground-glass opacity in both lungs, this effect is better described by texture-related features. By adding the shape-related features, we ended up incorporating unnecessary and redundant information.

However, we still achieved satisfactory performance in both scenarios. Several models showed accuracy above 90%, with kappa statistic between 0.80 and 0.90.

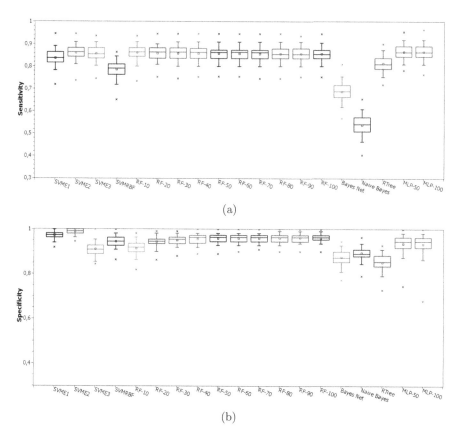

(a)

(b)

Fig. 7 Classification performance for detection of SARS-CoV-2 when combining Haralick and Zernike for feature extraction. In (**a**) are the results for sensitivity, while in (**b**) we show the specificity of each model

As for sensitivity, the best models achieved results around 0.95 using Haralick features and around 0.85 using both Haralick and Zernike features. Considering the specificity, we found values above 0.95 for Haralick alone and for the combination of feature extractors. Overall, the methods SVM, Random Forest, and MLP showed the better results. All tested configurations of Random Forest and MLP presented similar performances. As for SVM, the RBF kernel performed worse than the polynomial kernels. The worst results for both databases were achieved by the Bayesian methods, reaching accuracy between 70% and 80%, kappa around 0.50 and greater data dispersion. The Random Tree showed intermediate performance. These results indicate that although the problem is possibly easy to generalize, it still seems to be complex to solve. Furthermore, the low performance of Bayesian networks points to a dependency among the extracted features.

Such as mentioned before, Tables 4 and 5 particularly show the average and standard deviation results for all metrics used to assess classification performance.

Table 5 Classification performance for the best and the worst models using Haralick and Zernike for feature extraction

Model	Accuracy (%)		Kappa statistic		Recall/Sensitivity		Precision		Specificity		Area under ROC	
	Average	StdDev	Average	StdDev	Average	StdDev	Average	StdDev	Average	StdDev	Average	StdDev
SVM (E2)	92.5243	1.7025	0.8505	0.0340	0.8608	0.0305	0.9883	0.0131	0.9897	0.0116	0.9252	0.0170
Naive Bayes	71.6360	2.6524	0.4327	0.0531	0.5383	0.0466	0.8369	0.0384	0.8944	0.0283	0.8170	0.0277

Table 6 Confusion matrix for the classification using the best model and Haralick moments. This matrix confirms that it is more difficult to detect Covid-19 than non-Covid-19. Thirty-five instances of Covid-19 were classified as non-Covid-19, and seven instances of non-Covid-19 were classified as Covid-19

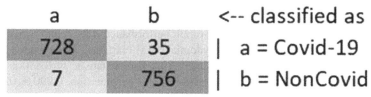

In these tables we present the results for the best and the worst models. For the dataset described by Haralick features, we considered polynomial SVM with exponent 3 as the best model, which achieved average accuracy of 96.994% ± 1.38, kappa statistic of 0.940 ± 0.028, sensitivity or recall of 0.952 ± 0.024, precision of 0.987 ± 0.014, specificity of 0.987 ± 0.014, and AUC of 0.970 ± 0.014. The worst performance was achieved by the Bayes Network; this performance was about 30% lower than the one for the best model.

Considering the second database, using features from Haralick and Zernike, the polynomial SVM model with exponent 2 performed better than the others. In this situation, the accuracy was 92.524% ± 1.702 and kappa was 0.850 ± 0.034, the results for the other metrics were also high, we found sensitivity or recall of 0.861 ± 0.030, precision of 0.988 ± 0.013, specificity of 0.990 ± 0.012, and AUC of 0.925 ± 0.017. For this database, Naive Bayes showed the worst performance.

Finally, since this SVM model using the Haralick features was the best one found in this study, we present the confusion matrix for this condition in Table 6. This matrix shows the amount of instances that were correctly and incorrectly classified in each group. From the matrix we may confirm that most instances were correctly classified in their respective group. Moreover, we found that Covid-19 class is harder to differentiate, since there is more confusion in this group (35 instances) than in Non-Covid group (7 instances).

6 Conclusion

The Covid-19 cases have already affected several countries. To date, more than three million people were infected with the new coronavirus and over 200 thousand people have died from the disease. With a high number of infected people, it is crucial to plan and implement health policies and measures to control the spread of the disease. In this context, it is essential to test the population quickly and accurately in order to provide the patients early medical attention and proper isolation. For this reason, it is important to have a robust tool to diagnose Covid-19 cases wherein the results are released in lower turnaround time than the ground-truth exam. In this context, our motivation is to provide a robust method to diagnose precisely Covid-19 cases.

Thus, we combined CT images and machine learning techniques to identify the confirmed cases of the disease. Our method shows high results using texture and shape features to represent CT images and classic intelligent algorithms for Covid-19 detection. The best overall performance was found when using Haralick as feature extractor and polynomial SVM classifier. This model achieved average values of accuracy of 92.542 ± 1.702, kappa statistic of 0.850 ± 0.034, sensitivity/recall of 0.861 ± 0.030, 0.988 ± 0.013 of precision, 0.990 ± 0.012 of specificity, and AUC of 0.925 ± 0.017.

These results indicate that an effective path for Covid-19 diagnosis may be found by combining CT image analysis by AI and clinical assessment. This combination might increase diagnosis assertiveness, thus improving treatment options and prognosis. However, system usage is compromised by the high cost and relatively low availability of CT imaging technique, specially in low-income regions. In addition, the examination process itself can become a risk factor for contamination, since the patient needs to be taken to the exam room. Moreover, extra care in room cleaning and control are necessary to avoid any type of contamination between patients.

Despite these disadvantages, CT images have good resolution and information that can be useful to monitor pulmonary changes caused by Covid-19. As pointed out in this work, the most commonly found abnormalities are consolidation, bilateral and peripheral disease, linear opacities, and the reverse halo sign. However, as these findings are also common to other types of respiratory diseases, and can be confusing factors, intelligent systems can play an important role.

Disclosure Statement

All authors declare they have no conflicts of interest.

Compliance with Ethical Standards

This study was funded by the Federal University of Pernambuco and the Brazilian research agencies FACEPE, CAPES, and CNPq.

All procedures performed in studies involving human participants were in accordance with the ethical standards of the institutional and/or national research committee and with the 1964 Helsinki declaration and its later amendments or comparable ethical standards.

Acknowledgments The authors are grateful to the Federal University of Pernambuco and the Brazilian research agencies FACEPE, CAPES, and CNPq, for the partial financial support of this research.

References

1. Ackermann, M., Verleden, S. E., Kuehnel, M., Haverich, A., Welte, T., Laenger, F., ...et al. (2020). Pulmonary vascular endothelialitis, thrombosis, and angiogenesis in covid-19. *New England Journal of Medicine, 383,* 120–128.
2. Apostolopoulos, I., Aznaouridis, S., & Tzani, M. (2020). Extracting possibly representative covid-19 biomarkers from X-ray images with deep learning approach and image data related to pulmonary diseases. Preprint. arXiv:2004.00338.
3. Apostolopoulos, I. D., & Mpesiana, T. A. (2020). Covid-19: automatic detection from x-ray images utilizing transfer learning with convolutional neural networks. *Physical and Engineering Sciences in Medicine, 43,* 635–640.
4. Azevedo, W. W., Lima, S. M. L., Fernandes, I. M. M., Rocha, A. D. D., Cordeiro, F. R., da Silva-Filho, A. G., & dos Santos, W. P. (2015). Morphological extreme learning machines applied to detect and classify masses in mammograms. In *2015 International Joint Conference on Neural Networks (IJCNN)* (pp. 1–8).
5. Bai, H. X., Hsieh, B., Xiong, Z., Halsey, K., Choi, J. W., Tran, T. M. L., ...others (2020). Performance of radiologists in differentiating covid-19 from viral pneumonia on chest CT. *Radiology, 296,* E46–E54.
6. Barbosa, V. A. F., Santana, M. A., Andrade, M. K. S., Lima, R. C. F., & Santos, W. P. (2020). Deep-wavelet neural networks for breast cancer early diagnosis using mammary termographies. In H. Das, C. Pradhan, & N. Dey (Eds.), *Deep learning for data analytics: Foundations, biomedical applications, and challenges* (1st ed.). Academic Press
7. Beeching, N. J., Fletcher, T. E., & Beadsworth, M. B. J. (2020). Covid-19: testing times. *BMJ, 369.* Retrieved from https://www.bmj.com/content/369/bmj.m1403
8. Bernheim, A., Mei, X., Huang, M., Yang, Y., Fayad, Z. A., Zhang, N., ...others (2020). Chest CT findings in coronavirus disease-19 (covid-19): relationship to duration of infection. *Radiology, 295*(3), 200463.
9. Bhandary, A., Prabhu, G. A., Rajinikanth, V., Thanaraj, K. P., Satapathy, S. C., Robbins, D. E., ...Raja, N. S. M. (2020). Deep-learning framework to detect lung abnormality–a study with chest x-ray and lung ct scan images. *Pattern Recognition Letters, 129,* 271–278.
10. Borges, L. (2016). Medidas de acurácia diagnóstica na pesquisa cardiovascular. *International Journal of Cardiovascular Science, 29*(3), 218–222.
11. Boser, B. E., Guyon, I. M., & Vapnik, V. N. (1992). A training algorithm for optimal margin classifiers. In *Proceedings of the Fifth Annual Workshop on Computational Learning Theory* (pp. 144–152).
12. Breiman, L. (2001). Random forests. *Machine Learning, 45*(1), 5–32.
13. Burog, A. I. L. D., Yacapin, C. P. R. C., Maglente, R. R. O., Macalalad-Josue, A. A., & Uy, E. J. B. (2020). Should IgM/IgG rapid test kit be used in the diagnosis of COVID-19? *Asia Pacific Center for Evidence Based Healthcare, 2020*(04), 1–12.
14. Caruso, D., Zerunian, M., Polici, M., Pucciarelli, F., Polidori, T., Rucci, C., ...Laghi, A. (2020). Chest CT features of COVID-19 in Rome, Italy. *Radiology, 296*(2), E79–E85.
15. Cheng, J., & Greiner, R. (2001). Learning bayesian belief network classifiers: Algorithms and system. *Advances in Artificial Intelligence, 2056*(1), 141–151.
16. Connors, J. M., & Levy, J. H. (2020). Covid-19 and its implications for thrombosis and anticoagulation. *Blood, The Journal of the American Society of Hematology, 135*(23), 2033–2040.

17. Cordeiro, F. R., Santos, W. P., & Silva-Filho, A. G. (2016). A semi-supervised fuzzy growcut algorithm to segment and classify regions of interest of mammographic images. *Expert Systems with Applications*, *65*, 116–126.
18. Cordeiro, F. R., Santos, W. P. d., & Silva-Filho, A. G. (2017). Analysis of supervised and semi-supervised growcut applied to segmentation of masses in mammography images. *Computer Methods in Biomechanics and Biomedical Engineering: Imaging & Visualization*, *5*(4), 297–315.
19. Coronavirus disease (covid-19) pandemic [Computer software manual]. (2020). Retrieved from www.who.int/emergencies/diseases/novel-coronavirus-2019. Last accessed: 28 April 2020.
20. Cortes, C., & Vapnik, V. (1995). Support-vector networks. *Machine Learning*, *20*(3), 273–297.
21. de Lima, S. M., da Silva-Filho, A. G., & dos Santos, W. P. (2014). A methodology for classification of lesions in mammographies using Zernike moments, ELM and SVM neural networks in a multi-kernel approach. In *2014 IEEE international conference on systems, man, and cybernetics (SMC)* (pp. 988–991).
22. de Lima, S. M., da Silva-Filho, A. G., & dos Santos, W. P. (2016). Detection and classification of masses in mammographic images in a multi-kernel approach. *Computer Methods and Programs in Biomedicine*, *134*, 11–29.
23. de Santana, M. A., Pereira, J. M. S., da Silva, F. L., Lima, N. M. d., de Sousa, F. N., de Arruda, G. M. S., ... Santos, W. P. (2018). Breast cancer diagnosis based on mammary thermography and extreme learning machines. *Research on Biomedical Engineering*, *34*, 45–53.
24. de Vasconcelos, J., dos Santos, W., & de Lima, R. (2018). Analysis of methods of classification of breast thermographic images to determine their viability in the early breast cancer detection. *IEEE Latin America Transactions*, *16*(6), 1631–1637.
25. Döhla, M., Boesecke, C., Schulte, B., Diegmann, C., Sib, E., Richter, E., ... others (2020). Rapid point-of-care testing for SARS-CoV-2 in a community screening setting shows low sensitivity. *Public Health*, *182*, 170–172.
26. Egner, W., Beck, D. C. C., Davis, G., Dayan, C., El-shanawany, T., Griffiths, E., ... others (2020). Statement from RCPath's immunology specialty advisory committee on COVID-19/SARS CoV2 antibody evaluation. *Institute of Clinical Epidemiology, National Institutes of Health-UP Manila and Asia-Pacific Center for Evidence Based Healthcare Inc.*
27. Fang, Y., Zhang, H., Xie, J., Lin, M., Ying, L., Pang, P., & Ji, W. (2020). Sensitivity of chest CT for COVID-19: comparison to RT-PCR. *Radiology*, *296*(2), E115–E117.
28. Geurts, P., Ernst, D., & Wehenkel, L. (2006). Extremely randomized trees. *Machine Learning*, *63*(1), 3–42.
29. Gozes, O., Frid-Adar, M., Greenspan, H., Browning, P., Zhang, H., Ji, W., ... Siegel, E. (2020). Rapid AI development cycle for the coronavirus (covid-19) pandemic: Initial results for automated detection & patient monitoring using deep learning ct image analysis. arXiv.
30. Guo, L., Ren, L., Yang, S., Xiao, M., Chang, D., Yang, F., ... Wang, J. (2020). Profiling early humoral response to diagnose novel coronavirus disease (COVID-19). *Clinical Infectious Diseases*, *71*(15), 778–785.
31. Hani, C., Trieu, N. H., Saab, I., Dangeard, S., Bennani, S., Chassagnon, G., & Revel, M.-P. (2020). Covid-19 pneumonia: a review of typical CT findings and differential diagnosis. *Diagnostic and Interventional Imaging*, *101*(5), 263–268.
32. Haralick, R. M., Shanmugam, K., & Dinstein, I. (1973). Textural features for image classification. *IEEE Transactions on Systems, Man, and Cybernetics*, *SMC-3*(6), 610–621.
33. Haykin, S. (2001). Neural networks: principles and practice. *Bookman*, *11*, 900.
34. Hoffman, T., Nissen, K., Krambrich, J., Rönnberg, B., Akaberi, D., Esmaeilzadeh, M., ... Lundkvist, Å. (2020). Evaluation of a covid-19 IGM and IGG rapid test; an efficient tool for assessment of past exposure to SARS-COV-2. *Infection Ecology & Epidemiology*, *10*(1), 1754538.
35. Kan, C., & Srinath, M. D. (2001). Combined features of cubic b-spline wavelet moments and zernike moments for invariant character recognition. In *Proceedings International Conference on Information Technology: Coding and Computing* (pp. 511–515).

36. Kanazawa, K., Niki, N., Satoh, H., Ohmatsu, H., & Moriyama, N. (1994). Computer assisted diagnosis of lung cancer using helical X-ray CT. In *Proceedings of IEEE Workshop on Biomedical Image Analysis* (pp. 261–267).

37. Leisman, D. E., Deutschman, C. S., & Legrand, M. (2020). Facing COVID-19 in the ICU: vascular dysfunction, thrombosis, and dysregulated inflammation. *Intensive Care Medicine, 46*(6), 1105–1108.

38. Lerner, B., Levinstein, M., Rosenberg, B., Guterman, H., Dinstein, L., & Romem, Y. (1994). Feature selection and chromosome classification using a multilayer perceptron neural network. In *Proceedings of 1994 IEEE International Conference on Neural Networks (ICNN'94)* (Vol. 6, pp. 3540–3545).

39. Li, K., Wu, J., Wu, F., Guo, D., Chen, L., Fang, Z., & Li, C. (2020a). The clinical and chest ct features associated with severe and critical covid-19 pneumonia. *Investigative Radiology, 55,* 327–331.

40. Li, L., Qin, L., Xu, Z., Yin, Y., Wang, X., Kong, B., ...Xia, J. (2020b). Using artificial intelligence to detect covid-19 and community-acquired pneumonia based on pulmonary CT: Evaluation of the diagnostic accuracy. *Radiology, 296*(2), E65–E71.

41. Li, M., Lei, P., Zeng, B., Li, Z., Yu, P., Fan, B., ...Liu, H. (2020c). Coronavirus disease (covid-19): Spectrum of ct findings and temporal progression of the disease. *Academic Radiology, 27*(5), 603–608.

42. Li, X., Geng, M., Peng, Y., Meng, L., & Lu, S. (2020d). Molecular immune pathogenesis and diagnosis of COVID-19. *Journal of Pharmaceutical Analysis, 10*(2), 102–108.

43. Li, Y., & Xia, L. (2020). Coronavirus disease 2019 (covid-19): role of chest CT in diagnosis and management. *American Journal of Roentgenology, 214*(6), 1280–1286.

44. Li, Z., Yi, Y., Luo, X., Xiong, N., Liu, Y., Li, S., ...others (2020e). Development and clinical application of a rapid IGM-IGG combined antibody test for SARS-COV-2 infection diagnosis. *Journal of Medical Virology, 92,* 1518–1524.

45. Lima, S., Azevedo, W., Cordeiro, F., Silva-Filho, A., & Santos, W. (2015). Feature extraction employing fuzzy-morphological decomposition for detection and classification of mass on mammograms. In *Conference Proceedings:...Annual International Conference of the IEEE Engineering in Medicine and Biology Society. IEEE Engineering in Medicine and Biology Society. Annual Conference* (Vol. 2015, pp. 801–804).

46. Lin, D., Liu, L., Zhang, M., Hu, Y., Yang, Q., Guo, J., ...et al. (2020). Evaluations of serological test in the diagnosis of 2019 novel coronavirus (SARS-CoV-2) infections during the COVID-19 outbreak. medRxiv.

47. Liu, Y., Liu, Y., Diao, B., Ren, F., Wang, Y., Ding, J., & Huang, Q. (2020). Diagnostic indexes of a rapid IGG/IGM combined antibody test for SARS-COV-2. medRxiv.

48. Magro, C., Mulvey, J. J., Berlin, D., Nuovo, G., Salvatore, S., Harp, J., ...Laurence, J. (2020). Complement associated microvascular injury and thrombosis in the pathogenesis of severe covid-19 infection: a report of five cases. *Translational Research, 220,* 1–13.

49. Marietta, M., Ageno, W., Artoni, A., De Candia, E., Gresele, P., Marchetti, M., ...Tripodi, A. (2020). Covid-19 and haemostasis: a position paper from Italian society on thrombosis and haemostasis (SISET). *Blood Transfusion, 18*(3), 167.

50. Narin, A., Kaya, C., & Pamuk, Z. (2020). Automatic detection of coronavirus disease (covid-19) using X-ray images and deep convolutional neural networks. Preprint. arXiv: 2003.10849.

51. Okba, N. M., Muller, M. A., Li, W., Wang, C., GeurtsvanKessel, C. H., Corman, V. M., ...et al. (2020). SARS-CoV-2 specific antibody responses in covid-19 patients. medRxiv.

52. Patel, R., Babady, E., Theel, E. S., Storch, G. A., Pinsky, B. A., St George, K., ...Bertuzzi, S. (2020). Report from the American Society for Microbiology COVID-19 International Summit, 23 March 2020: Value of diagnostic testing for SARS–CoV-2/COVID-19. *mBio, 11*(2), e00722-20.

53. Pereira, J. M. S., Santana, M. A., Lima, R. C. F., Lima, S. M. L., & Santos, W. P. (2020a). Method for classification of breast lesions in thermographic images using elm classifiers. In W. P. dos Santos, M. A. de Santana, & W. W. A. da Silva (Eds.), *Understanding a cancer diagnosis* (1st ed., pp. 117–132). Nova Science.

54. Pereira, J. M. S., Santana, M. A., Lima, R. C. F., & Santos, W. P. (2020b). Lesion detection in breast thermography using machine learning algorithms without previous segmentation. In W. P. dos Santos, M. A. de Santana, & W. W. A. da Silva (Eds.), *Understanding a cancer diagnosis* (1st ed., pp. 81–94). Nova Science.

55. Pereira, J. M. S., Santana, M. A., Silva, W. W. A., Lima, R. C. F., Lima, S. M. L., & Santos, W. P. (2020c). Dialectical optimization method as a feature selection tool for breast cancer diagnosis using thermographic images. In W. P. dos Santos, M. A. de Santana, & W. W. A. da Silva (Eds.), *Understanding a cancer diagnosis* (1st ed., pp. 95–118). Nova Science.

56. Phung, S. L., Bouzerdoum, A., & Chai, D. (2005). Skin segmentation using color pixel classification: analysis and comparison. *IEEE Transactions on Pattern Analysis and Machine Intelligence, 27*(1), 148–154.

57. Rodrigues, A. L., de Santana, M. A., Azevedo, W. W., Bezerra, R. S., Barbosa, V. A., de Lima, R. C., & dos Santos, W. P. (2019). Identification of mammary lesions in thermographic images: feature selection study using genetic algorithms and particle swarm optimization. *Research on Biomedical Engineering, 35*(3), 213–222.

58. Rosenblatt, F. (1958). The perceptron: a probabilistic model for information storage and organization in the brain. *Psychological Review, 65*(6), 386.

59. Santana, M. A., Pereira, J. M. S., Lima, R. C. F., & Santos, W. P. (2020). Breast lesions classification in frontal thermographic images using intelligent systems and moments of haralick and zernike. In W. P. dos Santos, M. A. de Santana, & W. W. A. da Silva (Eds.), *Understanding a cancer diagnosis* (1st ed., pp. 65–80). Nova Science.

60. Sethy, P. K., & Behera, S. K. (2020). Detection of coronavirus disease (Covid-19) based on deep features. Preprints, *2020030300*, 2020.

61. Silva, W. W. A., Santana, M. A., Silva Filho, A. G., Lima, S. M. L., & Santos, W. P. (2020). Morphological extreme learning machines applied to the detection and classification of mammary lesions. In T. K. Gandhi, S. Bhattacharyya, S. De, D. Konar, & S. Dey (Eds.), *Advanced machine vision paradigms for medical image analysis*. Elsevier.

62. Tang, Y. W., Schmitz, J. E., Persing, D. H., & Stratton, C. W. (2020). Laboratory diagnosis of COVID-19: current issues and challenges. *Journal of Clinical Microbiology, 58*(6), e00512–20.

63. Wang, S., Kang, B., Ma, J., Zeng, X., Xiao, M., Guo, J., ...Xu, B. (2020). A deep learning algorithm using ct images to screen for corona virus disease (COVID-19). medRxiv.

64. Yamamoto, S., Matsumoto, M., Tateno, Y., Iinuma, T., & Matsumoto, T. (1996). Quoit filter-a new filter based on mathematical morphology to extract the isolated shadow, and its application to automatic detection of lung cancer in X-ray ct. In *Proceedings of 13th International Conference on Pattern Recognition* (Vol. 2, pp. 3–7).

65. Zhao, J., Zhang, Y., He, X., & Xie, P. (2020). Covid-CT-dataset: a CT scan dataset about covid-19. arXiv preprint arXiv:2003.13865.

Analysis of Blockchain-Backed COVID-19 Data

Tadepalli Sarada Kiranmayee and Ruppa K. Thulasiram

1 Introduction

Coronavirus (COVID-19) is a new virus outbreak that is broadly believed to have started in December 2019 in Wuhan, China, and has rapidly spread to different countries around the world [1]. COVID-19 was declared as a pandemic by World Health Organization (WHO) on March 11, 2020, and by the end of March 2020 around 166 countries were affected [2]. This calls for the need of a unified platform, where governments, local authorities, and hospitals can share COVID-19 data with trust. This data would help in building related applications which provides accurate results for preventing further spread. These applications, developed on trusted and verifiable data, should also guarantee the data privacy of the patients [3].

All the above required properties for data analysis (trust, verifiability, and privacy) can be provided by blockchain technology. The blockchain, which maintains an immutable ledger of records in a decentralized way has attracted various researchers and businesses. Although this concept was initially meant to be used with cryptocurrencies, it is constantly being explored for use in different areas to ease the processes by acting as a source of trust or data validation. Many data analysts worldwide are helping in providing best insights from the COVID-19 data (confirmed cases, deaths, and recovered cases) to help governments to create best policies or measures for their citizens. This data was provided by many different agencies such as WHO, European Centre for Disease Prevention and Control (ECDC), and John Hopkins University. Initially, the COVID-19 Situation Report by WHO provided these statistical data, but unfortunately, on March 18, 2020, they shifted the cutoff time for Situation Report 58 leading to the data becoming

T. S. Kiranmayee · R. K. Thulasiram (✉)
Department of Computer Science, University of Manitoba, Winnipeg, MB, Canada
e-mail: tadepask@myumanitoba.ca; tulsi.thulasiram@umanitoba.ca

compromised. The data between the midnight and 9 am (CET) overlapped in that report [4]. It is evident that integrating worldwide COVID-19 related data and making it accessible to others would be convenient for building predictive models for decision-making. MiPasa by HACERA provided normalized and standardized information from all contributing data streams (e.g., for country codes using ISO 3166, or date and time of information using a unified time zone). Furthermore, it also used blockchain technology – Hyperledger fabric – for tamper-proof data consumption [5]. Hence, the WHO, partnered with government agencies, international health organizations, and tech giants on March 27, 2020, and announced Mipasa, blockchain-based control and communications system – aiming for precise and rapid detection of COVID-19 carriers and infection hotspots. It promises to share health and location data between different levels of organizations and authority taking privacy into consideration [6].

When the data is provided on the promising blockchain-backed platform, then the data analyst can perform productive models with such data. This chapter presents a state-of-the-art blockchain data analytics for COVID-19. The rest of this chapter is structured as follows: In Sect. 2, related work provides the recent research work blockchain data analysis and COVID-19 data analysis on centralized data. In Sect. 3, the motivation is described. In Sect. 4, methodology and implementation are elaborated with the sequential steps. In Sect. 5, the required results are showcased and Sect. 6 is conclusion.

2   Related Work

Bitcoin and blockchain concepts were first devised in 2008 by an unknown person or a group using the pseudonym Satoshi Nakamoto [7]. Bitcoin is a combination of open distributed ledger and cryptographic hash functions used together for digital currency applications. Many countries avoided the use of Bitcoin since it violates many government policies and due to its high price volatility. However, its underlying technology – blockchain, is increasingly attracting the attention of researchers. Few of the advantages of the blockchain are that it is decentralized, distributed ledger, information transparency, tamper-proof construction, and openness. The applications of blockchain technology is not only restricted to digital currency but also used for finance, supply chain management (SCM), healthcare system, monitoring market, smart energy, and copyright protection, etc. [8].

To the best of our knowledge, peer-reviewed literature on the topics of blockchain data are on these topics, and there are very little COVID-19 data analysis studies in the peer-reviewed literature. Hence, what we have in the following is the most about the use of blockchain technology in these sectors that form data network(s), similar to what we try to achieve for COVID-19 cluster.

Song et al. [9] discussed materials flowing throughout the supply chain from suppliers, manufacturing facilities, warehouses/distribution centers, to customers. This requires supply chain management (SCM) to improve the traceability, transparency, and auditability. Logistics, quality assurance, inventory management, and forecasting are some functions of a supply chain management (SCM) system. The paper provided details of the impact of blockchain in tracability systems.

There are different categories [10] of a blockchain network – permissionless (public) and permissioned (private and federated). Anyone can join as a new user or a miner in the network where all the participants can perform transactions in permissionless blockchain. However, in permissioned blockchains, only the allowed users who are defined with particular characteristics and permissions participate in the network operations. The blockchain analysis has been performed on both permissionless platforms such as Bitcoin and Ethereum as well as on data by Hyperledger fabric.

Ron and Shamir [11] discussed the bitcoin, a digital payment system where people perform transactions anonymously. They collected the full history of Bitcoin scheme, and many statistical properties were analyzed which were associated with the transaction graph. The paper provided insights on various questions related to the users, for example, the behavior of the users, how they spend and acquire bitcoins, and also balance in the accounts, movement of bitcoins between accounts to maintain their privacy. This analysis helped in additional discovery of a large transaction which occurred in November 2010 that the users tried to hide with fork-merge structures and strange long chains in the transaction graph.

Spagnuolo et al. [12] present the framework, BitIodine – a modular framework, parsing the blockchain and clustering the addresses that are likely to belong to a same user or group of users by classifying such users and adding labels to them, and finally visualizing the complex information extracted from the Bitcoin network. From all the openly available sources, the labeled users are scraped from website where the Bitcoin address information is provided. It also helps in manual exploration of the paths to or from an addresses or users. They evaluated this framework on many real-world scenarios and links were found to Silk Road Wallets and CryptoLocker ransomware where the details of the victims and the ransom paid by them were extracted. This prototype was used for Bitcoin forensic analysis tools. Oggier et al. [13] created a graph mining tool, BiVA, for analyzing and visualizing the Bitcoin network. Bitcoin data is used to explore as well as visualize the subgraphs. The algorithms include clustering algorithms and address aggregation mechanisms. The interface supports basic visualization algorithms; however, it is adaptable and new algorithms can be integrated. In this paper, they provided the Ashley Madison data hack as a case study. Battista et al. [14] coined a visual analysis which depicted the Bitcoin flow mixes with other currency flows in a transaction graph. It graphically tracks the transactions and stores at multiple UTXOs (unspent transaction outputs). Accumulation, distribution, and mixing are some applications for the analysts to reveal Bitcoin flow patterns. Chen et al. [15] collected all Ethereum transaction data and constructed three graphs, money flow graph (MFG), smart contract creation graph (CCG), and smart contract innovation graph (CIG), to characterize the major activities and to later draw insights from these graphs.

A new approach called cross-graph analysis was introduced to address two new security issues – attack forensics and anomaly detection. Donna Dillenberger et al. [16] discussed connecting the analytics engines to blockchains for providing convenient predictive models, provenance histories, configurable dashboards, and compliance checking. They also showed how the blockchain data can be combined with external data sources for secure and private analytics. This can enable creation of artificial intelligence model over geographically dispersed data, and enable provenance and lineage tracking for trusted artificial intelligence models.

Zhou et al. [17] proposed – Ledgerdata Refiner – a ledger data query platform. It provides an interface to retrieve data such as blocks and transactions with the help of ledger data analysis middleware. Historical data can be tracked which can be in any state. Ledger state schemas are analyzed and clustered which help the users to query the ledger data.

Dey et al. [18] discussed how the recent outbreak of pneumonia in Wuhan, China, caused by the SARS-CoV-2 has prioritized the analysis of the epidemiological data of this novel virus and focused on predicting the number of people getting infected all around the globe. The study presents a way to gather epidemiological outbreak data and to analyze on COVID-19. The Johns Hopkins University, World Health Organization, Chinese Center for Disease Control and Prevention, National Health Commission, and DXY (an online community for physicians and healthcare professionals in China) are few open datasets for 2019-nCoV. To understand the number of different cases reported, that is, confirmed, death, and recovered in China and around the world, an exploratory data analysis with visualizations is inevitable. Moreover, this type of evaluation is inevitable to the risks and begin containment activities at the earliest.

Hamzah et al. [19] developed an online platform called Corona Tracker, which provides latest and reliable news updates, as well as statistics and analysis on COVID-19. This paper is done by the research team in the CoronaTracker community and it aims to predict and forecast COVID-19 cases, deaths, and recoveries through predictive modeling. Interpreting public sentiment patterns on healthcare information and assessing political and economic influence toward the spread of the virus are the main significant features of this model.

3 Motivation

During this unprecedented time, it is inevitable to provide speedy analysis and solutions for the safety of the people. Mostly the data analysis has been done on the centralized data. This data will be limited and will not provide complete data for the analysis. However due to the discrepancies and laborious preprocessing of the centralized data, it calls for a reliable alternative solution. Mipasa [20] is the blockchain-based, distributed data platform launched by HACERA. The main goal is to simplify and to streamline the process of sharing and usage of data among multiple providers, so that the researchers, software developers, and decision-makers worldwide can access the data for decision-making. This paper showcases the data analytics on the blockchain-backed data for COVID-19.

4 Methodology and Implementation

In this section, the centralized datasets of ECDC COVID-19 [21], The COVID Tracking Project [22] and COVID-19 dataset in Japan [23] are compared with the blockchain-backed data Mipasa blockchain-backed dataset. This will showcase the importance and need for normalized data to perform easy and quick data analysis for unprecedented situation such as COVID-19. Later, analysis is performed to create a knowledge tree for comparing the USA and Japan's response to COVID-19 by analyzing the confirmed cases.

4.1 Data Collection

The datasets and their descriptions are presented below.

4.1.1 ECDC Covid-19 Data

The data is collected from the website [21]. The dataset consists of the columns are presented in Table 1.

Table 1 COVID-19 ECDC data description

Columns	Description
dateRep	Date in the dd/mm/yyyy format
Day	dd extracted from the date
month	mm extracted from date
year	yyyy extracted from the date
cases	Total number of COVID19 confirmed cases
deaths	Total number of COVID19 related deaths
countriesAndTerritories	Name of the country or Union Territory
geoId	Geographical ID is the Alpha-2 country code
countryterritoryCode	Alpha-3 country code
popData2019	total population of the country in 2019
continentExp	Name of the continent
Cumulativenumberfor14daysof COVID-19casesper100000	The 14-day notification rate of newCOVID-19 cases and the 14-day death rate are the main indicators displayed

Table 2 Mipasa's COVID-19 ECDC cases description

Columns	Description
dataId	It is a 32 alphanumeric value representation like Wallet ID
countryCode2	Country codes Alpha-2
date	Date when data collected (ISO 8601 time representation)
cases	Number of positive cases
New_cases	Number of new cases
Cumulativecases	Number of cumulative cases

4.1.2 Mipasa ECDC Covid-19 Blockchain Data

The main dataset used for this paper ECDC COVID-19 blockchain data collected from Mipasa Website [24]. The source.csv is the original data which is similar to the ECDC data. This data is normalized and presented in two files Output_ECDC_Cases.csv and Output_ECDC_Deaths.csv. The ECDC dataset is collected from the Mipasa and the blockchain data description is presented in Table 2.

4.1.3 The COVID Tracking Project

This dataset is collected from the COVID tracking website [22] and the data presented in Table 3.

4.1.4 The-Mipasa's-COVID-Tracking-Project

This dataset is collected from the Mipasa and the blockchain data is described in Table 4.

4.1.5 Mipasa's COVID-19 in Japan

Output_States.csv dataset is collected from the Mipasa and the blockchain data is described in Table 5.

4.1.6 COVID-19-Dataset-in-Japan-From-Kaggle

covid _jpn_prefecture.csv dataset from Kaggle [23] is derived from manually collecting data by using pdf-excel converter on the reports published by Ministry of Health, Labour and Welfare HP. This data is described in Table 6.

Note, that the positive cases are defined when a patient visits doctor and is tested for positive for COVID-19 at local level. However, a presumptive positive test is

Table 3 The COVID Tracking Project data description

Columns	Description
date	Date with yyyyddmm format
state	Alpha-2 state code
positive	Total number of positive cases
negative	Total number of negative cases
pending	Total number of viral tests that have not been completed
hospitalizedCurrently	Individuals who are currently hospitalized with COVID-19
hospitalizedCumulative	Total number of individuals who have ever been hospitalized with COVID-19
inIcuCurrently	Individuals who are currently hospitalized in the Intensive Care Unit with COVID-19
inIcuCumulative	Total number of individuals who have ever been hospitalized in the Intensive Care Unit with COVID-19
onVentilatorCurrently	Individuals who are currently hospitalized under advanced ventilation with COVID-19 Definitions vary by state/territory
onVentilatorCumulative	Total number of individuals who have ever been hospitalized under advanced ventilation with COVID-19
recovered	Total number of people that are identified as recovered from COVID-19
dataQualityGrade	The COVID Tracking Project grade of the completeness of the data reporting by a state
lastUpdateEt	The COVID Tracking Project compiles data once each day
dateModified	Modified date
checkTimeEt	Check time
death	Total fatalities with confirmed OR probable COVID-19 case diagnosis
hospitalized	Total number of people hospitalized
dateChecked	Date when checked
totalTestsViral	Total number of completed viral tests (or specimens tested)
positiveTestsViral	Total number of completed viral tests (or specimens tested) that return positive
negativeTestsViral	Total number of completed viral tests (or specimens tested) that return negative
positiveCasesViral	Total number of people with a completed viral test that return positive as reported by the state or territory
deathConfirmed	Total fatalities with confirmed COVID-19 case diagnosis
deathProbable	Total fatalities with probable COVID-19 case diagnosis
fips	Federal Information Processing Standards (FIPS) code for the state or territory
grade	The COVID Tracking Project grade of the completeness of the data reporting by a state

passed along to a national level for confirmation, if it is confirmed by National body then it is considered as confirmed case [25]. Here, ECDC dataset provides details for confirmed cases where The COVID Tracking Project provides positive cases.

Table 4 Mipasa's The COVID Tracking Project data description

Columns	Description
dataId	It is a 32 alphanumeric value representation like Wallet ID
date	Date when data collected (ISO 8601 time representation
stateId	US state Alpha-2
positive	Number of positive cases

Table 5 Mipasa's COVID-19 in Japan

Columns	Description
dataId	It is a 32 alphanumeric value representation like Wallet ID
date	Date when data collected (ISO 8601 time representation
Tests	Number of tests
Confirmed	Number of confirmed cases
Deaths	Number of death cases
Recovered	Number of recovered cases
Hospitalized	Number of hospitalized cases
Severe	Number of severe cases
Population	Population in the state
Administrative_area_level	Administrative division
administrative_area_level_2	The state name
JIS_Code	Japan state codes

Table 6 COVID-19 in Japan

Columns	Description
Date	YYYY-MM-DD
Prefecture	Prefecture name. Tokyo, Osaka etc.
Positive	PCR tested positive cases
Tested	PCR tested cases
Discharged	Discharged cases
Fatal	Fatal cases
Hosp_require	Requiring hospitalization
Hosp_severe	Positive and with severe symptoms

4.2 Analysis

The Mipasa ECDC COVID-19 Blockchain data and Mipasa's COVID Tracking Project datasets are already normalized. This makes the analysis of the data convenient as the data preprocessing time is minimal and data preprocessing is reduced for a data analyst. On the other hand, the centralized ECDC and COVID Tracking Project datasets presented in Tables 1 and 3 are different and there are number of fields which have to be understood first before analyzing the data. Moreover, if the data from different sources is provided on one platform and if the data is already normalized, then the data analyst can provide good analysis in less time.

4.3 Implementation

The ECDC and COVID Tracking Project which are blockchain-backed data feed are used to find the countries with COVID-19 confirmed patients and in US states, respectively, for all the months from January 2020 to July 2020. This analysis of the data is implemented using Python and visualization is done in R programming.

The Mipasa's ECDC confirmed cases dataset, that is, Output_ECDC_Cases.csv was downloaded. The data description of the dataset is given in Table 4. The aim of the analysis is to find the top 10 countries suffering from a greater number of COVID cases every month from beginning of January 2020 to July 2020. This aim can be achieved by firstly extracting the month from the ISO 8601 date format and defining the respective month. Then, finding the maximum confirmed cases each month. The countrycode2 is linked to "Unbounded Taxonomy Representation (UTR)" to avoid inconsistencies country codes in the dataset [26]. To attain the corresponding country name, the country code has to be linked to the country UTR [26].

Finally, knowledge tree is constructed to showcase the 10 countries which were adversely affected by COVID-19.

5 Results

Knowledge tree is constructed using Collapsible Tree [27] in R the Fig. 1 shows the 10 countries with maximum number of COVID-19 confirmed cases for months from January 2020 to July 2020. We can easily comprehend that Japan was having the maximum number of confirmed cases in January and February, but the government took important measures such as effective contact tracing for early detection and early response to clusters [28] for containment of COVID-19 cases along with required measures such as "social distance," "wearing mask," and "hand hygiene for example hand washing." We can see the efforts paid off as the number of cases decreased and the first wave ended early in Japan. However, the USA persistently had maximum number of cases. Hence, we can understand that February marked the end of the first wave in Japan but unfortunately, it was not the same with the USA.

Finally, the results from the above analysis provides 10 most affected US states with maximum number of COVID-19 confirmed cases each month.

Figure 2 shows number of cases in a particular month for Japan and the USA, Fig. 2 shows the top countries for the month of February with countries with maximum number of COVID-19 confirmed case numbers on the edges.

Now we would like to know within the USA and Japan which are the 10 states or provinces that are having maximum COVID-19 patients. Figure 3 shows the hierarchy tree structure for the cases. This tree hierarchy was possible to create by joining the COVID-19 Tracking Data for the USA and COVID19 in Japan datasets,

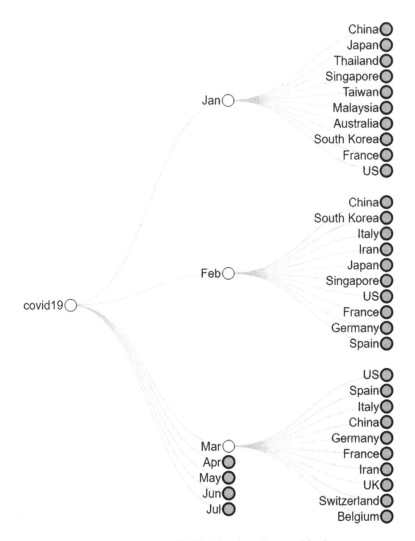

Fig. 1 Ten most affected countries with COVID-19 patients January–March

respectively. We can see the COVID-19 Tracking Data did not have the January COVID-19 data, whereas ECDC data shows 6 COVID-19 cases. This hierarchical tree structure shows the provinces are suffering the most and in turn the country is suffering of the pandemic. Moreover, there are more provinces suffering in Japan, whereas the USA had only Washington with COVID-19 cases.

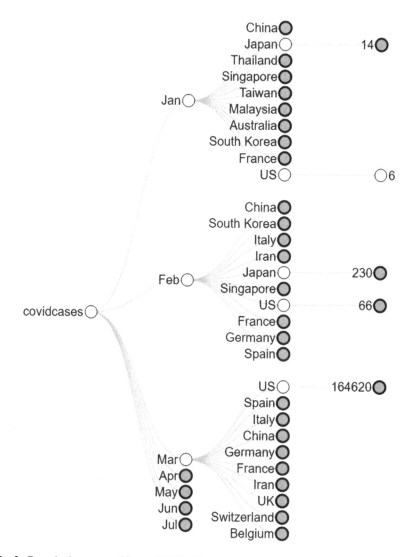

Fig. 2 Cases in the cases and Japan COVID-19 cases

We can also analyze the data corresponding to a particular state in the USA. Figures 3 and 4 show the monthly positive cases in New York and monthly confirmed cases in Tokyo, respectively. We can understand that the government has taken measures to reduce the cases in Tokyo since January 2020, but on the other hand, the containment measures of COVID-19 in New York showed affect only after April 2020.

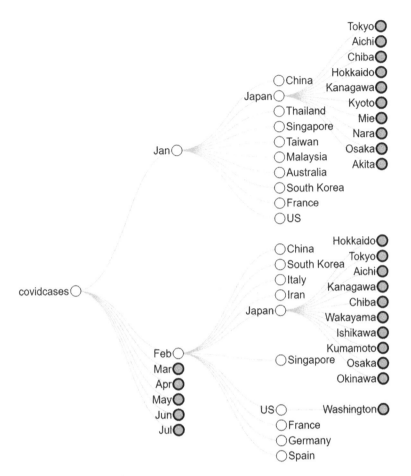

Fig. 3 COVID-19-affected provinces in Japan and states in the USA for the month of January and February

6 Conclusion

This research presents network analysis on the COVID-19 blockchain data. It also showcases the importance of normalized data provided by Mipasa, that is, data preprocessing was minimum for Mipasa data when compared to the data provided by the centralized system. Blockchain data was analyzed and insights related to first wave of COVID-19 in the USA and Japan were discussed by visualizing the data as a network.

This research can be further extended by understanding the timeline of the government measure which helped in containment of COVID-19.

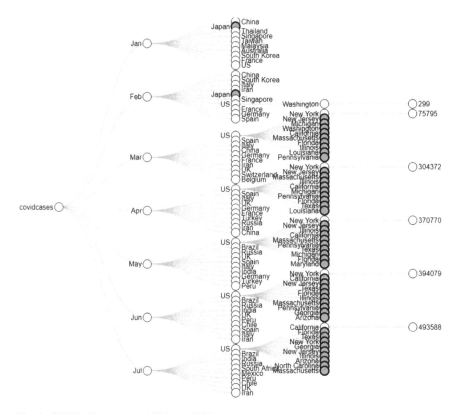

Fig. 4 COVID-19 epicenter drift in the USA

References

1. Huang, C., Wang, Y., Li, X., Ren, L., Zhao, J., Hu, Y., Zhang, L., Fan, G., Xu, J., Gu, X., Cheng, Z., Yu, T., Xia, J., Wei, Y., Wu, W., Xie, X., Yin, W., Li, H., Liu, M., Xiao, Y., Gao, H., Guo, L., Xie, J., Wang, G., Jiang, R., Gao, Z., Jin, Q., Wang, J., & Cao, B. (2020). Clinical features of patients infected with 2019 novel coronavirus in Wuhan, China. *Lancet, 395*(10223), 497–506. https://doi.org/10.1016/S0140-6736(20)30183-5. Epub 2020 Jan 24. Erratum in: Lancet. 2020 Jan 30; PMID: 31986264; PMCID: PMC7159299.
2. Khafaie, M., & Rahim, F. (2020, April). Article history: Cross-country comparison of case fatality rates of COVID-19/SARS-COV-2. *Osong Public Health and Research Perspectives, 11*, 74–80. https://doi.org/10.24171/j.phrp.2020.11.2.03.
3. Hamade, M. A., & Holdings, A. *COVID-19: How to fight disease outbreaks with data.* https://www.weforum.org/agenda/2020/03/covid-19-how-to-fight-disease-outbreaks-with-data/
4. Roser, M., Ritchie, H., Ortiz-Ospina, E., & Hasell, J. *COVID-19 deaths and cases: How do sources compare?* https://ourworldindata.org/covid-sources-comparison. Accessed: 2020-06-12.
5. Levi, J., & Stowell, P. *Utilize blockchain-backed COVID-19 data with MiPasa by HAC-ERA.* https://developer.ibm.com/callforcode/blogs/mipasa-open-data-hub-enables-developers-to-build-apps-to-fight-covid-19. Accessed: 2020-06-12.

6. Chamola, V., Hassija, V., Gupta, V., & Guizani, M. (2020). A comprehensive review of the COVID-19 pandemic and the role of IoT, drones, AI, blockchain and 5G in managing its impact. *IEEE Access*. https://doi.org/10.1109/ACCESS.2020.2992341.
7. Nakamoto, S. (2019). *Bitcoin: A peer-to-peer electronic cash system*. Tech. rep. Manubot.
8. Casino, F., Dasaklis, T., & Pat-sakis, C. (2018, November). A systematic literature review of blockchain-based applications: Current status, classification and open issues. *Telematics and Informatics*. https://doi.org/10.1016/j.tele.2018.11.006.
9. Song, J. M., Sung, J., & Park, T. (2019, January). Applications of blockchain to improve supply chain traceability. *Procedia Computer Science, 162*, 119–122. https://doi.org/10.1016/j.procs.2019.11.266.
10. Xu, M., Chen, X., & Kou, G. (2019, December). A systematic review of blockchain. *Financial Innovation, 5*. https://doi.org/10.1186/s40854-019-0147-z.
11. Ron, D., & Shamir, A. (2012, November). *Quantitative analysis of the full bitcoin transaction graph*. https://doi.org/10.1007/978-3-642-39884-12.
12. Spagnuolo, M., Maggi, F., & Zanero, S. (2014, March). *Bitiodine: Extracting intelligence from the Bitcoin network* (Vol. 8437, pp. 457–468). ISBN: 978-3-662-45471-8. https://doi.org/10.1007/978-3-662-45472-529.
13. Oggier, F., Phetsouvanh, S., & Datta, A. (2018). BiVA: Bit-coin network visualization analysis. In *2018 IEEE international conference on Data Mining Workshops (ICDMW)* (pp. 1469–1474).
14. Di Battista, G., Di Donato, V., Patrignani, M., Pizzonia, M., Roselli, V., & Tamassia, R. (2015). Bitconeview: Visualization of flows in the bitcoin transaction graph. In *2015 IEEE symposium on Visualization for Cyber Security (VizSec)* (pp. 1–8). Chicago, IL. https://doi.org/10.1109/VIZSEC.2015.7312773.
15. Chen, T., et al. (2018). Understanding ethereum via graph analysis. In *IEEE INFOCOM 2018-IEEE conference on Computer Communications* (pp. 1484–1492). Honolulu, HI. https://doi.org/10.1109/INFOCOM.2018.8486401.
16. Dillenberger, D., Novotny, P., Zhang, Q., Jayachandran, P., Gupta, H., Mehta, S., Hans, S., Chakraborty, S., Walli, M., Thomas, J., Vaculin, R., Sarpatwar, K., & Verma, D. (2019). Blockchain analytics and artificial intelligence. *IBM Journal of Research and Development*, 1. https://doi.org/10.1147/JRD.2019.2900638.
17. Zhou, E., et al. (2019). Ledgerdata refiner: A powerful ledger data query platform for hyperledger fabric. In *2019 sixth international conference on Internet of Things: Systems, Management and Security (IOTSMS)* (pp. 433–440).
18. Dey, S., Siddiqi, U., & Howlader, A. (2020, March). Analyzing the epidemiological outbreak of COVID-19: A visual exploratory data analysis (EDA) approach. *Journal of Medical Virology, 92*, 632–638. https://doi.org/10.1002/jmv.25743.
19. Hamzah, F. B., Lau, C., Nazri, H., Ligot, D. V., Lee, G., & Tan, C. L. (2020). CoronaTracker: Worldwide COVID-19 outbreak data analysis and prediction. *Bull World Health Organ, 1*, 32.
20. Rakhmilevich, M. *Oracle and HACERA team up on multi-cloud blockchain-based MiPasa Data Platform to strengthen COVID-19 response*. https://blogs.oracle.com/blockchain/oracle-and-hacera-team-up-on-multi-cloud-blockchain-based-mipasa-data-platform-to-strengthen-covid-19-response. Accessed: 2020-07-31.
21. European Centre for Disease Prevention and Control. https://www.ecdc.europa.eu/en/publications-data/download-todays-data-geographic-distribution-covid-19-cases-worldwide. Accessed: 2020-07-31.
22. The COVID19 Tracking Project. https://covidtracking.com/. Accessed: 2020-07-31.
23. COVID-19 dataset in Japan. https://www.kaggle.com/lisphilar/covid19-dataset-in-japan. Accessed: 2020-08-31.
24. Mipasa-Because I care. https://app.mipasa.org/. Accessed: 2020-07-31.
25. What is a confirmed case vs. a presumptive case of COVID-19?. https://regina.ctvnews.ca/just-curious/what-is-a-confirmed-case-vs-a-presumptive-case-of-covid-19-1.4864155. Accessed: 2020-07-31.

26. Mipasa-Finding, Viewing Using Datasets. https://mipasa.org/documentation/finding-viewing-using-datasets/. Accessed: 2020-07-31.
27. Collapsible Tree. https://github.com/AdeelK93/collapsibleTree. Accessed: 2020-08-31.
28. How Japan Has Responded to COVID-19 Pandemic. https://valdaiclub.com/a/highlights/how-japan-has-responded-to-covid-19-pandemic/. Accessed: 2020-07-31.

Intelligent Systems for Dengue, Chikungunya, and Zika Temporal and Spatio-Temporal Forecasting: A Contribution and a Brief Review

Clarisse Lins de Lima, Ana Clara Gomes da Silva, Cecilia Cordeiro da Silva,
Giselle Machado Magalhães Moreno, Abel Guilhermino da Silva Filho,
Anwar Musah, Aisha Aldosery, Livia Dutra, Tercio Ambrizzi, Iuri
Valério Graciano Borges, Merve Tunali, Selma Basibuyuk, Orhan Yenigün,
Tiago Lima Massoni, Kate Jones, Luiza Campos, Patty Kostkova,
and Wellington P. dos Santos ⓘ

1 Introduction

Neglected tropical diseases (NTDs) affect more than one billion people worldwide [102]. There are more than 20 different diseases that, disproportionately, leave their mark on vulnerable populations. NTDs have a direct impact in the 149, tropical and subtropical, countries where they are located. The direct impacts are related to the number of infected people and death cases, whereas the indirect impacts are related to socioeconomics impacts. This large group of conditions became known as "diseases of poverty" and generates an annual cost of billions of dollars [79].

C. L. de Lima
Graduate Program in Computer Engineering, Polytechnic School of the University of
Pernambuco, Recife, Brazil

A. C. G. da Silva · W. P. dos Santos (✉)
Department of Biomedical Engineering, Federal University of Pernambuco, Recife Pernambuco,
Brazil
e-mail: wellington.santos@ufpe.br

C. C. da Silva · A. G. da Silva Filho
Center for Informatics, Federal University of Pernambuco, CIn-UFPE, Recife, Brazil

G. M. M. Moreno · L. Dutra · T. Ambrizzia · I. V. Graciano Borges · P. Kostkova
Department of Atmospheric Sciences, IAG, University of São Paulo, USP, São Paulo, Brazil

A. Musah · A. Aldosery
Institute for Risk and Disaster Reduction, University College London, UCL-IRDR, London, UK

M. Tunali · S. Basibuyuk · O. Yenigün
Bogaziçi University, Institute of Environmental Sciences, Istanbul, Turkey

Caused by different pathogens, NTDs pose major challenges for public health sectors worldwide [9, 28]. According to the World Health Organization, they are responsible for 17% of infectious disease cases. Moreover, they cause more than 700,000 deaths per year [101].

An important group of diseases called arboviruses is part of these neglected diseases. Arboviruses (short for *arthropod-borne virus*), in turn, are diseases caused by viruses, which are maintained in nature by means of a vertebrate host and a transmitting vector, hematophagous arthropod. The biological transmission of an arbovirus infection commonly occurs when the hematophagous arthropod feeds on the blood of a viremic non-human host and deposits its infectious saliva in a human through a bite. However, some arboviruses have adapted to the urban environment in such a way that they are able to remain in nature only through the mosquito–human–mosquito cycle. Although mosquito bites are the most common form of transmission of arboviruses, other types of transmission have already been reported: during pregnancy, blood transfusion, and through sexual intercourse. Zika, dengue, and chikungunya viruses are important examples of arboviruses. All of them are transmitted by mosquitoes of the genus Aedes [49, 71, 72].

Dengue is a disease present in 100   countries and has experienced a staggering 300% increase in the number of cases over the past decade. For the year 2017, for example, it is estimated that there were 104 million cases of dengue worldwide [102]. Asymptomatic individuals, in most cases, tend to increase clinical severity when reinfection occurs. The most common symptoms are fever, muscle and joint pain, headache, vomiting, and cutaneous rash. The individual may present abdominal pain, persistent vomiting, and hemorrhage in the most severe cases, which can lead to death [102].

Chikungunya has symptoms similar to dengue. The most common clinical findings characteristic are lymphopenia and joint pain that can be disabling and have a prolonged duration from months to years. It is estimated that only about 10% of chikungunya cases are reported [102].

The Zika virus was responsible for the most recent outbreak of arbovirus in the Americas. In Brazil alone, it affected more than 1.5 million cases concentrated mainly in low-income areas [34]. Congenital deformities, such as microcephaly in newborns and acute paralysis in adults, are examples of the devastating consequences of the epidemic [72, 85].

T. L. Massoni
Department Systems and Computing, Federal University of Campina Grande, Campina Grande, Brazil

K. Jones
Centre for Biodiversity and Environment Research, Department of Genetics, Evolution and Environment, University College London, UCL, London, UK

L. Campos
Department of Civil Environmental and Geomatic Engineering, University College London, UCL, London, UK

The distribution and transmission of arboviruses are determined by complex processes that involve demographic aspects, social and environmental factors. Environmental and climatic factors, such as temperature, rainfall, favor the arbovirus pandemics. Local climatic conditions such as temperature, rainfall, and humidity interfere in the vector development, the hatching of mosquito eggs, the life span and dispersion of mosquitoes [28, 38]. In addition, the mobility of human populations, the swelling of overcrowded urban areas associated with difficulty in accessing basic sanitation, regular water supply, and the health system make it difficult to control the vector [28, 38, 66]. In this context, several research groups have concentrated their efforts to understand the dynamics of the distribution of these diseases through mathematical models and also through computational models [28, 46, 54, 78].

A widely used approach for combating arboviruses is the creation of case prediction systems based on the behavior of past disease events. Thus, in addition to the number of cases, the models also take into account the historical series of climatic variables for the prediction of future behaviors. Among the most common models, we can mention the seasonal auto-regressive integrated moving average model (SARIMA), the generalized linear model (GAM), and linear model with minimal absolute selection shrinkage operator (LASSO). On the other hand, the advancement of Digital Epidemiology, of the techniques of geoprocessing, data mining, and machine learning, has provided rapid monitoring, control, and dissemination of diseases [5, 39, 91].

In this chapter, we propose a review of the literature regarding the state-of-the-art methods for predicting arboviruses, focusing on dengue, chikungunya, and Zika fever. This chapter is organized in the following way: in Sect. 2, we present a review of compartment models; in Sect. 3, we detail the approaches based on statistical learning, the prediction models based on machine learning, and hybrid models. In Sect. 4, we present our concluding remarks and conclusions.

2 Forecasting by Statistical Learning and Compartment Models

Subramanian et al. [94] developed two lines of evidence to investigate the uncertainty and predictability of re-emergence timing for diseases with low basic reproduction numbers R_0. In the study the monthly estimates of dengue cases in the city of Rio de Janeiro, Brazil, between 1986 and 1990 were used. For this purpose, a deterministic SIR model described by ordinary differential equations (ODEs) was formulated. First an analytical approach was made to the proposed SIR system with seasonality and forced with intermittent outbreaks, and population turnover, to consider the general characteristics of re-emergence at low R_0. And for this, the approach used in the study by Stone et al. [93] was extended to take into account population growth and sinusoidal seasonality and case underreporting. After that, a stochastic SIR model fit was made using the cosine function as a simplification to represent the seasonal forcing that would be created by variation in climate through changes in infected mosquitoes. And then two approaches are taken: the first comparing the average seasonal R_0 resulting from the model with values of this reproductive number estimated directly from time-series data in the literature for DENV1 and DENV4 in Rio de Janeiro from 2010 to 2016, and a second one that considers a simple vector model based on temperature. A stochastic simulation is then performed using the data and numerical prediction of expected resurgence times.

For the Kaohsiung and Tainan regions of Taiwan, China, Musa et al. [70] proposed the epidemic models $S_h E_h A_h I_{h_1} I_{h_2} R_h$ and $S_m E_h I_h$ based on ODEs. Mathematical analysis of disease-free, endemic, and global equilibrium, system stability equilibrium, and bifurcation was performed. Sensitivity analysis of the model was done. This work also did numerical simulation with plug-and-play inference structures. The epidemic model is fitted to the time series of dengue cases in Kaohsiung and Tainan cities in 2014–2016. The analysis suggests that the reverse bifurcation can be removed when the basic reproduction number is equal to one. And the unsynchronized square wave approach has better fitting performance (in terms of AIC) than the cubic spline approach.

The researchers Abidemi et al. [1] bring a modified deterministic SIR model based on ODEs. The proposal is to describe the dynamics of vector–host interactions in the presence of two dengue virus strains in Madeira Island, Portugal. This model incorporates the vaccine and adulticide to study the control strategy. This model was explained mathematically, the existence of equilibria, the basic reproduction number, determination of the control constant, and sensitivity analysis are analyzed. And finally a numerical experiment in MATLAB with ode45 in the time interval of 365 days.

de Lima et al. [28] proposed the tool entitled DengueME (Dengue Modeling Environment) to study dengue transmission in Ilha do Governador, Rio de Janeiro, Brazil. DengueME is an open source tool that supports the development of spatio-temporal models to simulate dengue and its vector to Geographic Information Systems (GIS). Several models have been developed, including a basic SIR transmission model. From this model, implemented and accessible through VDE DengueME, it is possible to create and analyze scenarios. This model brings the use of successful mathematical–computational modeling to explain the spatio-temporal dynamics of vector ecology and dengue transmission. It has a user-friendly interface that allows users to create and customize scenarios from integrated models, and its extensible architecture allows new models to be added. Modelers can test, adapt, extend, integrate, and compare different models and modeling approaches for dengue and its vectors, and can be used by teachers as a teaching resource. In the present study, no solutions were found to be available dedicated to modeling dengue targeted at local dynamics at intra-urban scales. Kao and Eisenberg [50] discuss the importance of examining identifiability when linking models with data to make predictions and inferences. The study proposes a deterministic ASEI SEIR model, in which they include class A for larvae and pulps. The identifiability (or non-identifiability), which is the ability to provide insights into its unknown, structural, and practical parameters is analyzed and an estimate of the basic reproduction number R_0 is made. Through numerical implementation, various scenarios and conditions for the non-identifiability problem were analyzed. And for this, the Nelder–Mead of NumPy in Python 2.7.10 was used. But for the model to be structurally identifiable, i.e. consistent with the results of differential algebra and FIM approaches, it is practically unidentifiable. To understand the effects of dengue transmission in a spatio-temporal context taking into account human mobility, temperature, and mosquito control in the Pearl River Delta region was done by Zhu et al. [112]. To this end, a SEI SEIR compartmental modeling framework was proposed to infer spatio-temporal transmission patterns of Dengue. To estimate the uncertain parameters of the model, the Markov Chain Monte Carlo method was used, and thereby evaluate the model. And an uncertainty analysis was performed. The model was validated with real epidemic data from the 2014 outbreak. This study brings a new modeling framework with consideration of multi-scale factors and surveillance data to shed light on the hidden dynamics responsible for the spatio-temporal transmission of Dengue fever. The model is robust in exploring the transmission dynamics. This work has some limitations such as accounting for data only from symptomatic infected individuals as the data is based on the surveillance system of the Pearl Delta region. The biological parameters that were applied in the model were extracted from the literature, and therefore may apportion geographic disparities. In addition, the model does not include factors such as precipitation, social economics, and random factors that are potential factors. To study the impacts of temperature on the emergence of dengue fever in the USA, Robert et al. [84] proposed a mathematical model. The model used was the SEI SEIR model described by ODEs to analyze the influence of temperature on disease dynamics and the influences of temperature on viral dynamics within the mosquito. This study divided the mosquito population into five classes, of which the first four represent the larval stages, and the SEI model is for adult vectors only. The model fits were implemented numerically using the nonlinear least squares method, using the lsqnonlin function in MATLAB (MathWorks, 2018). In order to maintain some consistency when analyzing the impacts of temperature in the absence of other influential variables, the value of the density-dependent survival parameter was estimated using the dengue case data from San Juan. This value was used for the populations in each of the six cities in the study. Temperature data were obtained from the National Oceanic and Atmospheric Administration (NOAA)—National Centers for Environmental Information. The mortality time of adult female mosquitoes as a function of temperature was also analyzed, as well as larval development time, extrinsic incubation period, probability of mosquito survival to IPT, and the vector–host ratio. This study aims to understand all the impacts of temperature on the entire life cycle of the vector. The paper highlights the importance of temperature in the current and future patterns of Dengue transmission and emergence; however, temperature is one of many climatic factors that can be taken into consideration. The study also addresses the importance of understanding disease transmission with individual factors. And all the insights brought by the study can help in planning health intervention and vector control programs. For this work, some assumptions were made, among them: the impacts of mosquito fecundity and egg survival, larval survival, human–mosquito contact, and dengue transmission rates were left out. Another potential limitation of this study was that many dynamic processes were based on data that is collected in static settings; however, this approach was necessary in this study.

Researchers Suparit et al. [95], meanwhile, created a compartmental model to analyze the spread of ZIKV during the 2015–2016 outbreaks in the state of Bahia, Brazil. The model searched for a weekly bite rate that provides the best fit between the simulation result and the corresponding reported weekly data. The search process was repeated until the complete time series of weekly bite rates was obtained. The model was calibrated using the total number of reported cases of the outbreak in Bahia. And after May 2016 there were no state data, so the national level data between June and December 2016 was used. Finally, MATLAB was used to calculate the correlation coefficients between mosquito bite rates and temperature. With this, it was shown that the dynamics of ZIKV transmission are less sensitive to the change in mosquito population size than the change in mosquito bite rate. A prediction of weekly cases was made for Bahia. And, possibly, this was the first compartmental model that can accurately simulate two peak ZIKV outbreaks in Bahia. It should be noted that this model assumes that ZIKV behaves the same as, for example, dengue and chikungunya viruses. And it assumes spatial homogeneity in contact and transmission of the disease. It also does not consider sexual transmission of ZIKV. Described by system of ODEs, the SLI SRI model was proposed, mosquito and human, respectively, for Latin America and the Caribbean by Muñoz et al. [69]. In this model, the latent class of mosquitoes is introduced. The global coupled model ensemble participating in the North American Multi-Model Ensemble (NMME) project was selected. The model was coded and run in MATLAB on a monthly tempo scale for a total of 792 months when forcing it with observed data, and 348 months per members when using NMME hindcasts. Interestingly, this predictive tool can be used by health professionals and decision makers interested in Aedes mosquito-borne diseases in Latin America and the Caribbean. This model considers only the effect of climatic conditions, through temperature and precipitation, on disease transmission via vectors and viruses of interest. Transmissions via sexual intercourse and blood transfusion have not been considered. In addition, the current version of the model cannot simulate co-infections or mixed states.

In order to study the progression of Zika outbreak dynamics in El Salvador during 2015–2016, the authors Kumar et al. [55] developed a study. A SEI SEIR transmission model was proposed. In this work, an analysis of the disease-free equilibrium and reproduction number was performed. For numerical simulation of the model, MATLAB platform and Runge–Kutta fourth-order (RK4) method was used. As a result, a good correlation between the simulations and the actual number of cases in El Salvador was obtained. The model identified transition parameters that affect the spread of the outbreak. The most important parameter identified was the mosquito–human ratio that affected the disease transmission of a given population. And to predict the progression of Zika in different geographical regions, numerical implementation in MATLAB can be used. This will aid the prediction of outbreaks and formulation of control measures. However, the main drawback of this work is that it does not consider sexual and mother to fetus transmission. In addition, homogeneity of the infection was assumed, but in reality it is heterogeneous.

To model and predict Zika cases in Brazil, Ndaïrou et al. [75] proposed SEI SEIR. This model has 11 compartments described by a system of ODEs, with epidemiological parameter estimates and with basic reproduction number R_0. The partial rank correlation coefficient method was used for global sensitivity and analysis to identify the most relevant parameters. Using optimal control theory and Pontryagin's principle, a control model was proposed. The conditions for optimal control are determined for the proposed deterministic Zika virus model. Numerical simulations were performed in MATLAB, using the Runge–Kutta method, with data from 1 to 52 weeks of 2016, data from Brazil. The model is robust and multifaceted, making it very useful for government agencies, health managers, and researchers. To understand and determine the important parameters in Zika virus transmission Usman et al. [99] developed a compartmental SEI SVEIR model. This model described by differential equations introduces a class of vaccinated individuals to the classical SEI SEIR model. The disease-free equilibrium and the basic reproduction number R_0 were studied mathematically. The importance of each parameter of the Zika virus infection transmission model is studied by means of the stability index. And these parameters vary, increasing or decreasing, as new cases of infected individuals appear in the host population. Thus, this study sheds light on strategies to help control the spread of Zika virus using these parameters. In order to understand the Zika outbreak that happened between 2015 and 2016 in Colombia, El Salvador and Suriname, Shutt et al. [90] developed a compartmental SEI SEIR model described by differential equations. With data from these countries, the model was fitted using Approximate Bayesian Computation to estimate the uncertainty quantification in addition to estimating the reproduction number. The system solution was solved using MATLAB's ode45 function and the Markov Chain Monte Carlo method for numerical manipulation. The use of Bayesian Computation applied to the proposed model brought good results for El Salvador and Suriname, and provided additional information about the spread of the disease. However, the methods used were inaccurate in estimating the statistics for Colombia. Apparently, the problem was the appearance of a second peak in the epidemic. In the development of this study, several assumptions were made, such as homogeneity of the population and of disease transmission, invariant parameters in time and space, lifelong immunity, among others. The simulations were done with data from less than 1 year and with little data for El Salvador and Suriname.

In the present study, Bonyah et al. [11] proposed a SEI SEIR theoretical model to understand the transmission of Zika virus in China. To determine the conditions for optimal control of the disease, the Pontryagin Maximum Principle is used. Mathematically, a study of the (local) balance and free stability of the disease, endemic equilibrium and bifurcation analysis, existence of bifurcation, endemic balance of local and global stability, in addition to parameter sensitivity analysis such as R_0 and optimal control analysis using the Pontryagin Maximum Principle. The numerical simulations were performed on the MATLAB platform. In order to study the stability of the endemic balance, the central variety theory was used. And this balance is asymptotically stable. The effects of dynamic multi-pathway transmission of Zika virus in El Salvador was studied by Olawoyin and Kribs [77]. To this end, a deterministic SEI SEIAR model was proposed that incorporates four routes of ZIKV transmission: the sexual transmission, vertical transmission of ZIKV in mosquito populations, as well as sexual and human and vector transmission for human transmission. Analysis of the disease-free equilibrium and estimation of the baseline reproduction number were performed. Using baseline values, numerical simulations were generated in Mathematica software to visualize the difference between the dynamics of the vector-only model and the full model. To fit the full model to time-series data of ZIKV-infected individuals, Mathematica's FindFit function was used. This work used demographic data from the mosquito population and not from humans, unlike most studies. The parameter values were estimated from the literature. This model may be advantageous for health authorities to estimate the R_0 by means of incidence data. And the differential of this study is the sensitivity analysis of other important epidemiological quantities such as final size and peak outbreak. And this has not been analyzed in previous studies. In addition, this study brings out the time differential effects of control measures based on additional ZIKV transmission channels. However, this study omits vertical transmission of Zika virus in humans. In order to estimate parameters and characteristics of the disease and prediction, the researchers Rahman et al. [81] did a modeling by means of a mathematical model. The proposed model was the SEI SEIR model described by ODEs for French Polynesia and Yap Island. In this study, sensitivity analysis based on complex step derivatives was performed to identify the parameters that can be estimated from limited data. Nonlinear least squares (NLS) method was used to estimate the parameters. To ensure that the reduction of the free parameters is adequate the F-test was used. To solve numerically the system of differential equations describing the model, the Runge–Kutta fourth-order (RK4) method was used. All calculations were performed in MATLAB (The MathWorks, Inc.). The final parameters are robust in that they are less sensitive to the fixed values of the 200 ensembles that were used in the fit. The model predictions from these estimated final parameters provide excellent agreement with the weekly data from each of the seven islands considered. This study brings new insights into Zika virus-related parameters, as well as the infection dynamics of this virus and the effect of disease prevention programs, which may be useful for developing optimal prevention and control strategies. However, with the methodology used in this work for the seven islands (six French Polynesia and one Yap), it was realized that the data set does not contain enough information to estimate more than three parameters in the French Polynesia islands and more than two parameters in Yap Island with a reasonable degree of certainty attached to the estimates.

Doing a comparative analysis of Dengue and Zika outbreaks in Yap and Fais islands, the study by Funk et al. [35] proposed a SEI SEIR mathematical model with extensive sensitivity analysis. Then use was made of dengue transmission models to estimate the transmission dynamics of Zika, but with care when comparing outbreaks in different locations. In the present study, an estimation of the basic reproduction number was made. Dengue time series of the main islands of Yap (dengue and Zika) and Fais (dengue) from three outbreaks were used and the model was fitted to them, keeping the parameters constant across outbreaks, where they represent a common element. We also studied the equilibrium generation intervals, where p is mean time between infection of a primary case and its secondary cases and was 20 days (mean, 95% CI) and a standard deviation of 7.4 days (mean, 95% CI). To estimate the parameters a Bayesian framework was used for generating a posterior distribution sample using Markov Chain Monte Carlo. And for this procedure, the libbi software package was used, run with the R package using rbi and rbi.helpers packages. The result of this analysis is that the single serotype dengue model does not incorporate cross-reactivity between viruses or heterologous serotypes. Two alternative fitted models were also created. Thus, in this study it is necessary to measure mosquito population density and exposure to biting in different environments can provide information for expected attack rates. This work explores the overlap between environmental outbreaks and diseases, and parameter constraints. This work also uses this to investigate the dynamics of disease transmission. In addition the approach used made it possible to identify parameters that would not be identifiable by observing the outbreak in isolation. For example, the Zika model is the assumed equivalent of a single-serotype dengue model that does not incorporate cross-reactivity between heterologous viruses or serotypes and the probability of infection of a susceptible human when bitten by an infectious mosquito, fixed in multiple locations. However, this work has some limitations, such as the generation intervals were misidentified in the data; it left open questions such as how best to explain a rapidly growing epidemic that spreads to large parts of the population in a few generations without making everyone seropositive; the model is deterministic and all uncertainty is in the probability that encodes the reporting process; and the reproduction number is influenced by heterogeneity, and is proportional to it, so it should be interpreted with caution. Thus, this work has produced good results, but care should be taken when extrapolating the results to another environment. Bates et al. [6] proposed a compartmental approach and SEI SEIR theoretical-math model to study Zika virus. This research provided a rigorous global stability analysis of the models. The global stability of the equilibrium point of the model in terms of R0, the disease-free equilibrium, was studied. LaSalle's Invariance Principle was used and shown that the equilibrium is globally and asymptotically stable, so the solution is unique. The transcriptional bifurcation was also calculated using Matcont software. The basic reproduction number R_0 was also calculated using the spectral radius of the next generation matrix. In the present study, it was seen that there was no endemic equilibrium, so the model was modified for such an equilibrium to exist, but most of the equations were kept. And with this new system in addition to showing the existence of an endemic equilibrium point, theorems were also established that provide conditions for the global stability of both the disease-free and the endemic point. To construct the Lyapunov functions, a theory-matrix method and a theory-graph method were used. And with phase portraits numerical evidence is given that there is a transcritical bifurcation, this bifurcation was calculated using Matcont as well, and mosquito bite rate was used as parameters.

Tang et al. [96] studied the impacts of Dengue and Zika coinfection. A compartmentalized SI SIRJ model was formulated to describe such transmission dynamics with a focus on the effects of Zika outbreak by dengue vaccination among human hosts. This study concludes that interventions such as vaccination against one virus can be detrimental to control the other. The proposed model is age-structured dual and incorporated age-specific compartments for Dengue and Zika infected individuals. For the basic reproduction number to be calculated, the model was simplified with stage-specific homogenetic assumptions about virus dynamics. And this study aims to determine the conditions under which dengue vaccination can contribute to dengue prevention and control without inducing a significant increase in Zika infection.

Dodero-Rojas et al. [31] developed a modeling of strategies for Chikungunya outbreak for Rio de Janeiro city, Brazil. A SEIR model was proposed and the analysis done in the years 2016, 2017, and 2019. The basic reproduction number (R_0) was estimated using the next generation matrix method, the reproduction number R_T was also obtained, and a sensitivity analysis was performed. In this analysis, 100,000 Monte Carlo simulations were performed, showing a variation of 5% from the best-fit value of the 2018 Chikungunya epidemic data. And thus, indicate the importance of each parameter for the epidemic profile of the disease in the different phases of the outbreak in a quantitative way. The study further expanded and included Mayaro virus, which was reported as an emerging disease in Latin America. Assuming that Mayaro and Chikungunya have similar dynamics. And has as a result the potential possibility of the virus becoming an epidemic disease in Rio de Janeiro.

Yamana and Shaman [104] built a SEI SEIRA model. The accuracy of operational disease predictions, however, depends on the quality of the available system optimization observations. In doing synthetic testing, it was seen that predictions improve as the observation error decreases. The present study applies a synthetic testing framework to an infectious disease system and the effects of data availability and quality on the prediction of vector-borne diseases using a process-based model are explored. This disease transmission model was coupled with the EAKF Ensemble Fit Filter, which is a data assimilation method that uses observations to iteratively optimize model parameters and state variables over the course of the disease outbreak in a forecast update cycle. This filter-model system was used for parameter estimation and prediction of outbreak characteristics for each "synthetic" observation set. It is shown that the addition of mosquito observations produced better classification predictions, implying a significant improvement in average classifications for most predictions. The predictions improve with decreasing observation error. Prediction with respect to human observations alone improves as observations of vector infection rates are included. Predictive ability is improved by reducing the uncertainty in the parameter values of the model. With these data it is possible to quantify some parameters and use them to formulate measures to better control and manage outbreaks. Furthermore, this framework can be adapted to other disease systems and models for predicting and controlling disease transmission. It is assumed that the dynamics of disease transmission is perfect, but this is not true, since, for example, effects of seasonality, interactions between dengue serotypes, and heterogeneous population mixing are not considered. And the dynamics of disease transmission is dependent on the disease in question, which in turn depends on a number of factors, e.g., for a certain region models based on rainfall and temperature would be more appropriate than those used in this paper.

The increase in threats of arboviruses outbreaks and its dispersion around the world are the main reasons for researchers to seek to develop diverse and effective tools for the prediction of diseases [91]. In this sense, besides the compartment methods mentioned in the works above, several other methods have been explored.

In order to develop a dengue prediction model to help the disease prevention plans, Jing et al. [47] studied auto-regressive models of moving average to fulfill this goal. For this, the authors used data from monthly dengue cases from 2001 to 2016 in the city of Guangzhou, China. Among the confirmed cases, there were local cases (97.41% of cases) and imported cases (2.59%). In addition to the confirmed cases of dengue, data were obtained regarding the following climatic variables: temperature (average, maximum, and minimum) and precipitation (total, average, and minimum). The best model was evaluated according to the Akaike Information Criterion (AIC) and according to the root mean squared error (RMSE). Among the models evaluated by Jing et al. [47], the one with the best fit was ARIMA (0.1.1) (0.0.2) with an AIC $= 400.8313$ and RMSE equal to 0.7762. For this model, the RMSE associated with the prediction was 0.6820.

Jing et al. [47] found that the number of imported cases, as well as the minimum temperature, were important variables associated with local transmission of dengue in Guangzhou, China. However, despite the good results obtained, the model in question only identifies the correlation between dengue cases and risk variables. It does not identify causal relationships. Moreover, the proposed model does not take into account socioeconomic variables and the effects of under reporting cases.

In the work of Zhu et al. [111], the authors used a linear probit regression to predict dengue cases in Guandong, China. Data on dengue cases were collected from the China National Notifiable Infectious Disease Reporting Information System. The data was collected for the period from 2008 to 2016. Climatic factors were gathered from the China Integrated Meteorological Information Sharing System. The data collected were: temperature (average, maximum, and minimum), atmospheric pressure (average, maximum, and minimum), average relative humidity, and wind speed (maximum and extreme). Before building the models, the authors investigated the ideal minimum temperature using the Ross–Macdonald model. Also, Zhu et al. [111] investigated the correlation between dengue cases and covariants using the Spearman's rank correlation and multivariate analysis of variance. Data from the years 2008 to 2016 were used to train the probit regression model for the prediction of cases in the first 41 weeks of 2017.

Zhu et al. [111] observed that, for the studied area, the minimum temperature suitable for the transmission of dengue is above 18 °C. Another point noted by the authors was that dengue epidemics are related to temperature (average, maximum, and minimum) and atmospheric pressure (average), and relative humidity (average). The model proposed by Zhu et al. [111] presented $R^2 = 0.6$ and a correlation coefficient of 0.8104. Despite the good correlation coefficient, the prediction model did not have a good fit with the observed cases. In addition, the authors did not present any metrics to assess prediction errors. Another point which is important to highlight is that the proposed model does not take into account socioeconomic factors, as well as spatial factors.

One of the important factors for the dengue transmission dynamics is the population dynamics. Individuals who travel to endemic regions can be infected and, when returning to their place of origin, can contribute to local transmission [86]. Considering this scenario, Sang et al. [86] studied a prediction model taking into account infected cases outside the study region. Data on dengue cases were obtained from the Notifiable Disease Surveillance System in China from 2006 to 2014. The data collected from the variables of the climatic variables were obtained for March 2006 to January 2016. In this work, Sang et al. [86] performed the time-series decomposition based on loess. The Poisson distribution was used to investigate the relationship between climatic factors and the incidence of dengue. Only the minimum temperature and the accumulated precipitation were used to build the model. The predictor was modeled using smooth cubic spline functions, with 3 degrees of freedom. As a result, the proposed model obtained a correlation coefficient of 0.98 as the methods proposed by Jing et al. [47], Zhu et al. [111], and Sang et al. [86] did not use geographic information for the elaboration of a prediction model. Therefore, it does not return information regarding the spatial distribution of cases. Despite the high correlation coefficient, the authors also did not present a metric to access the error related to the prediction.

Zhang et al. [109] in their proposal, presented a method based on the negative binomial model with a log link function to perform the forecast of weekly cases in the city of Zhongshan from dengue cases in Guangzhou. The confirmed dengue cases for both cities were collected in the Chinese National Disease Surveillance Reporting System for the period from January 2005 to December 2014. Besides the confirmed cases, data related to climatic variables and demographic data were also collected. The proposed model showed that weekly dengue cases in the city of Zhongshan were strongly associated with dengue cases in Guangzhou. The findings in Zhang et al. [109] work suggest that the dengue outbreak in a region may impact its surroundings. The model developed by Zhang et al. [109] presented an AUC of 0.969 for a limit of 3 cases per week, with a sensitivity and specificity of 78.83% and 92.48%, respectively. For a limit of 2 cases per week, the model presented an AUC of 0.957 and a sensitivity and specificity of, respectively, 91.17% and 91.39%. For a limit of 3 cases per week, the model presented an AUC of 0.938, with a specificity sensitivity of 85.16% and 87.25%. Like previous works, the model did not return any information regarding the geographic information on the distribution of the forecasted cases.

Chan et al. [18] proposed a method for predicting the local risk of dengue cases. According to the authors, this method can assist with early warnings, resource allocation, risk communication, and disease control. For the construction of the model, Chan et al. [18] used confirmed cases of dengue, collected in the National Notifiable Diseases Surveillance System, from 2009 to 2012. They also collected information regarding population density and environmental variables. The predictor was constructed using logistic regression to predict the likelihood of the incidence of dengue outbreaks in the given region. The proposed model presented an average sensitivity of 83% for the presence of the outbreak, with an average of false positive of 23%. With a simple and robust model, the authors performed well to predict the risk of dengue cases in a given area.

The work of Monaghan et al. [67] presents an approach for the prediction of the abundance of the mosquito *Aedes aegypti* with a month in advance. In this work, the authors used data on the abundance of the transmitting vector in Phoenix (AZ) from January 2006 to December 2015. The abundance of mosquito eggs was also obtained in the city of Miami (FL) in the period from June 2006 to June 2008. The climatic variables used were related to the historical series of rainfall, relative humidity, vapor pressure, and temperature (maximum and minimum). As a predictor, Monaghan et al. [67] used mixed models with Gaussian effect. The abundance potential was correctly predicted in 73% of the months, for the city of Phoenix (for a period of 10 years). In the case of the city of Miami, the model was able to correctly predict the mosquito abundance in 63% of the months (for a period of 2 years). Although the results presented are considerably good, the models do not return any information regarding the geographic distribution of the mosquito.

Nur Aida et al. [76] presented an approach to predict mosquito breeding sites. In this work, the authors investigated the use of auto-regressive moving average models to be able to predict the population of eggs of Aedes aegypti. The ovitraps data were obtained for the period from February 2008 to March 2010. The predictors evaluated were ARIMA models with a horizon of 1–11 weeks. The evaluated models performed well, however, as did the works by Monaghan et al. [67], Zhang et al. [109], and Jing et al. [47].

Nowadays, new data sources have been used to build predictors to assist in disease surveillance and monitoring [65, 68, 91]. An example of this is the work developed by Marques-Toledo et al. [64]. In this work, the authors investigated tweets as a source of data to feed an early epidemic monitoring system at local and national level. Data related to confirmed dengue cases were obtained in two ways. The official records were collected in the Information System for Notifiable Diseases of the Ministry of Health of Brazil. Initiatives of dengue cases were obtained by Twitter, Google Trends, and Wikipedia. In addition to data on dengue cases, socio-demographic data found on the platform of the Brazilian Institute of Geography and Statistics (IBGE) were also collected. The authors used the following socio-demographic indicators: population, GDP per capita, HDI (Human Development Index), human income development index (HDI-Income), human education development index (HDI-education). The percentage of households with access to the Internet and computers was also assessed. In total, 238 cities (with a population greater than 40,000 inhabitants) were investigated. Marques-Toledo et al. [64] evaluated the relationship between data from Google Trends ($r = 0.92$, $p < 0.001$), Twitter ($r = 0.87$, $p < 0.001$), and Wikipedia ($r = 0.71$, $p <$

0.01). However, only Twitter data was considered because it contained geographic information. The model proposed by the authors were generalized additive models (GAM), where 170 weeks were used for training and 39 weeks were used for validation. The findings of Marques-Toledo et al. [64] showed that Twitter can be a good tool for real-time surveillance of dengue cases. Mainly, for regions where the disease is endemic, especially where health surveillance services are weak or case reporting is slow. On the other hand, the best models generated were generated in cities with a larger population and better socioeconomic indexes with access to the internet and computers. In this way, the method can underestimate cases of disease in the poorest populations.

The spatio-temporal approach can be very useful to assist health managers in planning strategies to fight diseases. Thus, Chen et al. [19] proposed the development of a framework for the spatio-temporal prediction of dengue cases in Singapore at the neighborhood level. In the proposed model, the authors used (1) confirmed cases of dengue in Singapore, containing the address and date of the diagnosis, (2) weekly meteorological variables such as temperature (average, maximum, and minimum), (3) the vegetation index referring to the Normalized Difference Vegetation Index (NVDI), (4) travel history. This last factor was used to calculate how each neighborhood is connected to other regions of the country through public transport, and finally (5) the age of the buildings. Individual-level data were aggregated to weekly cases in 315 space units with a size of 1×1 km. For the modeling of dengue cases, Chen et al. [19] built submodels with the LASSO method for each prediction window, where the 315 neighborhoods were included in the submodels. For dengue cases, the authors used an 8-week lag, while for climatic variables, a 5-week lag was used. The models were evaluated at prediction from 1 to 12 weeks ahead. As a result, neighborhoods were ordered according to the predicted number of dengue cases at each point in time. The neighborhoods present in the upper decile were classified as "high risk," otherwise, they were classified as "low risk."

The model proposed by Chen et al. [19] showed a better performance in the prediction of cases for a week ahead. In this situation, the area under the ROC curve was 0.88. On the other hand, the worst performance of the model was for predictions for 12 weeks ahead, with an AUC of 0.76. The results presented by Chen et al. [19] are very promising because they are able to access the risk of the disease in a good resolution in environments that are urbanized and, therefore, conducive to the proliferation of the transmitting vector. On the other hand, the authors did not present other metrics to corroborate the good performance of the system. In addition, for the reproduction of this method, it is necessary to obtain a database with a georeference of each case in real time.

3 Forecasting by Machine Learning and Hybrid Models

According to Siriyasatien et al. [91], new data on dengue is constantly being generated and must be incorporated into existing data to ensure that predictive models have a complete set of new data from which to learn, making predictive models current and relevant. This is also true for Zika and Chikungunya. However, ensuring that new observations are incorporated into the existing body of data is essentially a manual task. Thus, it can be a time-consuming and inconvenient task, and it may not yet be comprehensive. All of this can result in ineffective forecasting models. This is a fundamental issue to develop automatic data update mechanisms on an ongoing basis. This would optimize the effectiveness and efficiency of the forecasting model. This problem is aggravated by the need to update the forecasting models frequently, which would impose very high overheads (manual import of data from the databases). Infrequent updating can decrease the effectiveness and efficiency of the forecasting model and make the planning and management of vector control policies ineffective. Since manual data collection is a laborious task, it is desirable to automatically collect information using mobile applications and the Internet of Things (IoT).

IoT is the network of mobile devices, portable devices, and sensors of all types [91]. The existence of millions of connected devices, which gather and transmit data, makes IoT an important new technology that should be of great interest to epidemiological researchers. A significant application of this technology is in the field of medicine, using what is now called the Internet of Medical Things (IoMT). IoMT is a fusion of medical devices and applications that can connect to health information technology systems using network technologies [91]. It is possible, for example, to build biosensors and connected devices for dengue identification and patient monitoring to measure various patient parameters to identify and diagnose dengue infection, patient immunity levels, blood pressure, heart rate, and respiration [62, 63, 100].

For Siriyasatien et al. [91], a limitation of the epidemic prediction models is that the interpretation of the forecast results may not be easy to understand by users who have no experience, knowledge, or professional experience. Medical information is often difficult to understand even by medical professionals. Machine Learning techniques are difficult for most people to understand without the necessary training. The more complex the forecasting model, the more essential it is to have the forecasting results displayed in a way that is easily interpreted and understood by users, including a necessary customization [2]. In any case, the presentation of data must be easily understood as being of maximum use for decision-making.

According to Bhunia and Shit [10], the use of geographic information systems is very useful in supporting the control of epidemics and endemics, organizing the variables of interest in the form of information plans, providing also qualitative data for public health managers.

Kesorn et al. [52] proposed a system for predicting the mortality rate of hemorrhagic dengue based on support vector machines. In the proposal of Kesorn et al. [52], instead of modeling the problem using regression and thus seeking to directly predict the mortality rate, the problem was transformed into a classification problem. Thus, the problem is to classify whether the mortality rate is high or low. As evaluation metrics, sensitivity, specificity, and accuracy are used. As areas of study, the central region of Thailand was considered, composed of three provinces: Nakhon Pathom, Ratchaburi, and Samut Sakhon. Data were collected from 2007 to 2013. This region was chosen for the following reasons: (1) high mortality rate from hemorrhagic dengue; (2) high mosquito density; (3) similar climatic factors. For prediction four databases were used: population density distribution, provided by the Ministry of Interior of Thailand; infection rate by Aedes aegypti (larvae, male and female mosquito population, in percentage), provided by the University of Chulalongkorn; climatic factors (temperature, humidity, wind speed, and rainfall), provided by the Thai Meteorological Department; and spatio-temporal distribution of dengue cases, provided by the Bureau of Epidemiology. From the distribution of dengue cases, the mortality rate was calculated and the threshold from which a rate should be considered high was defined. Each of these variables of interest was organized in the form of a time series, without considering geographic location information. The resulting database has two classes and 648 instances with 9 attributes. Health data are for monthly notifications.

Kesorn et al. [52] investigated the following classifiers and configurations: Decision Tree (DT), k-Nearest Neighbors (kNN), multilayer perceptron neural network (MLP), and support vector machines (SVMs). There is no information about which neural network was used in the study. Given that the authors used the Weka library and the standard supervised neural network of that library is MLP with a standard configuration of a single hidden layer with a number of neurons equal to the sum of the number of inputs and the number of outputs, it is believed that 11 neurons were used in the single hidden layer. Regarding SVMs, the following kernel configurations were investigated: linear, polynomial (polynomial degree not specified), and Radial Basis Function (RBF). All tests were done with 10 parts cross-validation. The best results were obtained with SVM-RBF: sensitivity of 0.882, specificity of 0.789, and accuracy of 90.740%.

Despite the good prediction results, the transformation of a problem where a regression would better fit into a classification problem by Kesorn et al. [52] carries inaccuracies. For example: Why classify only between high or not high mortality rates? Is the specific fee not important? Could not the approximate rate be more useful to public health managers and would it not have a greater epidemiological interest? In addition, considering that geographic positioning information is not considered, the proposed method does not return an approximate spatial distribution of the mortality rate or of the criterion (high or not), which makes it difficult to use the result in local planning, in this case, at the level of the provinces and their cities. Kesorn et al. [52] also do not analyze the weight of each factor in the prediction.

Although Ch et al. [16] present research on prediction of malaria and not dengue, we decided to present here a critical summary of this work for its relevance and potential application also in the prediction of arboviruses transmitted by Aedes aegypti, since the Malaria prevention is also achieved by controlling the vector, and the malaria vector is also the distribution of mosquitoes, in this case, mosquitoes of the Anopheles species. Ch et al. [16] proposed a method to predict malaria transmission based on the optimization of an SVM regressor using the artificial Firefly Algorithm (FFA). As a study area, they investigated the regions of Jodhpur and Bikaner, in the State of Rajasthan, India. The authors used databases obtained from monthly notifications, from January 1998 to December 2002, corresponding to rainfall, temperature, humidity, and cases of malaria. The FFA was used to find the best values of the gap parameter C and the parameters γ and ϵ of an SVM-RBF for the largest possible value of Normalized Mean Square Error (NMSE). The results were compared with classic methods, such as Auto-Regressive Models (AR), Auto-Regressive Moving Average (ARM), and Auto-Regressive Integrated Moving Average (ARIMA). The results were also compared with MLP (single layer, 19 and 17 neurons in the hidden layer for Bikaner and Jodhpur regressors, respectively) and SVM with linear kernel. The experiments were performed using percentage division, with 80% of the data set for training and 20% for testing. The proposed model, SVM-FFA, proved to be much superior to the others, as it had a much smaller average error and was more stable, that is, it varied less, which was true for both regions. As the geographic positioning information was not considered, the proposed method does not return an approximate spatial distribution of cases of malaria, which makes it difficult to use the result in local planning, in this case, at the level of the provinces and their cities. As Kesorn et al. [52] and Ch et al. [16] also does not analyze the weight of each factor in the prediction.

Shaukat et al. [89] propose a method for predicting dengue outbreaks by mining data from clinical information provided by health facilities. The information used is the most well-known dengue symptoms, i.e. fever, fatigue, myalgia (muscle pain), very strong flu, hemorrhage, other symptoms, and the diagnosis of the presence or absence of dengue. The base is composed by vectors of seven attributes and a single class. The information is binary and the problem is modeled as a classification. The database used was provided by the Public Health System of Pakistan and corresponds to data from the municipality of Jhelum, and corresponds to records of 95 patients. Bayes' naive classifier, decision trees (J48, Random Tree, and Rep Tree), and linear SVM were evaluated. As evaluation metrics, accuracy, false positive rate, false negative rate, precision, recall, F-measure, and area of the ROC curve were considered. This work does not take into account patients' geographic location information or climatic and environmental factors. It is much more of a diagnostic support system than a system for predicting outbreaks. The best results were obtained with the naive Bayes classifier, which shows that the modeling can be too simple, given that this classifier considers that the attributes are statistically independent, which is not true for most of the real problems.

Laureano-Rosario et al. [56] applied artificial neural networks to predict occur-rences of dengue outbreaks in San Juan, Puerto Rico, and in several municipalities in the State of Yucatán, Mexico. The models were trained with 19 years of dengue data for Puerto Rico and 6 years for Mexico. The climatic and environmental data and the demographic information included in the predictive models were: Sea Surface Temperature (SST), precipitation, air temperature (minimum, maximum and average), humidity, previous cases of dengue, and population size. Two models were built to predict the occurrence of dengue: one aimed at the population at risk, that is, young people under 24 years old; and another model, aimed at the most vulnerable population, that is, children under 6 years old and elderly people over 65 years old. The data were collected weekly. For Puerto Rico, 986 cases of population at risk and 986 cases of the most vulnerable population were collected each week. For Mexico, 310 cases of population at risk and 310 cases of the most vulnerable population were collected each week. The problem was modeled as a classification problem, with thresholds being assumed:

- Puerto Rico: population at risk:

  - Pre-Epidemic: 2 cases per week;
  - Epidemic: 6 cases per week;
  - Post-Epidemic: 5 cases per week.

- Puerto Rico: most vulnerable population:

  - Pre-Epidemic: 1 case per week;
  - Epidemic: 2 cases per week;
  - Post-Epidemic: 1 case per week.

- Mexico: population at risk:

  - Pre-Epidemic: 2 cases per week;
  - Epidemic: 6 cases per week;
  - Post-Epidemic: 5 cases per week.

- Mexico: most vulnerable population:

  - Pre-Epidemic: 1 case per week;
  - Epidemic: 6 cases per week;
  - Post-Epidemic: 4 cases per week.

Laureano-Rosario et al. [56] trained MLP networks using the NSGA-II optimization algorithm (Non-dominated Sorting Genetic Algorithm II) [24, 25, 45], aiming both false positives and false negatives. The probability of crossing was 0.1. The probability of mutation was 0.2. These values were chosen low to avoid destructive diversity, i.e. the destruction of good solutions due to excessive crossings or very aggressive mutations. Laureano-Rosario et al. [56] organized the training as follows: Data from Puerto Rico on the number of cases of population at risk (N_{PR24}) and most vulnerable population (N_{PR5-65}), were separated into training ($N_{PR24} = 1950$ and $N_{PR5-65} = 437$), validation ($N_{PR24} = 609$ and $N_{PR5-65} = 140$), and testing ($N_{PR24} = 1950$ and $N_{PR5-65} = 437$); for Mexico, data on the number of cases of the population at risk (N_{MX24}) and most vulnerable (N_{MX5-65}) were also organized in training ($N_{MX24} = 2169$ and $N_{MX5-65} = 187$), validation ($N_{MX24} = 1643$ and $N_{MX5-65} = 89$), and testing ($N_{MX24} = 4096$ and $N_{MX5-65} = 459$).

Despite the sophistication of the Laureano-Rosario et al. [56] model, that is, the use of an MLP optimized by a multiobjective genetic algorithm, and a very large database, the results cannot be considered good: for Puerto Rico, accuracy was obtained 47% and 58% for the prediction of the population at risk and the most vulnerable, respectively; for Mexico, the accuracy was 51% and 66% for the prediction of population at risk and more vulnerable, in that order. The area under the ROC curve and F-measure were also used as metrics for evaluating the results.

As the geographic positioning information was not considered, the method proposed by Laureano-Rosario et al. [56] does not return an approximate spatial distribution of the cases, which makes it difficult to use the result in local planning. However, unlike Kesorn et al. [52] and Ch et al. [16], the relevant environmental factors for the prediction were assessed using the NPSFS algorithm (Neural Pathway Strength Feature Selection) [33, 51, 57]. There was a considerable influence on the size of the population and the occurrence of previous cases of dengue. Of the climatic and environmental factors, the most influential was the air temperature.

Lee et al. [58] proposed two models for predicting the number of mosquitoes: one is based on artificial neural network, and the other on multiple linear regressions, MLR (Multiple Linear Regression). The Yeongdeungpo-gu district of Seoul, South Korea was chosen as the study area. Daily data on the number of mosquitoes collected in the years 2011 and 2012 were used. Mosquitoes are counted using a system that performs emissions CO_2 to attract them. The counts are then performed and made available on the Digital Mosquito Monitoring System (DSM). As climatic and environmental variables, temperature, wind speed, humidity, and rainfall were used. The objective was to forecast the mosquito population from May to October 2011 and 2012. The base consists of 317 instances. Of these, 220 instances were used for training and 97 for testing. To build the model based on artificial neural network, a single layer MLP was used. MLP configurations were investigated, varying the number of hidden layer neurons from 2 to 23. Both models are regression oriented. As metrics for evaluating quality, we used the Root Mean Square Error (RMSE), the Index of Agreement (IA), and the Correlation Coefficient (R). The best results were achieved with the model based on the MLP neural network, with $RMSE = 14.38$, $IA = 0.75$, and $R = 0.61$. The weight of climatic factors was also analyzed using relevance analysis by attribute selection. The relevance of humidity was only 20%, while temperature, wind speed, and pluviometry had a relevance of 65% each. As the geographic positioning information was not considered, the proposed method does not return an approximate spatial distribution of the cases, which makes it difficult to use the result in local planning.

Iqbal and Islam [44] did a bibliographic review on studies on dengue prediction methods, from 1995 to 2013. Thirty articles were searched. Iqbal and Islam [44] concluded that, to predict outbreaks of dengue, it is necessary to consider historical series of symptoms (fever, severe headache, nausea, vomiting, diarrhea, hypotension, gastrointestinal bleeding, excess appetite or anorexia, pain behind the eyes, abdominal pain, pleural effusion, bleeding from the mouth or nose, abdominal pain, etc.) [12, 15, 83], climatic and environmental factors (temperature, rainfall, wind speed, and humidity) [21, 61, 103], in addition to genetic factors [12, 15, 26, 74, 98] and proteins [40, 82]. Iqbal and Islam [44] conclude that the best models for predicting dengue outbreaks should be based on: decision trees; Artificial neural networks; evolutionary classifiers optimized by meta-heuristic optimization algorithms, such as genetic algorithms; and ensemble classifiers. As the geographic positioning information was not considered, the methods studied and proposed by Iqbal and Islam [44] do not return an approximate spatial distribution, which makes it difficult to use the result in local planning, in this case, at the level of the provinces and their cities. Iqbal and Islam [44] also do not analyze the weight of each factor in the prediction.

Hamlet et al. [41] described a way to parameterize the effect of temperature on yellow fever transmission using statistical models to predict geographic and temporal heterogeneities in the distribution of the disease, while demonstrating its robustness compared to models that simply predict geographic distribution. Hamlet et al. [41] claim that this quantification of seasonality could lead to more accurate applications of vaccination campaigns and vector control programs. This in turn would help to maximize the impact of these campaigns and rationalize resources, in addition to helping to reduce the risk of large-scale outbreaks. The authors believe that the methods described can be applied to other diseases transmitted by the Aedes mosquitoes, providing a useful tool for understanding and combating several other important diseases, such as dengue and Zika.

Hamlet et al. [41] used linear regression models over the historical series of each of the 167 selected provinces of countries on the African continent, to approximate the probabilities of an outbreak of yellow fever. The data were collected weekly, from 1971 to 2015. Probability thresholds were adopted to evaluate the regression models as a classification. The quality metric used was the area of the ROC curve, which ranged from 0.79 to 0.84. Unlike other works, Hamlet et al. [41] combine climatic and environmental information (temperature and vegetation index), dengue outbreak probability and geographic location, with data organized as historical series of geospatial distributions, also allowing for qualitative assessment of results. Despite the evident advantages of the adopted spatio-temporal analysis, more metrics to evaluate the quality of the regression and the classification should have been adopted, since the area under the ROC curve tends to overestimate the quality of the classification.

Cortes et al. [23] used data from 2001 to 2014 to forecast 2015 for the cities of Recife and Goiânia. The data were collected monthly and correspond to the number of dengue cases reported by the National Notification System (SINAN), of the Unified Health System. The prediction problem is approached as the adjustment of two time series using the regression methods ARIMA and SARIMA (Season Auto-Regressive Integrated Moving Average). The results were not considered adequate for the prediction of the cases in Recife. The research lacked consideration of other factors important to the prediction than the case history, such as climatic and environmental variables and other information of interest. In addition, there was no type of refinement that would allow prediction at the level of neighborhoods and districts, since information on the geographic positioning of the cases was not collected.

Buczak et al. [14] built two models to predict dengue outbreaks in the cities of Iquitos, Peru, and San Juán, Puerto Rico, based on sets of ARIMA and SARIMA regressors. Data on dengue cases were collected weekly from the respective public health systems of the two countries, from 2009 to 2010 and from 2012 to 2013. Data on rainfall, temperature, and vegetation were collected daily by satellite. Geographic positioning information was not considered. Therefore, the methods proposed by Buczak et al. [14] do not return an approximate spatial distribution, which makes it difficult to use the result in local planning, in this case, at the level of neighborhoods in each city. Buczak et al. [14] also did not analyze the weight of each factor in the prediction.

For Albrieu-Llinás et al. [4], remote sensing systems and geographic information offer valuable tools for mapping the distribution of species in a given area. However, the prediction of species occurrences by means of probability distribution maps based on entomological research[1] cross-cutting has limited utility for local authorities. Albrieu-Llinás et al. [4] aimed to examine the temporal evolution of the number of houses infested with immature stages of Aedes aegypti in each individual neighborhood and to investigate the environmental clusters generated with information provided by variables of remote sensing to explain the behavior observed over time. Entomological surveys were carried out between 2011 and 2013 in the city of Clorinda, Argentina, recording the number of homes with breeding sites with Aedes aegypti larvae. 10,981 houses were visited, chosen at random. Data were organized by neighborhood and collected monthly. Clorinda has 32 neighborhoods. A SPOT 5 satellite image was used to obtain seven land cover variables: bare soil, surface water, wetlands, low vegetation (grass), tall vegetation (shrubs and trees), urban buildings, and pastures or crops. These variables were subjected to partitioning using the k-means algorithm for grouping neighborhoods into four environmental clusters. The problem of prediction of sites infected by Aedes aegypti was modeled as a regression problem. A regressor based on a generalized linear model was used. The results were also presented in the form of geospatial distributions, using heat maps, that is, pseudo-color maps, assembled after interpolation of the results. As a tool for visualizing spatial distribution and qualitative analysis, the Quantum GIS geographic information system, or QGIS, was used. The method proposed by Albrieu-Llinás et al. [4] showed great potential for local planning. However, other regression methods could have been tested.

Similar to Albrieu-Llinás et al. [4], Scavuzzo et al. [87] proposed an approach to predict sites infected by eggs and larvae of Aedes aegypti using satellite images and position data from infected breeding sites. The study area was the city of Tartagal, in the province of Salta, northwest of Argentina, close to the border of Bolivia. Breeding data were collected weekly, from August 2012 to July 2016, always covering 50 properties chosen at random. To collect the mosquito's eggs, ovitraps were used, traps made of plastic that attract the females to deposit the eggs and leave them trapped [13]. The climatic and environmental variables were obtained through remote sensing: vegetation index (Normalized Difference Vegetation Index, NDVI)

---

[1]Entomological research is understood as research involving insects and their relationship with humans, with other living beings and with the environment.

and water and humidity index (Normalized Difference Water Index, NDWI), from the MODIS satellite MOD13Q1; temperature distribution, obtained from JAXA's TRRM satellite (Tropical Rainfall Measuring Mission—NASA/Japan Aerospace Exploration Agency). The prediction problem was modeled as a prediction of time series using regression. Linear regression methods were tested, SVM with RBF kernel, MLP with three hidden layers of three neurons each, k closest neighbors (kNN) and decision trees. As quality indexes, the correlation index and the MSE were used. The best results were obtained with kNN, MLP, and SVM, with a correlation of 0.888, 0.875, and 0.837, in this order, against linear regression and decision tree, with 0.774 and 0.679, respectively. The results were not presented in the form of geospatial distributions, since the approach did not use the positioning information of each breeding site. The approach of Scavuzzo et al. [87] is similar to that of Scavuzzo et al. [88], which focuses on artificial neural networks.

Ahmad et al. [3] studied the warning signs of the Malaysian Clinical Practice Guideline in predicting severe dengue and its associated factors among confirmed cases presented at a teaching hospital in northeastern Malaysia in 2014. The authors conducted a cross-sectional study in February 2015, using data obtained from hospital records. There were 2607 confirmed cases of dengue presented to Sains Malaysia University Hospital (HUSM) in 2014. 700 patients were selected after proportional stratified random sampling (i.e. respecting the proportionality of gender, ethnicity, geographical origin of the patient and age group population) performed according to the number of cases in 12 different months in 2014. The base consisted of 700 patients, corresponding to 700 instances, with the following characteristics: 75.4% over 15 years old, 24.6% aged less than or equal to 15 years; 49.7% men and 50.3% women; 94.6% Malay and 5.4% from other ethnicities; 83.7% from the Kota region, 8.9% from Bachok, and 7.4% from other regions. The results of severe dengue represented 4.9% of the cases. The prevalence of any of the warning signs in severe dengue was 91.2%. The most common warning signs before severe dengue were persistent vomiting (55.9%) and abdominal pain or tenderness (52.9%). The most sensitive warning sign for detecting severe dengue was abdominal pain (59%). The individual warning signs proved to be very relevant for the prediction of severe dengue, especially the accumulation of clinical fluid (99%), hepatomegaly (98%), and bleeding of the mucosa (93%). Other factors associated with severe dengue were persistent vomiting (2.41%), bleeding of the mucosa (4.73%), and increased hematocrit with rapid drop in platelets (2.74%). The problem was modeled as a classification. Thus, a classifier based on a single class was used, detecting the presence or absence of severe dengue. The classifier was based on logistic regression. The result reached an accuracy of approximately 80% and an area under the ROC curve of 70%. Unlike the other proposals, Ahmad et al. [3] proposes a much more supportive system for diagnosis, with assessment of the relevance of attributes and considering geographic information at a certain level, than a predictor system to support the control of the distribution of the disease. There is no spatio-temporal analysis. In addition, other methods could have been used in the classification, such as artificial neural networks and support vector machines, which could eventually contribute to a higher classification performance.

For Beltrán et al. [8], the devastating consequences of newborns infected with the Zika virus make it necessary to combat and stop the spread of this virus and its vectors: the mosquitoes Aedes aegypti. An essential part of the fight against mosquitoes is the use of mobile technology to support routine surveillance and risk assessment by Endemic Control Agents (ACEs). In addition, to improve early warning systems, public health officials need to more accurately predict where an outbreak of the virus and its vector is likely to occur. The ZIKA system, proposed by Beltrán et al. [8], aims to develop a comprehensive framework that combines e-learning to empower ACEs, and provide community-based participatory surveillance and prediction of occurrences and distribution of the Zika virus and its vectors in real time. Currently, this system is being implemented in Brazil, in the cities of Campina Grande, Recife, Jaboatão dos Guararapes and Olinda, in the State of Pernambuco and Paraíba, with the highest prevalence of Zika virus disease. The ZIKA system also aims to help ACEs to learn new techniques and good practices to improve virus surveillance and offer a real-time forecast of the virus and vector. The proposed forecasting model can be recalibrated in real time with information from ACEs, government institutions, and weather stations to predict the areas most at risk of an outbreak of Zika virus and other arboviruses transmitted by Aedes aegypti in a interactive map. This mapping and alerting system has the potential to help government institutions make quick decisions and use their resources more efficiently to prevent the spread of the Zika virus. Although they propose the use of Random Forest regressors to make predictions, Beltrán et al. [8] did not carry out experiments and focused on the proposal of the mobile application to support participatory surveillance, the proposal being only a theoretical model.

Zhao et al. [110] developed a national pooled model to predict counts of dengue cases across different departments of Colombia. The authors used the assumption that precipitation, air temperature, and land cover type have been shown to be three important determinants of Aedes mosquito abundance and are often used as predictors in dengue forecasting [7, 32, 36, 87]. Precipitation data was obtained from the CMORPH (Climate Prediction Center morphing method) daily estimated precipitation data set [48]. The land surface temperatures were extracted from the MODIS Terra Land Surface Temperature 8-day image products (MOD11C2.006). Enhanced vegetation index (EVI) estimates were obtained from the MODIS Terra Vegetation Indices 16-Day image products (MOD13C1.006). Several studies have shown that socio-demographic factors may influence dengue transmission and incidence as significantly as environmental factors [41–43]. Considering the role of social injustice in epidemics, the authors also included population, education coverage, and the Gini Index (a measure of income inequity) as potential predictors, which were retrieved from the Colombian National Administrative Department of Statistics. The dengue case surveillance data were extracted from an electronic platform, SIVIGILA, created by the Colombia national surveillance program and was available at the department level. The national surveillance program receives weekly reports from all public health facilities that provide services to cases of dengue. The dengue cases reported were a mixture of probable and laboratory confirmed cases without distinguishing between the two different case definitions.

Laboratory confirmation for dengue is based on a positive result from antigen, antibody, or virus detection and/or isolation [38]. Probable cases are based on clinical diagnosis plus at least one serological positive immunoglobulin M test or an epidemiological link to a confirmed case within 14 days prior to symptom onset. Cases are typically reported within a week with severe cases usually being reported immediately.

Zhao et al. [110] found that for the majority of Colombia departments, the national model more accurately forecasted future dengue cases at the department level compared to the local model. This indicates the similarity in importance of dengue vectors across different administrative regions of Colombia. Pooling data from individual departments creates a training data set with larger ranges of variables, increasing the extrapolating capacity of Random Forest models. Results with Random Forests were superior to the ones obtained with MLPs and Deep Convolutional Neural Networks. The national pooled model trained by a larger data set had higher prediction accuracy compared to the local models. The authors also discovered that the meteorological and environmental variables were more important for prediction accuracy at smaller forecasting horizons compared to the socio-demographic variables, with socio-demographics being more important at larger forecasting horizons. Poor quality housing and sanitation management with high population density are key risk factors for dengue transmission, closely related to education and poverty. These results demonstrate the complementary nature of these different groups of predictor variables and the importance of their inclusion in dengue forecasting models.

According to Chakraborty et al. [17], dengue data sets are neither purely linear nor nonlinear. They usually contain both linear and nonlinear patterns. If this is the case, then the individual ARIMA or Artificial Neural Network (ANN) is not adequate to model this situations. Consequently, the combination of linear and nonlinear models can be well suited for accurately modeling such complex autocorrelation structures. Hybrid ARIMA-ANN models have become more popular due to its capacity to forecast complex time series accurately [108]. Neural Network Auto-Regression (NNAR) corresponds to a feed-forward neural network model with only one hidden layer with a time series with lagged values of the series as inputs. Differently from pure ANNs, SVMs, and LSTMs, NNAR is a nonlinear auto-regressive model. Popular hybrid models are hybrid ARIMA-ANN [53, 107], hybrid ARIMA-SVM model [80], and hybrid ARIMA-LSTM [20]. These models try to fit both linear and nonlinear patterns of the time-series data.

Chakraborty et al. [17] proposes a hybrid ARIMA-NNAR model to capture complex data structures and linear plus nonlinear behavior of dengue data sets. In the first phase, ARIMA catches the linear patterns of the data set. Then the NNAR model is employed to capture the nonlinear patterns in the data using residual values obtained from the base ARIMA model. Three popular open-access dengue data sets, namely San Juan, Iquitos, and the Philippines data are used to determine the effectiveness of the proposed model. Different linear and nonlinear models have been studied on these data sets that show highly nonlinear patterns in these regions. Mean absolute error (MAE), root mean square error (RMSE), and symmetric Mean Absolute Percent Error (SMAPE) are used as evaluation metrics. For the endemic regions San Juan and Iquitos, weekly laboratory-confirmed cases for the time periods from May 1990 through October 2011 and from July 2000 through December 2011, respectively are considered in this study. The Philippines data set contains the monthly recorded cases of dengue per 100,000 population in the Philippines. Monthly incidence of dengue is available for the time period January 2008 through December 2016. The Philippines monthly data set contains a total of 108 monthly observations and we use the total cases reported from all regions in the Philippines in this study. San Juan weekly data set contains a total of 1144 observations whereas Iquitos data set contains only 520 observations. The authors organized the three dengue data sets into training and test sets. They studied ARIMA, ANN, SVM, LSTM, NNAR model for these data. The data set is divided into two samples of training and testing to assess the forecasting performance of the proposed model. The proposed hybrid ARIMA-NNAR model was able to predict nonlinear tendencies in comparison with the other models, i.e. ARIMA, ANN, SVM, NNAR, LSTM, ARIMA-SVM, ARIMA-ANN, and ARIMA-LSTM.

Lima and Laporta [60] tried to evaluate the accuracy of deterministic and stochastic statistical models by using a protocol for monthly time-series analysis of the incidence of confirmed dengue cases in the federative units and federal district of Brazil from January 2000 to December 2017. A statistical and computational approach was applied for evaluating and validating the accuracy of 10 statistical models to predict the time series of new dengue cases: ARIMA, Exponential Smoothing Model (ETS) [29], BATS [29], TBATS [37, 73], Long-Short Term Memory (LSTM), Structural Model (StructTS) [59], Neural Network Auto-Regression (NNETAR), Extreme Learning Machine (ELM) [22, 27, 30, 42, 43, 97, 105, 106], and Multilayer Perceptron (MLP). In BATS models, B stands for Box–Cox transformation, A—ARMA residuals, T—trend component, S—seasonal component, which implies the prior Box–Cox transformation of the original data [37]. De Livera et al. [29] introduced a trigonometric representation of the seasonal components based on the Fourier series, providing the TBATS model. Official data on the monthly dengue cases from January 2000 to December 2016 were used to train the statistical methods, while those for the period January–December 2017 were utilized to test the predictive capacity of each model by considering three forecasting horizons (12, 6, and 3 months). Results were evaluated using Mean Absolute Percentage Error (MAPE). According to the authors, the failure to consider the external factors (climatic, environmental, social, or immunological) and the building infestation index could be viewed as a possible limitation of this study. According to the authors, the forecasting models based on ARIMA, TBATS, and ELM could be used to predict dengue in Brazilian states.

Stolerman et al. [92] developed machine learning algorithms to analyze climate time series and their connection to the occurrence of dengue epidemic years for seven Brazilian state capitals. The authors focused on the impact of two key variables: frequency of precipitation and average temperature during a determined range of time windows in the annual cycle. The authors used the publicly available data sets of the Brazilian Notifiable Diseases Information System (SINAN), specially the total number of dengue cases per year, from 2002 to 2017, for all Brazilian state capitals. The authors assumed that the numbers reported were sufficient to identify dengue epidemic years. A year is considered epidemic if, for a given city, the incidence of dengue is above 100 cases per 100,000 inhabitants in the period January–December. To find critical climate signatures, the research was restricted to seven state capitals with at least three epidemic years and three non-epidemic years in the period 2002–2012. Climate data used in this work was acquired from the National Institute of Meteorology (INMET), including average temperature time series and precipitation for the following cities: Aracajú, Belo Horizonte, Manaus, Recife, Salvador, and São Luís (from 1/1/2001 to 12/31/2012) and for Rio de Janeiro (from 1/1/2002 to 12/31/2013). Instead of regression, the authors reduced the dengue forecasting problem to temporal classification. They used RBF and linear kernel SVMs. To evaluate their results, they employed accuracy, a metric consistent with a classification approach. The authors used the model trained with data from earlier years 2002–2012 to forecast dengue outcomes from 2013 to 2017. The state capital of São Luís exhibited the higher accuracy (100% corresponding to 3 correct predictions from a total of 3 test years), followed by Manaus and Salvador (80% accuracy corresponding to 4 correct predictions from a total of 5 test years). For Rio de Janeiro, Aracajú, Belo Horizonte, and Recife, the authors reached accuracies below 70%. The authors obtained an overall accuracy of 74% considering all seven capital cities. Therefore, the proposed method correctly predicted the outcome of 23 out of 31 experiments. Their results indicate that each Brazilian state capital considered has its own climate signatures that correlate with the overall number of human dengue cases. The immediate winter before an epidemic year is a strong factor in epidemic year predictions. However, the authors recognize that their approach to reduce dengue forecasting to classification could be considered somewhat arbitrary, since it is not the canonical way to forecast arboviruses according to the Brazilian Ministry of Health.

4 Conclusion

The potential for damage caused by arboviruses such as dengue, Zika, and chikungunya has already been felt by several populations in seasonal outbreaks that leave marks in the form of suffering, mortality, and morbidities that will still be present when the next outbreak arrives. The possibility of preventing and/or controlling the occurrence of these diseases is concrete and depends largely on the location of breeding sites, real or potential, of the mosquito. This is the aspect that gives priceless value to the models of temporal and spatial prediction of the occurrence of these arboviruses.

The temporal approach allows an understanding of the epidemiological curve of the disease to assist in quick decision-making in the control and prevention of outbreaks of infectious diseases. The spatio-temporal approach provides a better assessment of regions where there is a greater or lesser concentration of cases of a particular disease. Disease distribution maps provide relevant information for public health managers and epidemiologists. Therefore, the development of tools that provide predictions based on the two types of approach allows health managers to analyze the forecast of the quantities the disease cases, as well as their distribution in a given territory. In this way, it can assist in the elaboration of public policies, and preventive and corrective measures in public health. Enabling human and financial resources to be better targeted, especially in regions where they are scarce.

Compartmental models are designed from real problems, and then an abstraction is made, where the parameters to be studied are analyzed. From this, a model based on a system of differential equations is formulated that well describes the interaction of these parameters. Then a quantitative analysis (analytical and numerical study) of this system is made and then the solution to the problem is found. With this type of approach, it is possible to understand the disease and how it is transmitted. And based on this you can predict how the disease will behave over time. And therefore, this information can help to control and thereby help to develop control and decision-making strategies by health managers. However, to use this approach it is necessary to select the parameters used and analyze them. When doing this, the model is limited, and even, assumptions are made to develop the analytical analysis, for example not taking into account the demographic effects of the population as vital dynamics, for example, and this limits the predictive power of mathematical models.

The advancement of Artificial Intelligence and machine learning techniques has enabled the development of applications in several fields of science. In the area of Epidemiology, Artificial Intelligence has enabled the development of tools that provide for rapid monitoring, control and simulation, and dissemination of diseases. However, machine learning based models are black box models, mainly artificial neural networks and support vector machines. That is, these models are intelligible to specialists, because for them the important thing is the correct mapping of the model's inputs and outputs. In addition to the AI-based models, hybrid models, obtained from the combination of sanding machines and statistical methods, can combine the advantages of the two approaches. The hybrid model can also be composed of a combination of a learning machine and a compartmental model. For, despite not obtaining predictions with the same precision as a learning machine, compartmental models can assist in the human understanding of epidemiological aspects as phenomena.

References

1. Abidemi, A., Abd Aziz, M., & Ahmad, R. (2020). Vaccination and vector control effect on dengue virus transmission dynamics: Modelling and simulation. *Chaos, Solitons & Fractals, 133*, 109648.
2. Abreu, P. H., Santos, M. S., Abreu, M. H., Andrade, B., & Silva, D. C. (2016). Predicting breast cancer recurrence using machine learning techniques: a systematic review. *ACM Computing Surveys (CSUR), 49*(3), 52.

3. Ahmad, M., Ibrahim, M., Mohamed, Z., Ismail, N., Abdullah, M., Shueb, R., & Shafei, M. (2018). The sensitivity, specificity and accuracy of warning signs in predicting severe dengue, the severe dengue prevalence and its associated factors. *International Journal of Environmental Research and Public Health, 15*(9), 1–12.

4. Albrieu-Llinás, G., Espinosa, M. O., Quaglia, A., Abril, M., & Scavuzzo, C. M. (2018). Urban environmental clustering to assess the spatial dynamics of Aedes aegypti breeding sites. *Geospatial Health, 13*(1), 135–142.

5. Baquero, O. S., Santana, L. M. R., & Chiaravalloti-Neto, F. (2018). Dengue forecasting in São Paulo city with generalized additive models, artificial neural networks and seasonal autoregressive integrated moving average models. *PLoS One, 13*(4), 1–12. Retrieved from https://doi.org/10.1371/journal.pone.0195065

6. Bates, S., Hutson, H., & Rebaza, J. (2017). Global stability of zika virus dynamics. *Differential Equations and Dynamical Systems, 29*, 657–672.

7. Beketov, M. A., Yurchenko, Y. A., Belevich, O. E., & Liess, M. (2014). What environmental factors are important determinants of structure, species richness, and abundance of mosquito assemblages? *Journal of Medical Entomology, 47*(2), 129–139.

8. Beltrán, J. D., Boscor, A., dos Santos, W. P., Massoni, T., & Kostkova, P. (2018). ZIKA: A New System to Empower Health Workers and Local Communities to Improve Surveillance Protocols by E-learning and to Forecast Zika Virus in Real Time in Brazil. In *Proceedings of the 2018 International Conference on Digital Health* (pp. 90–94).

9. Bhatt, S., Gething, P. W., Brady, O. J., Messina, J. P., Farlow, A. W., Moyes, C. L., ... et al. (2013). The global distribution and burden of dengue. *Nature, 496*(7446), 504.

10. Bhunia, G. S., & Shit, P. K. (2019). *Geospatial analysis of public health*. Springer International Publishing.

11. Bonyah, E., Khan, M. A., Okosun, K., & Islam, S. (2017). A theoretical model for zika virus transmission. *PLoS One, 12*(10), e0185540.

12. Brasier, A. R., Ju, H., Garcia, J., Spratt, H. M., Victor, S. S., Forshey, B. M., ... Kochel, T. J. (2012). A three-component biomarker panel for prediction of dengue hemorrhagic fever. *The American Journal of Tropical Medicine and Hygiene, 86*(2), 341–348.

13. BRASIL, M. d. S. (2012). *Levantamento rápido de índices para Aedes aegypti LIRAa para vigilância entomológica do Aedes aegypti no Brasil: Metodologia para avaliação dos índices de Breteau e predial e tipos de recipientes* (1st ed.; G. Coelho, P. C. Silva, & R. L. Frutuoso, Eds.). Author.

14. Buczak, A. L., Baugher, B., Moniz, L. J., Bagley, T., Babin, S. M., & Guven, E. (2018). Ensemble method for dengue prediction. *PloS One, 13*(1), e0189988.

15. Butt, N., Abbassi, A., Munir, S., Ahmad, S. M., & Sheikh, Q. H. (2008). Haematological and biochemical indicators for the early diagnosis of dengue viral infection. *Journal of College of Physicians and Surgeons Pakistan, 18*(5), 282–285.

16. Ch, S., Sohani, S., Kumar, D., Malik, A., Chahar, B., Nema, A., ... Dhiman, R. (2014). A support vector machine-firefly algorithm based forecasting model to determine malaria transmission. *Neurocomputing, 129*, 279–288.

17. Chakraborty, T., Chattopadhyay, S., & Ghosh, I. (2019). Forecasting dengue epidemics using a hybrid methodology. *Physica A: Statistical Mechanics and its Applications, 527*, 121266.

18. Chan, T.-C., Hu, T.-H., & Hwang, J.-S. (2015). Daily forecast of dengue fever incidents for urban villages in a city. *International Journal of Health Geographics, 14*, 1–11.

19. Chen, Y., Ong, J. H. Y., Rajarethinam, J., Yap, G., Ng, L. C., & Cook, A. R. (2018). Neighbourhood level real-time forecasting of dengue cases in tropical urban Singapore. *BCM Medicine, 16*, 1–13.

20. Choi, H. K. (2018). Stock price correlation coefficient prediction with ARIMA-LSTM hybrid model. Preprint. arXiv:1808.01560.

21. Choudhury, Z. M., Banu, S., & Islam, A. M. (2008). *Forecasting dengue incidence in Dhaka, Bangladesh: A time series analysis*. WHO Regional Office for South-East Asia.

22. Christou, V., Tsipouras, M. G., Giannakeas, N., Tzallas, A. T., & Brown, G. (2019). Hybrid extreme learning machine approach for heterogeneous neural networks. *Neurocomputing, 361*, 137–150.

23. Cortes, F., Martelli, C. M. T., de Alencar Ximenes, R. A., Montarroyos, U. R., Junior, J. B. S., Cruz, O. G., ... de Souza, W. V. (2018). Time series analysis of dengue surveillance data in two Brazilian cities. *Acta Tropica, 182*, 190–197.

24. Deb, K., Agrawal, S., Pratap, A., & Meyarivan, T. (2000). A fast elitist non-dominated sorting genetic algorithm for multi-objective optimization: NSGA-II. In *International Conference on Parallel Problem Solving from Nature* (pp. 849–858).

25. Deb, K., Pratap, A., Agarwal, S., & Meyarivan, T. (2002). A fast and elitist multiobjective genetic algorithm: NSGA-II. *IEEE Transactions on Evolutionary Computation, 6*(2), 182–197.

26. de Kruif, M. D., Setiati, T. E., Mairuhu, A. T. A., Koraka, P., Aberson, H. A., Spek, C. A., ... van Gorp, E. C. M. (2008). Differential gene expression changes in children with severe dengue virus infections. *PLoS Neglected Tropical Diseases, 2*(4), e215.

27. de Lima, S. M., da Silva-Filho, A. G., & Dos Santos, W. P. (2016a). Detection and classification of masses in mammographic images in a multi-kernel approach. *Computer Methods and Programs in Biomedicine, 134*, 11–29.

28. de Lima, T. F. M., Lana, R. M., de Senna Carneiro, T. G., Codeço, C. T., Machado, G. S., Ferreira, L. S., ... Davis Junior, C. A. (2016b). Dengueme: A tool for the modeling and simulation of dengue spatiotemporal dynamics. *International Journal of Environmental Research and Public Health, 13*(9), 920.

29. De Livera, A. M., Hyndman, R. J., & Snyder, R. D. (2011). Forecasting time series with complex seasonal patterns using exponential smoothing. *Journal of the American Statistical Association, 106*(496), 1513–1527.

30. de Santana, M. A., Pereira, J. M. S., da Silva, F. L., de Lima, N. M., de Sousa, F. N., de Arruda, G. M. S., ... dos Santos, W. P. (2018). Breast cancer diagnosis based on mammary thermography and extreme learning machines. *Research on Biomedical Engineering, 34*(1), 45–53.

31. Dodero-Rojas, E., Ferreira, L. G., Leite, V. B., Onuchic, J. N., & Contessoto, V. G. (2020). Modeling chikungunya control strategies and mayaro potential outbreak in the city of rio de janeiro. *PLoS One, 15*(1), e0222900.

32. Dom, N. C., Hassan, A. A., Abd Latif, Z., & Ismail, R. (2013). Generating temporal model using climate variables for the prediction of dengue cases in Subang Jaya, Malaysia. *Asian Pacific Journal of Tropical Disease, 3*(5), 352–361.

33. Duncan, A. P. (2014). *The Analysis and Application of Artificial Neural Networks for Early Warning Systems in Hydrology and the Environment* (Unpublished doctoral dissertation). University of Exeter.

34. Fauci, A. S., & Morens, D. M. (2016). Zika virus in the americas – Yet another arbovirus threat. *New England Journal of Medicine, 374*(7), 601–604. Retrieved from https://doi.org/10.1056/NEJMp1600297 (PMID: 26761185)

35. Funk, S., Kucharski, A. J., Camacho, A., Eggo, R. M., Yakob, L., Murray, L. M., & Edmunds, W. J. (2016). Comparative analysis of dengue and zika outbreaks reveals differences by setting and virus. *PLoS Neglected Tropical Diseases, 10*(12), e0005173.

36. Gharbi, M., Quenel, P., Gustave, J., Cassadou, S., La Ruche, G., Girdary, L., & Marrama, L. (2011). Time series analysis of dengue incidence in Guadeloupe, French West Indies: forecasting models using climate variables as predictors. *BMC Infectious Diseases, 11*(1), 1–13.

37. Gos, M., Krzyszczak, J., Baranowski, P., Murat, M., & Malinowska, I. (2020). Combined TBATS and SVM model of minimum and maximum air temperatures applied to wheat yield prediction at different locations in Europe. *Agricultural and Forest Meteorology, 281*, 107827.

38. Gubler, D. J. (2011). Dengue, urbanization and globalization: the unholy trinity of the 21st century. *Tropical medicine and health, 39*(4SUPPLEMENT), S3–S11.

39. Guo, P., Liu, T., Zhang, Q., Wang, L., Xiao, J., Zhang, Q., ... Ma, W. (2017, 10). Developing a dengue forecast model using machine learning: A case study in China. *PLOS Neglected Tropical Diseases, 11*(10), 1–22. Retrieved from https://doi.org/10.1371/journal.pntd.0005973

40. Guzman, M. G., Halstead, S. B., Artsob, H., Buchy, P., Farrar, J., Gubler, D. J., ...et al. (2010). Dengue: a continuing global threat. *Nature Reviews Microbiology, 8*(12supp), S7.
41. Hamlet, A., Jean, K., Perea, W., Yactayo, S., Biey, J., Van Kerkhove, M., ...Garske, T. (2018). The seasonal influence of climate and environment on yellow fever transmission across Africa. *PLoS Neglected Tropical Diseases, 12*(3), e0006284.
42. Huang, G. B., Wang, D. H., & Lan, Y. (2011). Extreme learning machines: a survey. *International Journal of Machine Learning and Cybernetics, 2*(2), 107–122.
43. Huang, G. B., Zhu, Q. Y., & Siew, C. K. (2006). Extreme learning machine: theory and applications. *Neurocomputing, 70*(1–3), 489–501.
44. Iqbal, N., & Islam, M. (2017). Machine learning for dengue outbreak prediction: An outlook. *International Journal of Advanced Research in Computer Science, 8*(1), 93–102.
45. Jensen, M. T. (2003). Reducing the run-time complexity of multiobjective EAs: The NSGA-II and other algorithms. *IEEE Transactions on Evolutionary Computation, 7*(5), 503–515.
46. Jindal, A., & Rao, S. (2017). Agent-based modeling and simulation of mosquito-borne disease transmission. In *Proceedings of the 16th Conference on Autonomous Agents and Multiagent Systems* (pp. 426–435).
47. Jing, Q. L., Cheng, Q., Marshall, J. M., Hu, W. B., Yang, Z. C., & Lu, J. H. (2018). Imported cases and minimum temperature drive dengue transmission in Guangzhou, China: Evidence from arimax model. *Epidemiology and Infection, 146*, 1226–1235.
48. Joyce, R. J., Janowiak, J. E., Arkin, P. A., & Xie, P. (2004). Cmorph: A method that produces global precipitation estimates from passive microwave and infrared data at high spatial and temporal resolution. *Journal of Hydrometeorology, 5*(3), 487–503.
49. Kamal, M., Kenawy, M. A., Rady, M. H., Khaled, A. S., & Samy, A. M. (2019, 12). Mapping the global potential distributions of two arboviral vectors aedes aegypti and ae. albopictus under changing climate. *PLOS ONE, 13*(12), 1–21. Retrieved from https://doi.org/10.1371/journal.pone.0210122
50. Kao, Y.-H., & Eisenberg, M. C. (2018). Practical unidentifiability of a simple vector-borne disease model: Implications for parameter estimation and intervention assessment. *Epidemics, 25*, 89–100.
51. Keshtegar, B., Heddam, S., & Hosseinabadi, H. (2019). The employment of polynomial chaos expansion approach for modeling dissolved oxygen concentration in river. *Environmental Earth Sciences, 78*(1), 34.
52. Kesorn, K., Ongruk, P., Chompoosri, J., Phumee, A., Thavara, U., Tawatsin, A., & Siriyasatien, P. (2015). Morbidity rate prediction of dengue hemorrhagic fever (DHF) using the support vector machine and the Aedes aegypti infection rate in similar climates and geographical areas. *PloS One, 10*(5), e0125049.
53. Khashei, M., & Bijari, M. (2011). A novel hybridization of artificial neural networks and arima models for time series forecasting. *Applied Soft Computing, 11*(2), 2664–2675.
54. Kraemer, M. U., Sinka, M. E., Duda, K. A., Mylne, A. Q., Shearer, F. M., Barker, C. M., ...et al. (2015). The global distribution of the arbovirus vectors Aedes aegypti and Ae. albopictus. *eLife, 4*, e08347.
55. Kumar, N., Abdullah, M., Faizan, M. I., Ahmed, A., Alsenaidy, H. A., Dohare, R., & Parveen, S. (2017). Progression dynamics of zika fever outbreak in el salvador during 2015–2016: a mathematical modeling approach. *Future Virology, 12*(5), 271–281.
56. Laureano-Rosario, A. E., Duncan, A. P., Mendez-Lazaro, P. A., Garcia-Rejon, J. E., Gomez-Carro, S., Farfan-Ale, J., ...Muller-Karger, F. E. (2018a). Application of Artificial Neural Networks for Dengue Fever Outbreak Predictions in the Northwest Coast of Yucatan, Mexico and San Juan, Puerto Rico. *Tropical Medicine and Infectious Disease, 3*(1), 5.
57. Laureano-Rosario, A. E., Duncan, A. P., Symonds, E. M., Savic, D. A., & Muller-Karger, F. E. (2018b). Predicting culturable enterococci exceedances at Escambron Beach, San Juan, Puerto Rico using satellite remote sensing and artificial neural networks. *Journal of Water and Health, 17*(1), 137–148.
58. Lee, K. Y., Chung, N., & Hwang, S. (2016). Application of an artificial neural network (ANN) model for predicting mosquito abundances in urban areas. *Ecological Informatics, 36*, 172–180.

59. Lendek, Z., Guerra, T. M., Babuška, R., & Schutter, B. (2011). *Stability analysis and nonlinear observer design using takagi-sugeno fuzzy models*. Springer.
60. Lima, M. V. M. d., & Laporta, G. Z. (2020). Evaluation of the models for forecasting dengue in Brazil from 2000 to 2017: An ecological time-series study. *Insects, 11*(11), 794.
61. Lu, L., Lin, H., Tian, L., Yang, W., Sun, J., & Liu, Q. (2009). Time series analysis of dengue fever and weather in Guangzhou, China. *BMC Public Health, 9*(1), 395.
62. Manogaran, G., Lopez, D., Thota, C., Abbas, K. M., Pyne, S., & Sundarasekar, R. (2017). Big data analytics in healthcare Internet of Things. In *Innovative healthcare systems for the 21st century* (pp. 263–284). Springer.
63. Manogaran, G., Varatharajan, R., Lopez, D., Kumar, P. M., Sundarasekar, R., & Thota, C. (2018). A new architecture of Internet of Things and big data ecosystem for secured smart healthcare monitoring and alerting system. *Future Generation Computer Systems, 82*, 375–387.
64. Marques-Toledo, C. d. A., Degener, C. M., Vinhal, L., Coelho, G., Meira, W., Codeço, C. T., & Teixeira, M. M. (2017, 07). Dengue prediction by the web: Tweets are a useful tool for estimating and forecasting dengue at country and city level. *PLOS Neglected Tropical Diseases, 11*(7), 1–20. Retrieved from https://doi.org/10.1371/journal.pntd.0005729
65. Masri, S., Jia, J., Li, C., Zhou, G., Lee, M.-C., Yan, G., & Wu, J. (2019). Use of twitter data to improve zika virus surveillance in the united states during the 2016 epidemic. *BCM Public Health, 19*, 1–14.
66. Mohammed, A., & Chadee, D. D. (2011). Effects of different temperature regimens on the development of aedes aegypti (l.)(diptera: Culicidae) mosquitoes. *Acta Tropica, 119*(1), 38–43.
67. Monaghan, A. J., Schmidt, C. A., Hayden, M. H., Smith, K. A., Reiskind, M. H., Cabell, R., & Ernst, K. C. (2019). A simple model to predict the potential abundance of aedes aegypti mosquitoes one month. *American Journal of Tropical Medicine and Hygene, 100*, 434–437.
68. Morsy, S., Dang, T., Kamel, M., Zayan, A., Makram, O., Elhady, M., ... Huy, N. (2018). Prediction of zika-confirmed cases in Brazil and colombia using Google trends. *Epidemiology and Infection, 146*(13), 1625–1627.
69. Muñoz, Á. G., Thomson, M. C., Stewart-Ibarra, A. M., Vecchi, G. A., Chourio, X., Nájera, P., ... Yang, X. (2017). Could the recent zika epidemic have been predicted? *Frontiers in Microbiology, 8*, 1291.
70. Musa, S. S., Zhao, S., Chan, H.-S., Jin, Z., He, D., et al. (2019). A mathematical model to study the 2014–2015 large-scale dengue epidemics in Kaohsiung and Tainan cities in Taiwan, China. *Mathematical Biosciences and Engineering, 16*(5), 3841–3863.
71. Musso, D., & Gubler, D. J. (2016). Zika virus. *Clinical Microbiology Rewies, 29*, 487–524.
72. Musso, D., Stramer, S. L., & Busch, M. P. (2016). Zika virus: A new challenge for blood transfusion. *The Lancet, 387*, 1993–1994.
73. Naim, I., Mahara, T., & Idrisi, A. R. (2018). Effective short-term forecasting for daily time series with complex seasonal patterns. *Procedia Computer Science, 132*, 1832–1841.
74. Nasirudeen, A., Wong, H. H., Thien, P., Xu, S., Lam, K.-P., & Liu, D. X. (2011). RIG-I, MDA5 and TLR3 synergistically play an important role in restriction of dengue virus infection. *PLoS Neglected Tropical Diseases, 5*(1), e926.
75. Ndaïrou, F., Area, I., Nieto, J. J., Silva, C. J., & Torres, D. F. (2018). Mathematical modeling of zika disease in pregnant women and newborns with microcephaly in Brazil. *Mathematical Methods in the Applied Sciences, 41*(18), 8929–8941.
76. Nur Aida, H., Abu Hassan, A., Anita, T., Nurita, A. T., Dieng, H., Suhaila, A. H., ... Farida, A. (2017). Developing time-based model for the prediction of breeding activities of dengue vectors using early life cycle variables and epidemiological information in northern malaysia. *Tropical Biomedicine, 34*, 691–707.
77. Olawoyin, O., & Kribs, C. (2018). Effects of multiple transmission pathways on zika dynamics. *Infectious Disease Modelling, 3*, 331–344.
78. Padmanabhan, P., Seshaiyer, P., & Castillo-Chavez, C. (2017). Mathematical modeling, analysis and simulation of the spread of zika with influence of sexual transmission and preventive measures. *Letters in Biomathematics, 4*(1), 148–166.

79. PAHO. (2019). Vector-borne diseases [Computer software manual]. Retrieved from https://www.paho.org/bra/index.php?option=com_content&view=article&id=5796:doencas-transmissiveis-analise-de-situacao-de-saude&Itemid=0. Last accessed: 06 Apr 2021.

80. Pai, P.-F., & Lin, C.-S. (2005). A hybrid ARIMA and support vector machines model in stock price forecasting. *Omega, 33*(6), 497–505.

81. Rahman, M., Bekele-Maxwell, K., Cates, L. L., Banks, H., & Vaidya, N. K. (2019). Modeling zika virus transmission dynamics: Parameter estimates, disease characteristics, and prevention. *Scientific Reports, 9*(1), 1–13.

82. Rey, F. A. (2003). Dengue virus envelope glycoprotein structure: new insight into its interactions during viral entry. *Proceedings of the National Academy of Sciences, 100*(12), 6899–6901.

83. Rissino, S., & Lambert-Torres, G. (2009). Rough set theory—fundamental concepts, principals, data extraction, and applications. In *Data mining and knowledge discovery in real life applications*. InTech.

84. Robert, M. A., Christofferson, R. C., Weber, P. D., & Wearing, H. J. (2019). Temperature impacts on dengue emergence in the united states: investigating the role of seasonality and climate change. *Epidemics, 28*, 100344.

85. Sakkas, H., Bozidis, P., Giannakopoulos, X., Sofikitis, N., & Papadopoulou, C. (2018). An update on sexual transmission of zika virus. *Pathogens, 7*(3). Retrieved from https://www.mdpi.com/2076-0817/7/3/66

86. Sang, S., Gu, S., Bi, P., Yang, W., Yang, Z., Xu, L., ...Liu, Q. (2015, 05). Predicting unprecedented dengue outbreak using imported cases and climatic factors in Guangzhou, 2014. *PLOS Neglected Tropical Diseases, 9*(5), 1–12.

87. Scavuzzo, J. M., Trucco, F., Espinosa, M., Tauro, C. B., Abril, M., Scavuzzo, C. M., & Frery, A. C. (2018). Modeling Dengue vector population using remotely sensed data and machine learning. *Acta Tropica, 185*, 167–175.

88. Scavuzzo, J. M., Trucco, F. C., Tauro, C. B., German, A., Espinosa, M., & Abril, M. (2017). Modeling the temporal pattern of Dengue, Chicungunya and Zika vector using satellite data and neural networks. In *Information processing and control (RPIC), 2017 xvii workshop on* (pp. 1–6).

89. Shaukat, K., Masood, N., Mehreen, S., & Azmeen, U. (2015). Dengue fever prediction: A data mining problem. *Journal of Data Mining in Genomics & Proteomics, 2015*, 1–5.

90. Shutt, D. P., Manore, C. A., Pankavich, S., Porter, A. T., & Del Valle, S. Y. (2017). Estimating the reproductive number, total outbreak size, and reporting rates for zika epidemics in South and Central America. *Epidemics, 21*, 63–79.

91. Siriyasatien, P., Chadsuthi, S., Jampachaisri, K., & Kesorn, K. (2018). Dengue epidemics prediction: A survey of the state-of-the-art based on data science processes. *IEEE Access, 6*, 53757–53795.

92. Stolerman, L. M., Maia, P. D., & Kutz, J. N. (2019). Forecasting dengue fever in Brazil: An assessment of climate conditions. *PLoS One, 14*(8), e0220106.

93. Stone, L., Olinky, R., & Huppert, A. (2007). Seasonal dynamics of recurrent epidemics. *Nature, 446*(7135), 533–536.

94. Subramanian, R., Romeo-Aznar, V., Ionides, E., Codeço, C. T., & Pascual, M. (2020). Predicting re-emergence times of dengue epidemics at low reproductive numbers: Denv1 in Rio de Janeiro, 1986–1990. *Journal of the Royal Society Interface, 17*(167), 20200273.

95. Suparit, P., Wiratsudakul, A., & Modchang, C. (2018). A mathematical model for zika virus transmission dynamics with a time-dependent mosquito biting rate. *Theoretical Biology and Medical Modelling, 15*(1), 1–11.

96. Tang, B., Xiao, Y., & Wu, J. (2016). Implication of vaccination against dengue for zika outbreak. *Scientific Reports, 6*(1), 1–14.

97. Tang, J., Deng, C., & Huang, G.-B. (2015). Extreme learning machine for multilayer perceptron. *IEEE Transactions on Neural Networks and Learning systems, 27*(4), 809–821.

98. Tanner, L., Schreiber, M., Low, J. G., Ong, A., Tolfvenstam, T., Lai, Y. L., ...et al. (2008). Decision tree algorithms predict the diagnosis and outcome of dengue fever in the early phase of illness. *PLoS Neglected Tropical Diseases, 2*(3), e196.

99. Usman, S., Adamu, I. I., & Babando, H. A. (2017). Mathematical model for the transmission dynamics of Zika virus infection with combined vaccination and treatment interventions. *Journal of Applied Mathematics and Physics, 5*(10), 1964.

100. Wang, L., & Ranjan, R. (2015). Processing distributed Internet of Things data in clouds. *IEEE Cloud Computing, 2*(1), 76–80.

101. WHO. (2020). Vector-borne diseases [Computer software manual]. Retrieved from https://www.who.int/news-room/fact-sheets/detail/vector-borne-diseases. Last accessed: 06 Apr 2021.

102. WHO. (2021). Ending the neglect to attain the Sustainable Development Goals: a road map for neglected tropical diseases 2021–2030 [Computer software manual]. Retrieved from https://www.who.int/neglected_diseases/resources/who-ucn-ntd-2020.01/en/. Last accessed: 06 Apr 2021.

103. Wongkoon, S., Jaroensutasinee, M., & Jaroensutasinee, K. (2012). Development of temporal modeling for prediction of dengue infection in Northeastern Thailand. *Asian Pacific Journal of Tropical Medicine, 5*(3), 249–252.

104. Yamana, T. K., & Shaman, J. (2020). A framework for evaluating the effects of observational type and quality on vector-borne disease forecast. *Epidemics, 30*, 100359.

105. Yaseen, Z. M., Sulaiman, S. O., Deo, R. C., & Chau, K.-W. (2019). An enhanced extreme learning machine model for river flow forecasting: State-of-the-art, practical applications in water resource engineering area and future research direction. *Journal of Hydrology, 569*, 387–408.

106. Zhang, D., Peng, X., Pan, K., & Liu, Y. (2019). A novel wind speed forecasting based on hybrid decomposition and online sequential outlier robust extreme learning machine. *Energy Conversion and Management, 180*, 338–357.

107. Zhang, G., Patuwo, B. E., & Hu, M. Y. (1998). Forecasting with artificial neural networks:: The state of the art. *International Journal of Forecasting, 14*(1), 35–62.

108. Zhang, G. P. (2003). Time series forecasting using a hybrid ARIMA and neural network model. *Neurocomputing, 50*, 159–175.

109. Zhang, Y., Wang, T., Liu, K., Xia, Y., Lu, Y., Jing, Q., ...Lu, J. (2016). Developing a time series predictive model for dengue in Zhongshan, China based on weather and Guangzhou dengue surveillance data. *PLOS Neglected Tropical Diseases, 10*(2), 1–17.

110. Zhao, N., Charland, K., Carabali, M., Nsoesie, E. O., Maheu-Giroux, M., Rees, E., ...Zinszer, K. (2020). Machine learning and dengue forecasting: Comparing random forests and artificial neural networks for predicting dengue burden at national and sub-national scales in colombia. *PLoS Neglected Tropical Diseases, 14*(9), e0008056.

111. Zhu, B., Wang, L., Wang, H., Cao, Z., Zha, L., Li, Z., ...Sun, Y. (2019). Prediction model for dengue fever based on interactive effects between multiple meteorological factors in Guangdong, China (2008–2016). *PLoS One, 14*, 1–12.

112. Zhu, G., Liu, T., Xiao, J., Zhang, B., Song, T., Zhang, Y., ...et al. (2019). Effects of human mobility, temperature and mosquito control on the spatiotemporal transmission of dengue. *Science of the Total Environment, 651*, 969–978.

Machine Learning Approaches for Temporal and Spatio-Temporal Covid-19 Forecasting: A Brief Review and a Contribution

Ana Clara Gomes da Silva, Clarisse Lins de Lima, Cecilia Cordeiro da Silva, Giselle Machado Magalhães Moreno, Eduardo Luiz Silva, Gabriel Souza Marques, Lucas Job Brito de Araújo, Luiz Antônio Albuquerque Júnior, Samuel Barbosa Jatobá de Souza, Maíra Araújo de Santana, Juliana Carneiro Gomes, Valter Augusto de Freitas Barbosa, Anwar Musah, Patty Kostkova, Abel Guilhermino da Silva Filho, and Wellington P. dos Santos ⒾⒹ

1 Introduction

In the last month of 2019, in the city of Wuhan, China, a local outbreak of an illness was found that was cursing with symptoms such as cough, fever, sore throat, shortness of breath, fatigue, pneumonia evolving into severe acute respiratory syndrome, possibly being fatal. It was soon discovered that the disease was caused by a new coronavirus, SARS-CoV2 (severe acute respiratory syndrome coronavirus), and that the contagiousness and course of the disease would make

A. C. G. da Silva · W. P. dos Santos (✉)
Department of Biomedical Engineering, Federal University of Pernambuco, Recife, Pernambuco, Brazil
e-mail: wellington.santos@ufpe.br

V. A. de Freitas Barbosa
Academic Unit of Serra Talhada, Rural Federal University of Pernambuco, Serra Talhada, Brazil

C. L. de Lima · M. A. de Santana · J. C. Gomes
Graduate Program in Computer Engineering, Polytechnic School of the University of Pernambuco, Recife, Brazil

C. C. da Silva · E. L. Silva · G. S. Marques · L. J. B. de Araújo · L. A. A. Júnior
S. B. J. de Souza · A. G. da Silva Filho
Center for Informatics, Federal University of Pernambuco, CIn-UFPE, Recife, Brazil

G. M. M. Moreno · P. Kostkova
Department of Atmospheric Sciences, IAG, University of São Paulo, USP, São Paulo, Brazil

A. Musah
Institute for Risk and Disaster Reduction, University College London, UCL-IRDR, London, UK

© The Author(s), under exclusive license to Springer Nature Switzerland AG 2022
S. K. Pani et al. (eds.), *Assessing COVID-19 and Other Pandemics and Epidemics using Computational Modelling and Data Analysis*,
https://doi.org/10.1007/978-3-030-79753-9_18

it a threat to the world's different health systems. The disease, named COVID-19, quickly spread across four continents, until on March 11, 2020 it was declared a pandemic by the World Health Organization [20]. In little more than a year after the pandemic was recognized, covid-19 has now reached more than 130,000,000 people, adding up to more than 2,800,000 fatalities [98]. These numbers continue to grow. The biggest public health crisis in decades had begun. If on one hand COVID-19 exposed the epidemiological fragility of an extremely connected world, using tourist and commercial routes to reach the most diverse populations; on the other hand, the global connection proved its value once again, with scientists from various fields of knowledge and from the most varied countries establishing collaborations in the urgent effort to know, detail, prevent, detect, and contain the virus. Thus, a large amount of research has begun on epidemiological and pathophysiological aspects, drug development, virus detection tests, vaccines, and case prediction and control. Science has advanced by leaps and bounds and made the distance between the emergence of a new disease, the identification of the causative agent, the sequencing of its genetic material, and the appearance of the first viable vaccines seem shorter in just 1 year. Despite these achievements, one year after the WHO recognized the pandemic, the disease continues to spread, presenting an exuberance of possible clinical manifestations. The virus has new and even more transmissible variants (REF). The most viable form of control since the beginning of the pandemic continues to be: case identification, tracking and contact isolation. The ability to identify the presence of the pathogen plays an important role both in preventing the spread of the disease and in adequately combating it. Delay in diagnosis can delay proper patient care, hindering recovery, and especially allowing undiagnosed infected people to circulate in society, spreading the virus. The most well accepted test for diagnosing COVID-19 is RT-PCR (Reverse Transcription Polymerase Chain Reaction); however, the procedures for this test take several hours [24] and the result can take days to be available. In addition, there is the possibility of virus presence and transmission even if the RT-PCR test is negative, depending on the time of contamination at which the test was performed. Understanding more about the behavior of the virus in populations (identifying risk groups, more vulnerable social groups) or about its spatial and temporal spread in a region was, since the beginning of the pandemic, a factor that reduced the impact of the virus. And this has allowed greater assertiveness in the measures of isolation, protection, and vaccination, besides being determinant for economic, social, and administrative decisions of governments that have the intention to contain the pandemic by COVID-19. In this context, a relatively new area of Public Health, digital epidemiology, has gained space and recognition, providing effective monitoring of confirmed cases, accumulations, and excess deaths. Moreover, the possibility of using machine learning to make temporal and spatial predictions about the occurrence of COVID-19 has definitely brought artificial intelligence into the healthcare field. This chapter is dedicated to exploring some of the major studies that have been done on the use of forecasting by compartmental, statistical, machine learning, and hybrid approaches.

This chapter is organized as follows: in Sect. 2, we present the theoretical basis and a review of compartment forecasting models; in Sect. 3, we detail the forecasting approaches based on statistical learning and present the basis of the main machine learning methods applied to Covid-19 forecast, as well as state-of-the-art works selected taking into account academic relevance, i.e. the number of citations and the impact factor of journals and books; finally, in Sect. 4 we present our final considerations and general conclusions.

2 Forecasting by Statistical Learning and Compartment Models

With the outbreak of the 2019 coronavirus disease many researchers have become interested in mathematically modeling this new disease. Many have done these studies using compartmental models based on differential equations. These models can be described by two types of equations, ordinary differential equations (ODEs) and partial differential equations (PDEs). The techniques for solving each of these models and methods for doing numerical simulations are different.

The following are some studies that have used mathematical modeling to understand how disease dynamics work and even make predictions using computational techniques associated with these models.

Among the ODE-based compartmental models the researchers Sarkar et al. [85] developed a 6-compartment model that extends the classical SEIR to predict Covid-19 dynamics, where a sensitivity analysis was conducted to recognize the most influential parameters with respect to the infected population. For this purpose, the partial rank correlation coefficient (PRCC) technique was used for all input parameters with respect to variable I(infected or symptomatic individuals). And the numerical implementation was done in the FORTRAN program with the method of least squares (MMQ) to adjust the diary cases of the disease.

The researchers Suba et al. [89] developed a model based on ODEs and also used implementation by means of the method of least squares. In this work, seven models were developed, and to find the parameters of the model, excel spreadsheet and MMQ and plotted graphs in MATLAB were used. This study did the sensitivity analysis using real data from Tamil Nadu. Good results with simple methods, but the system is sensitive to the change of the basic reproduction number R_0, which changes the whole system automatically.

Some other studies follow the line of numerical implementation with MATLAB. This same software was used in the study developed by Zhong et al. [105], to perform the numerical calculation of the created differential equation system. And real data was used to predict the number of infected. This study brought predictions of the epidemic in different scenarios and with different levels of anti-epidemic measure and medical care represented by beta rate and gamma rate, with unreliable data through objective analysis. But this study has a prediction limited by the data and their reliability, because data before January 18, 2020 should be used with caution. Mandal et al. [59] also used MATLAB software to solve the system of differential equations that describes the proposed SEIQR model. The method used was the fourth-order Runge–Kutta (RK4). In this study, a theoretical analysis and numerical simulation are performed, as well as a stability analysis and estimation of R_0. The prediction made is sensitive to some parametric conditions, and since human behavior is uncertain there are changes in the parametric space corroborating to the change in the graphs of the COVID-19 cases. Therefore, the prediction made is short term. MATLAB was also used by Jiang [37]. Initially this work used the simulation repository built into the Netlogo software to create a SIR model to simulate virus transmission. The simulation took place in a closed environment (Small World) and assumed that there were no vital dynamics, i.e., no one died or was born naturally. To optimize the parameters of the proposed model, the MATLAB function fmincon was used. To find the numerical solution of the ODE system and adjust the curves, MATLAB's ode45 function was used. By using this function the values obtained were quite consistent with the real data as well as the simulation curves. This model was done for USA and for Hubei, China. For USA a model without vital dynamics was used, due to lack of data. The parameters definitely change with time in the real situation. The data from asymptomatic individuals is late, which makes it difficult to establish a SEIR-based model for fitting and prediction. As for Hubei, the prediction does not match the real situation. And finally, none of the models divides infected people into isolated and non-isolated infected individuals, or whether they received effective treatment. Massonis et al. [60] did a multi-state review using SIR and SEIR models described by systems of ODEs in which it evaluates structural identifiability, i.e., ability to provide insights into their unknown parameters, and observability (unmeasured states). A total of 255 articles were evaluated, 98 with SIR models and 157 with SEIR models. And a list of 36 model structures was made. The ability to provide reliable information was evaluated, and theoretical concepts of structural identifiability and reliability control were used for this. STRIKE-GOLDD, an open source toolbox and GenSSI2 MATLAB were used as analysis tools, and for some models the Observability Test code in Maple, Identifiability Analysis in Mathematica, SIAN in Maple, and others were used. Most models found in the literature have identifiable parameters. Often allowing for variability in an unknown parameter improves the observability and/or the identifiability of the model. This work has contributed to providing a detailed analysis of the structural identifiability and observability of a large set of compartmentalized COVID-19 models presented in the recent literature. To model and make prediction of COVID-19 evolution in Brazil, Bastos and Cajueiro [11] proposed two models SIRD and SIRASD described by ODEs. And to find the numerical solution of the ODE system and fit the curves, ode45 function, also from MATLAB, was used. And although this method controls the error by assuming fourth-order precision, it uses a precise fifth order formula to perform the steps. As a starting condition we used data from the Brazilian Institute of Geography and Statistics (IBGE). And the data used were from the Brazilian Ministry of Health (February 25 to March 30, 2020). For the estimation procedure, we minimized the loss functions using the method "$optimize.least_squares$" also from the scipy Python 34 library using the Cauchy loss with scaling parameter. It is notable that although the SIRASD model predicts that the number of infected is higher than the SIRD model estimates, it also manages to predict a lower peak for those infected with symptoms, which are those who require medical attention. This model is advantageous for short-term prediction for Brazil. The methodology of this study was able to estimate the asymptomatic individuals, who may not be entirely present in the data. But because the study was done at the beginning of the pandemic, there was little data and there were cases of underreporting of the actual number of infected people. In addition, this study did not take demographic effects into account, and it was assumed that there was no reinfection. The SIRASD model proved sensitive to the initial condition of asymptomatic individuals. Because the number of tests is small to map the entire population, it is necessary to work with assumptions.

A modeling of the spread of coronavirus taking into account the cases of undetected infections in China was done by Ivorra et al. [34]. In this paper, a deterministic SEIHRD model was made, which has low computational complexity and possibility of using ODE theory to analyze and interpret properly. This model is solved numerically via fourth-order Runge–Kutta (RK4) with 4 h time interval to approximate the solution of the system. Both Runge–Kutta and the WASF-GA algorithm have been implemented in Java. It is advantageous to use a deterministic model when you have little data, but the methodology used aimed precisely at solving this limitation. And a robust approach for overfitting the model parameters with respect to the reported data was created. However, the results are unsatisfactory because the estimation was done at the early stage of the epidemic.

Ambikapathy and Krishnamurthy [6] developed and validated a mathematical model to assess the impact of various scenarios on COVID-19 transmission in India. A compartmentalized ODE model incorporating the actual cases from 14 countries, China, Italy, Germany, France, USA, UK, Sweden, Netherlands, Austria, Canada, Australia, Malaysia, Singapore, and India, was proposed. The model was applied to predict transmission in India and the highest exposure situations, such as transit stations and shopping malls, were evaluated. It was validated using the infections reported in the adopted period and was used to predict future infected cases in the above countries, considering a 65-day period (IndiaSim implementation). Different intervention strategies were used with blocking periods of 4, 14, 21, 42, and 60 days. The model developed can capture the infection dynamics in each country to a considerable extent and predict future cases. The use of an ODE system to describe the models is advantageous because it is possible to apply controls to the model and find results. Nevertheless, the model suffers from numerical errors because at the beginning of the disease the S-compartment has a high value and the I and R-compartments have very low values. In addition, the model proposed in India assumes no spread of the disease in the community until the first week of March 2020, and the dynamic prediction interval is limited (110 days). The model will need to be updated. Also in order to do a predictive analysis of COVID-19 in China, Italy, and France, developed a SIRD model to predict the position of the epidemic peak of the disease. For stochastic evolution, the Python-Scipy package was used. For Italy, the prediction with nonlinear fit strategy for the endemic peak is robust.

With simulations it was shown that the recovery rate is the same for China as for Italy, but the infection and mortality rates seem to be different. This model showed that cultural factors influence the infection rate, varying from one country to another. The model has the limitation of data sensitivity, so it changes from one country to another. And when making numerical solution adjustments, it was found that the data reported for the outbreak in France is still too preliminary to justify a significant adjustment of this kind. The researchers [48] introduced a SIRD model described by a system of ODEs to analyze the behavior of COVID-19 disease in the USA, Germany, UK, and Russia and solved using numerical methods, and the data agree well with the model. The model predicts the peak of the epidemic in each country and compares the results obtained. Germany's prediction was the optimistic one. The authors Khajanchi et al. [42] proposed another paper in which two mathematical models were developed to describe the dynamics of the virus in China described by systems of ODEs and curves constructed for the number of infected, recovered, and dead. The optimal values of the model parameters, which accurately describe the statistical data, were found. World Health Organization (WHO) data was used to obtain the model parameters, obtaining good agreement between the statistical data and the model curves. Thus, it is shown that there was a broad fit of the proposed mathematical model. This indicates a high adequacy of the mathematical model for coronavirus infection. Hamzah et al. [31] have developed a framework to manage and track COVID-19 data called CoronaTracker. This framework is based on a SEIR predictive compartmental model to predict the outbreak of COVID-19 inside and outside China based on daily observations, analyzing the influence of news of people's behavior both positively and economically. John Hopkins University (UJH), World Health Organization (WHO), and Ding Xiang Yuan databases were used as data sources. The data collected in CoronaTracker is available on the data lakes platform. For numerical simulation, the Scipy implementation was used, and for numerical integration, odeint was used. The study showed that the spread of the outbreak is influenced by the social policy of each country. The developed platform has an easy interface in which citizens can register their feelings and express their opinions about news articles. CoronaTracker can assist the government and authorities to disseminate articles, provide updates on the situation, and advocate good personal hygiene. This study has the limitation that when using data from John Hopkins University (UJH), an initial number of exposed individuals was missing. A decision-making system for COVID-19 (CDMS) was created by Varotsos and Krapivin [93] for USA, Brazil, Russia, and Greece. For the model with deterministic components, a compartmental SPRD model was created, similar to the classic SIRD in the literature, described by ODEs and with parameters determined by the reported data. And for the model with stochastic compartments, the classes were represented with a stability indicator that characterizes the COVID-19 propagation trend. Numerical evaluations were done by the SARD block using stochastic reports on the state of disease effects. This study showed that temperature and humidity slowly affect the effects of the pandemic. The analysis of the spread of the disease and the loss of income due to the pandemic has different impacts for each country. The analysis of official data from Russia and Greece showed the results of the pandemic. The risk of infection and mortality increases with increasing population density. What limits this study is that there is not enough data to make the study reliable. In practice, it was impossible to coordinate measures to contain the COVID-19 pandemic under conditions of high uncertainty. In the study of Sadun [81] a compartmentalized SEIR model was developed. In this study strategies are developed to try to estimate the reproduction number R_0, and come to the conclusion that there is no direct way to measure it. The estimated value of R_0 depends on the length of the latency period for three versions of the classical SEIR model. The estimates of the reproduction number that have been published should be viewed with skepticism, one needs to understand the latency of COVID-19. However, there is no direct way to measure R_0, so what one can do is measure the time scale of the exponential growth of the pandemic and try to estimate R from it. The SEIPAHRF model was created by Ndaïrou et al. [67] to understand the transmission dynamics of COVID-19 in Wuhan. This model introduced a modification of the classical SEIR model by introducing asymptomatic (A), hospitalized (H), and fatality (F) infectious class. To study the basic reproduction number, a generation matrix was used in a sensitivity analysis. The local stability of the model was also studied. In this study, the theoretical findings and numerical results fit well with the actual results and reflect reality in Wuhan, China. This model can be used to study the reality in other countries whose outbreaks are increasing. However, the limited data at the beginning of the study, since it was early in the disease, was limited. Also to model and predict the dynamics of the COVID-19 pandemic in India, Sarkar et al. [85] created a 6-compartment model that extends the standard SEIR. And it divides the coronavirus-infected population from the susceptible individuals before the progression of clinical symptoms. It was also proven that quarantine decreases contact between uninfected and infected, and thus there is a reduction in the contact rate and can effectively reduce R_0. A sensitivity analysis was performed to recognize the most influential parameters with respect to the clinically infected population. And this sensitivity analysis was done by evaluating the technique of partial rank correlation coefficients (PRCC) for all input parameters in relation to variable I. The indices were evaluated at six time points: 30, 45, 60, 75, 90, and 100 days before steady state. The model variable was selected for sensitivity analysis I (infected or symptomatic individuals), generating six more influential parameters out of nine. And the actual daily COVID-19 data are fitted using least squares method (MMQ), which locally minimizes the sum of squares of errors. The numerical implementation was done in FORTRAN program. This model provides an important tool for assessing the consequences of possible policies, incorporating social distancing and blocking. Unfortunately, because of the short time scale, demographic effects are not considered. The Abou-Ismail [1] researchers focused on explaining and mathematically simplifying three models: SIR, SEIR, and SUQC (susceptible, unquarantined, quarantined, and confirmed). The goal was to understand the nature of the pandemic and to measure the impacts of social distancing through mathematical models. Making use of a system analysis of ODEs that describe the disease.

To analyze mathematically and do a numerical study the authors Viguerie et al. [95] created a new framework for understanding compartmental models by means of equilibrium equations similar to those found in Continuum Mechanics (Lotka–Volterra type). The model is SEIRD and made use of differential equation models to derive and analyze R_0. For models based on ODEs it has the concept of the basic reproduction number R_0 well defined, but the extension to a model based on PDEs is not clear due to the influence of diffusion. Therefore, in this work the EDO version of the EDP model was derived and its efficiency was evaluated with numerical tests. For the numerical tests either implicit second-order Backward Euler (BDF2) or implicit first-order Backward Euler was used. Picard linearization was performed at each time step. And the iterative Generalized Minimum Residual method (GMRES) with Jacobi preconditioning was used to solve all linear systems. PDE models are advantageous in that they allow a continuous space description of the relevant dynamics, allowing the dynamics to be described in time and space at all scales. Since models described by EDOs are limited for describing spatial information, implicit models are effective in describing the temporal dynamics of the system. In this model deaths other than by COVID-19 and births are not considered. The study developed by the researchers Khoshnaw et al. [47] used MATLAB's System Biology Tool (SBedit) package to compute the class dynamics of the model, and thus obtained a better understanding and identification of the key critical model parameters. And thus it was possible to understand the impacts of transmission rate and contact for New York. However, having several different models implies that one needs to create or identify the critical elements of each of the models. Furthermore, the model cannot simply be extrapolated to conditions in another country. Its parameters must be estimated from the new conditions. Making use of the same MATLAB package to obtain numerical solutions and calculate local sensitivity, Khoshnaw et al. [46] developed a model. Sensitivity analysis was done with the dynamics of the biological system modeled with law of mass action. This study also concluded that the most effective factors for the spread of coronavirus are: (1) the rate of person-to-person transmission, (2) the rate of quarantined exposure, and (3) the rate of transition from exposed individuals to individuals infected. MATLAB was used to numerically solve the compartmental model described by nonlinear differential equations proposed by Ahmed et al. [3]. And for the logistic model, the fitVirus function was used. The union of mathematical models and computer simulations is an effective tool that provides us with more understanding and good numerical predictions of the model states. However, in this study it is noticed that the number of people exposed to quarantine becomes stable after 40 days but the number of recovered people increases rapidly and becomes stable slowly. Having many approaches to identifying the estimates and understanding the disease makes the issue murky. However, this study has brought the identification of critical parameters of the model, helping to understand the overall issue more effectively and broadly. Using a code in MATLAB, Shao et al. [87] performed the numerical simulations. In this study, two time-delayed dynamic models were used to track COVID-19. The time-delayed dynamic coronavirus pneumonia model (TDD-NCP) introduced the delay process into the differential equations to describe the latent period of the epidemic and can be used to predict the trend of coronavirus outbreak. Whereas the Fudan-Chinese Center for Disease Control and Prevention (CCDC) model was established to determine the kernel functions in the TDD-NCP model by the public data of CDCC, this model is suggested to use the time delay model to adjust the real data. The advantage of the Fudan-Chinese model is that it can track the initial date of the epidemic, when provided the $I(t_0)$. Moreover, this model can reconstruct parameters such as the growth rate and the "isolation rage," and predict the cumulative number of confirmed cases in some cities in China. However, because this work was done in early March, there was still little knowledge about the disease and little data on confirmed cases. Rajagopal et al. [75] have developed a SEIRD model with integer and fractional differential equations to describe coronavirus in Italy. The fractional model is of the Caputo type, the most popular and most widely used for real problems. To find the optimal parameters, the model parameters are estimated. The number of infected, the number of deaths, and the associated mean square error (RMSE) are also considered. The fractional model gives more realistic predictions and has fewer modeling errors. And with that, the proposed model agrees with the actual data from Italy better than the classical model. A SCEAQHR model for predicting cases in Cameroon has been proposed by Nabi et al. [65]. This model integrates a new class for individuals who have made imperfect quarantine and disregarded blocking policies. The model parameters were estimated with real-time data, followed by a projection of the disease evolution. The model is described by Caputo fractional differential equations, and the existence and uniqueness of the solutions are presented. The optimization algorithm is based on the reliable-region-reflective (TRR) algorithm, which is the evolution of the Levenberg–Marquardt algorithm. The numerical implementation is done using the lsqcurve fit function of MATLAB. The Partial Rank Correlation Coefficient (PRCC) method was used to quantify the dominant mechanisms. The optimization is robust to solving nonlinear least squares problems.

The researchers Roda et al. [78] used the Akaike Information Criterion (AIC) to select the model. And performed an analysis of the predictions of the SIR and SEIR models. The SIR model outperformed the SEIR model in representing the information contained in the confirmed case data. The calibration of the model was done using the Monte Carlo Markov Chain algorithm, and the calibration was done with data from January 21 to February 04 from Wuhan city in China. The authors state that data before January 23 is unreliable and there is a lack of data. There is no identifiability because a group of model parameters cannot be determined solely from the data provided during model calibration. This impacts the reliability of the model.

Din et al. [22] brought out a new three-compartment model (PIQ) described by EDPs for COVID-19 transmission. To study the stability, the Atangana, Baleanu, and Caputo (ABC) model with arbitrary order was used. Banach's fixed point theorem and Guo–Krasnoselskii were used to prove the existence of the model. And the numerical simulations were done using the Adams–Bashforth (AB) method with fractional differentiation. Using this method is a sophisticated and powerful tool for investigating nonlinear problems. The model proves mathematically that it is well defined.

Through a system of ordinary differential equations, the disease is contextualized through social parameters to understand how the spread works and how it is possible to control the epidemics that affect society and thereby create preventive measures. Examples of this type of model are the modified SEIR models proposed by Yang et al. [101] as well as the SEIR (Susceptible, Exposed, Infectious, Recovered) model with age-structured quarantine class with the two types of control measures used to analyze the effects of policy control for the coronavirus epidemic in Brazil [15], and the SEIRQ (Susceptible, Exposed, Infectious, Recovered, Quarantine) model with age structure, proposed by Gondim and Machado [29]. This model aims to analyze optimal quarantine strategies in order to help in decision-making through health managers.

Regarding statistical epidemiological models, Sarkar et al. [85] propose a mathematical model to monitor the dynamics of six compartments: Susceptible (S), Asymptomatic (A), Recovered (R), Infected (I), Isolated Infected (Iq), and Quarantined Susceptible (Sq), collectively expressed SARIIqSq. The authors applied their proposal to real data on the COVID-19 pandemic in India. Starting from the date of first COVID-19 case reported in India, the authors have simulated the SARIIqSq model for 260 days for each states and for whole India to study the dynamics of the SARS-CoV-2 disease. They statistically confirmed that a reduction in the contact rate between uninfected and infected individuals by quarantined the susceptible individuals can effectively reduce the basic reproduction number. They also demonstrate that the elimination of ongoing SARS-CoV-2 pandemic is possible by combining the restrictive social distancing and contact tracing. However, the authors also emphasize the uncertainty of accessible authentic data, specially concerning to the accurate baseline number of infected individuals due to subnotifications, which may guide to equivocal outcomes and inappropriate predictions by orders of size.

Ndaïrou et al. [67] propose a novel epidemiological compartment model that takes into account the super-spreading phenomenon of some individuals. They consider a fatality compartment, related to death due to the virus infection. The constant total population size N is subdivided into eight epidemiological classes: Susceptible class (S), Exposed class (E), Symptomatic and Infectious class (I), Super-Spreaders class (P), Infectious but Asymptomatic class (A), Hospitalized (H), Recovery class (R), and Fatality class (F). This model reached a reasonably good approximation of the reality of the Wuhan outbreak, predicting a diminishing on the daily number of confirmed cases of the disease. The model also fits well the real data of daily confirmed deaths. The model can be considered useful for other realities than Wuhan, China, since the amount of hospitalized individuals is relevant as an estimate of the Intensive Care Units (ICU) needed.

Khajanchi and Sarkar [43] developed a new compartmental model to explain the transmission dynamics of Covid-19. They calibrated their model with daily Covid-19 data for four Indian states: Jharkhand, Gujarat, Andhra Pradesh, and Chandigarh. They studied the feasible equilibria of the proposed model and their stability with respect to the basic reproduction number R_0. The disease-free equilibrium becomes stable and the endemic equilibrium becomes unstable when the recovery rate of infected individuals increases, but if the disease transmission rate remains higher, then the endemic equilibrium always remains stable. The proposed model obtained $R_0 > 1$ for all studied Indian states, suggesting a significant outbreak. The model is able to provide short-time Covid-19 forecasting as well.

Samui et al. [84] proposed a deterministic ordinary differential equation model able to represent the overall dynamics of SARS-CoV-2. They stratified the total human population into four compartments: susceptible individuals (uninfected), asymptomatic individuals (pauci-symptomatic or clinically undetected), reported symptomatic infected individuals (symptomatic infectious individuals are reported by the public health service), and unreported symptomatic infected individuals (clinically ill but not reported) to formulate the SAIU (susceptible or uninfected (S), asymptomatic (A), reported symptomatic infectious (I), unreported symptomatic infectious (U)) model. This model assumes that infected individuals informed will no longer be associated with infections, as they are isolated or transferred to Intensive Care Units (ICU). Thus, only infectious individuals belonging to I(t) or U(t) spread or transmit the diseases. The authors designed the SAIU model to study the transmission dynamics of COVID-19 based on the accessible data for India during the time period January 30, 2020 to April 30, 2020. Based on the estimated data, the SAIU model predicts the outbreak of COVID-19 and computes the basic reproduction number R_0. The authors assessed the sensitivity indices of the basic reproductive number R_0, given that R_0 expresses the initial disease transmission and the sensitivity indices describe the relative importance of various parameters in coronavirus transmission. The SAIU model showed the persistence of diseases for $R_0 > 1$. The endemic equilibrium point $E*$, for this study, was locally asymptotically stable for $R_0 > 1$.

Khajanchi et al. [44] extended the classical deterministic Susceptible–exposed–infectious–removed (SEIR) compartmental model refined by introducing contact tracing-hospitalization strategies to study the epidemiological properties of Covid-19. They calibrated their mathematical model using data of confirmed cases in India and estimated the basic reproduction number for the disease transmission. The authors have their calibrated epidemic model for the short term prediction in the four provinces and the Republic of India. The simulation of the calibrated model was able to capture the increasing growth patterns for three different provinces, namely Delhi, Maharashtra, West Bengal and the Republic of India, whereas in case of the province Kerala, the model fitting is not good compared to other states and overall India. Model simulation and prediction suggest that Covid-19 has a potential to exhibit oscillatory but controllable dynamics in the near future by maintaining social distancing and effectiveness of home isolation and hospitalization. The proposed model forecasts that isolation or hospitalization of the symptomatic population, under stringent hygiene safeguards and social distancing, is considerably effective. Finally, Khajanchi et al. [44] give evidences that the size and duration of an epidemic can be considerably affected by timely implementation of the hospitalization or isolation program.

The classic mathematical models of epidemiological prediction are quite useful, but deterministic, demonstrating only the average behavior of the epidemic, which makes it difficult to quantify uncertainty. Wang et al. [97] proposed an analysis of the spatial structure and dynamics of the spread of Covid-19, providing a spatio-temporal prediction of the Covid-19 outbreak in the USA. Kapoor et al. [39] investigated large-scale spatio-temporal prediction using neural network graphs and human mobility data in US counties. Through this method and space-time information, the model learns the epidemiological dynamics. Tomar and Gupta [92] proposed a space-time approach to control and monitor Covid-19 using LSTM (Long Short-Term Memory) neural networks and adjusting curves to predict chaos. Ren et al. [76] used Ecological Niche Models (ENM) to gather epidemiological and socioeconomic data, aiming to accurately predict the risk areas for Covid-19 infection. Yesilkanat [102] made a study with space-time approach for 190 countries in the world and compared it with the number of real cases of the disease using the Random Forest method. Also using a space-time approach, Pourghasemi et al. [70] did a risk mapping, change detection and trend analysis of the Covid-19 spread in Iran using regression and machine learning. Roy et al. [79] developed a short-term prediction model for the new Coronavirus using canonical ARIMA (Autoregressive Integrated Moving Average) and disease risk analysis done using weighted overlap analysis in geographic information systems.

3   Forecasting by Machine Learning and Hybrid Approaches

Several efforts to aid Covid-19 screening and monitoring can be perused in the works of Dong et al. [25]. In this work, Dong et al. [25] created an online interactive panel to visualize Covid-19 infected cases and deaths in real time, providing researchers, health authorities, and the general public a tool to track cases as the disease progresses. Due to the rapid development of the coronavirus, the need to classify infected patients and analyze which individuals were more vulnerable to the disease also grew. Therefore, Xie et al. [100] proposed a model of clinical prediction for patient mortality based on multivariable logistic regression, to improve the use of limited healthcare resources and calculate the patient's survival rate. Furthermore, in order to aid the diagnosis, Feng et al. [26] developed the online calculator S-COVID-19-P based on Lasso regression, for early identification of suspected Covid-19 pneumonia in the admission of adult patients with fever. Jin et al. [38] proposed a system based on deep learning for the rapid diagnosis of Covid-19 with precision comparable to experienced radiologists, and can accurately classify pneumonia, CAP (Community-Acquired Pneumonia), influenza A and B, and Covid-19. They used LASSO to find the 12 most discriminating characteristics in the distinction between Covid-19 and other pneumonias. Gomes et al. [28] proposed a system to support the diagnosis of Covid-19 by analyzing chest X-ray images, capable of differentiating Covid-19 from bacterial and viral pneumonias using texture-based image representation and classification by Random Forests.

Different from other more complex Covid-19 X-ray feature extraction approaches [7, 8, 12, 19, 33, 35, 45, 53, 54, 63, 66, 96], Gomes et al. [28] avoided deep learning based solutions and adopted texture and shape features to provide the users a low-cost computational web-based computational environment able to deal with several simultaneous users without overcharging network resources.

In order to find a new way to perform early, efficient, and accurate control and screening of suspected individuals, Meng et al. [62] created the Covid-19 Diagnostic Aid APP to calculate the probability of infection through simple and easy laboratory test results. Screening a large number of suspicious people could optimize the diagnostic process and save medical resources. Barbosa et al. [10] considered the fact that, in many regions of the world, RNA testing is not always available due to the scarcity of inputs, created HegIA, an intelligent system based on Bayes Networks and Random Forests to aid at the diagnosis of Covid-19 based on blood tests from 24 blood tests. The performance is close to RT-PCR (Reverse Transcription Polymerase Chain Reaction) for symptomatic individuals, though coronavirus RNA is not searched [10]. HegIA is a fully functional system, available for free use, to provide low-cost rapid testing.

Several works have used Evolutionary Computing and Swarm Intelligence Methods to automatically adjust compartmental models [61, 71, 73, 83]. Putra and Khozin Mu'tamar [71] automatically estimated parameters in the Susceptible, Infected, Recovered (SIR) model using the Particle Swarm Optimization (PSO) algorithm. Their results suggest that the proposed method is able to tune SIR models precisely compared to other analytical approaches. Similarly, Mbuvha and Marwala [61] calibrated a SIR model to South Africa's Covid-19 reported cases taking into account several scenarios of the reproduction number R_0 for reporting infections and healthcare resource estimations. They assumed that the reported confirmed cases represent between 0.2% and 1% of the total infected population. The authors also assumed that SIR model parameters are fixed albeit at multiple ranges. However, they detected the uncertainty around SIR parameters and propose a Bayesian treatment using Markov Chain Monte Carlo techniques in the near future.

Qi et al. [73] investigated the influence of daily temperature (AT) and relative humidity (ARH) on the occurrence of Covid-19 in 31 Chinese provinces, mainly in Hubei. The authors collected daily counts of laboratory-confirmed cases in all provinces in China from the official reports of the National Health Commission of People's Republic of China from December 1, 2019 to February 11, 2020 for Hubei province and from January 20, 2020 to February 11, 2020 for other provinces. Tibet was not included in the following model since only one case was reported during the 23-day cited period. The meteorological data, including daily average temperature (AT) and daily average relative humidity (ARH) of each provincial capital, were retrieved from Weather Underground. Although this study suggests that both daily temperature and relative humidity influenced the occurrence of COVID-19 in Hubei province and in some other provinces, the association between COVID-19 and AT and ARH across the provinces was not considered consistent. The authors found spatial heterogeneity of COVID-19 incidence, as well as its relationship with daily AT and ARH, among provinces in Mainland China.

Salgotra et al. [83] propose prediction models based on genetic programming (GP) for confirmed cases and death cases across the three most affected states in India: Maharashtra, Gujarat, and Delhi. The authors also applied the model to forecast Covid-19 cases in whole India. The proposed prediction models are presented using explicit formula. The authors studied the impotence of prediction variables as well. Statistical parameters and metrics have been used to evaluate and validate the evolved models. Genetic evolutionary programming models have proven to be highly reliable for Covid-19 cases in India.

Rahimi et al. [74] present a systematic review on Computational Intelligence algorithms for Covid-19 forecasting. They searched on Web of Science (WoS) and Scopus for publications in accordance with the following keywords: forecasting, prediction, Covid-19, and coronavirus. The authors selected 920 technical research articles presenting just algorithmic descriptions, review articles, conference papers, case studies, and able to provide managerial insights, published until October 10, 2020. The authors focused on papers indexed by the Web of Science. Rahimi et al. [74] categorized the main forecasting works according to the following classification regarding the algorithms:

- Simple Moving Average [16] as defined by Maleki and Arellano-Valle [55], Maleki and Nematollahi [58], Zarrin et al. [103], Maleki et al. [56], and Hajrajabi and Maleki [30];
- Auto-Regressive Integrated Moving Average (ARIMA) [5, 50, 64, 80, 88];
- Two-piece distributions based on the scale [57];
- Logistic functions: S-shaped functions to model epidemiological curves [17, 52, 72];
- Regression Methods [4, 36, 77, 90, 94];
- Canonical neural networks [27, 64, 91];
- Deep learning methods based on Convolutional Neural Networks (CNNs) [13, 51, 86];
- Deep learning methods based on Long-Short Term Memory (LSTM) neural networks [9, 18];
- Genetic programming [82, 83]);
- Classical and modified compartment models: SIR, SEIR, and SIRD [2, 14, 41, 69].

Tamang et al. [91] used artificial neural network-based curve fitting techniques to predict and forecast Covid-19 infected and death cases in India, USA, France, and United Kingdom, considering the progressive trends of China and South Korea. The authors considered three cases to analyze the Covid-19 outbreak: (1) forecasting as per the present trend of rising cases of different countries; (2) one-week forecasting following up with the improvement trends as per China and South Korea; and (3) forecasting if followed up the progressive trends as per China and South Korea before a week. According to the authors, to reduce infection rates and achieve leveling of trends in epidemiological curves, these countries will require fewer days according to the forecast with the trend in China and more days with steady progress are seen with the South Korea's trend. In addition, it can also be concluded that, with the trend of China, countries with a greater number of cases could be better in fewer days with possibly stricter measures of social isolation, detachment, and confinement. Considering that South Korea's trend is toward slower and more constant control, which could be more effective in the initial stage with lower reported cases. All conclusions were made in accordance with the predictions obtained with the application of the multilayer perceptron artificial neural network technique. Although the case data used in the study are based on reliable sources, the predictions are in accordance with the conditions and techniques applied. Consequently, their experimental results suggest that artificial neural networks are able to forecast the future cases of COVID 19 outbreak of practically any country at low error rates.

Huang et al. [32] propose a new model of CNN deep neural network with multiple inputs to predict the cumulative number of confirmed cases of Covid-19. The cumulative number of confirmed cases on the following day is predicted according to the total number of confirmed cases from the previous 5 days, total new confirmed cases, total cured cases, total new cured cases, total deaths, and total new deaths. Datasets from seven Chinese cities in the provinces of Hubei, Guangdong, and Zhejiang were used with confirmed serious cases for the training and forecasting of the models. Data on confirmed cases of COVID-19 from January 23, 2020 to March 2, 2020, and from January 23, 2020 to March 2, 2020, were obtained from the media outlet Surging News Network and from the World Health Organization, respectively. The two evaluation indexes of the mean absolute error (MAE) and root mean square error (RMSE) were used. According to the authors, the proposed algorithm can quickly use small datasets to establish models with high predictive precision. This is a considerable advantage of this model over other models with similar characteristics. Through the proposed algorithm, a prediction model was established for the number of confirmed cases of COVID-19. Verification and comparison were conducted between different deep learning algorithms. The accuracy and reliability of the deep learning algorithm have been verified by predicting the future trend of Covid-19. In addition, experiments for several cities with more serious confirmed cases in China indicated that the prediction model in this study had the lowest error rate among its tested equivalents. As future work, the authors envisage using deep learning networks with a mixed structure, seeking to build more accurate models, which can be applied to more countries.

Distante et al. [23] modeled spreading of Covid-19 using Chinese data and used the model to predict epidemic curve in each Italian region, allowing to gain better information on the new daily cases peaks with the predicted epidemiological curve. According to the authors, the forecast portion of the curve allows to have a better prediction of active cases with the SEIR model, by computing the position of the peaks of active cases for each Italian region. Interestingly, the process of training on Chinese data and using the knowledge to forecast Italian spreading of Covid-19 has resulted in good forecasting results, considering the mean average precision between official Italian data and the forecast. SEIR models may fit better than other compartment models since they are based on the complete curve dynamic. Therefore, the proposed approach is valid since the predictive model learns from the dynamics of Covid-19 in China and exploits its knowledge to predict future daily cases in Italy.

Wieczorek et al. [99] proposes a predictive model based on a deep 7-layer neural network trained by the NAdam method to predict the number of infected cases. The authors used a dataset provided by the Center for Systems Science and Engineering (CSSE) at Johns Hopkins University on their github page. This dataset is composed of the following sources: (a) World Health Organization (WHO); (b) European Center for Disease Prevention and Control (ECDC); (c) DXY.cn. Pneumonia. 2020; (d) COVID Tracking Project; (e) National Health Commission of the People's Republic of China (NHC); (f) China CDC (CCDC); (g) Washington State Department of Health; (h) other smaller, regional US health departments. The predictive model was able to predict new cases with very high efficiency, above 99% in some geographic regions. However, the authors noticed that analysts should take into account several factors able to influence the epidemiological curves: behavior of the population in a given region, behavior of governments of given countries as well as access to knowledge and medical equipment. The neural network-based predictor employs a unified architecture. According to their experimental results, the authors do not need to change the architecture in dependence with each region or country. Accuracy for most of regions is around 87.70%. However, the authors believe that dedicated architectures should be used to contemplate differences among countries, like population and government behaviors.

Kırbaş et al. [49] modeled confirmed COVID-19 cases of Denmark, Belgium, Germany, France, United Kingdom, Finland, Switzerland, and Turkey using Auto-Regressive Integrated Moving Average (ARIMA), Nonlinear Autoregression Neural Network (NARNN) and Long-Short Term Memory (LSTM) approaches. They tested six model performance metrics: MSE, PSNR, RMSE, NRMSE, MAPE, and SMAPE. Cumulative confirmed case data of eight different European countries were used for modeling: Denmark, Belgium, Germany, France, United Kingdom, Finland, Switzerland, and Turkey. The datasets were acquired from the European Center for Disease Prevention and Control. Data were taken from the day the first case was seen, and the number of data for each country varies. The data covers 67, 90, 97, 100, 94, 90, 68, and 55 days, respectively, and ends on 3 May 2020. The data from cumulative confirmed cases in some European countries are modeled using three different approaches. According to the results, it was determined that LSTM approach has much higher success compared to ARIMA and NARNN. The lowest number of cases was observed in Finland during the epidemic, while the highest rate of increase was observed in the United Kingdom. According to the 2-week prospective estimation study, in many countries, the total case increase rate is expected to decrease slightly. Since the work was carried out entirely by considering statistical data and methodologies, the effects of social distancing and other similar measures, compliance with hygiene rules or lockdown were ignored. However, according to the results on real data, the authors considered the predictions satisfactory.

Pal et al. [68] have proposed to use the local data trend with a shallow Long Short-Term Memory (LSTM) based neural network combined with a fuzzy rule based system to predict long term risk of a country. The country-specific neural networks are optimized using Bayesian optimization. The authors used the dataset (https://github.com/datasets/covid-19) that included date, country, the number of confirmed cases, the number of recovered cases, and the total number of deaths. This data was combined with weather data (https://darksky.net/): humidity, dew, ozone, perception, maximum temperature, minimum temperature, and UV for analyzing the effect of weather. The authors considered mean and standard deviation over different cities of a country. The data spanned the duration 22-01-2020 to 02-08-2020. The authors propose to use country-specific optimized networks for accurate prediction, since this approach seems suitable for small and uncertain dataset. Combining the overall optimized LSTMs, they noticed that a shallow networks perform better compared to deep neural networks. The authors also noticed that the weather data does not affect the forecasting accuracy.

Zeroual et al. [104] performed a comparative study of five deep learning methods to forecast the number of new cases and recovered cases: simple Recurrent Neural Network (RNN), Long short-term memory (LSTM), Bidirectional LSTM (BiLSTM), Gated recurrent units (GRUs), and Variational AutoEncoder (VAE). These methods were applied for global forecasting of Covid-19 cases based on a small volume of data. This study is based on daily confirmed and recovered cases collected from six countries namely Italy, Spain, France, China, USA, and Australia. The values of parameters of deep learning models are selected such that the loss function is minimized during the training. The authors adopted the Adam optimizer. In the testing stage, the previously constructed models with the selected parameters are used to forecast the number of COVID cases. The accuracy of the model was verified by comparing the measured data with real data via different statistical indicators including RMSE, MAE, MAPE, and RMSLE (Root Mean Squared Log Error). The research was based on daily figures of confirmed and recovered cases collected from six highly impacted countries namely Italy, Spain, Italy, China, the USA, and Australia. The considered datasets are gathered from the starting of COVID-19 for the respective countries, i.e. 22 January 2020, till June 17th, 2020. These datasets are made publically by the Center for Systems Science and Engineering (CSSE) at Johns Hopkins University (https://github.com/ CSSEGISandData). Results demonstrate that the Variational AutoEncoder achieved the best forecasting performance in comparison to the other models.

Kapoor et al. [40] propose a novel spatio-temporal forecasting approach for Covid-19 case prediction based on Graph Neural Networks and mobility data. Differently from time series forecasting models, the proposed model learns from a single large-scale spatio-temporal graph, where nodes represent the region-level human mobility, spatial edges represent the human mobility based inter-region connectivity, and temporal edges represent node features through time. The authors applied their method to the US county level COVID-19 dataset. They perceived that the spatial and temporal information leveraged by the graph neural network allows the model to learn considerably complex dynamics. It is noticed a 6%

reduction of RMSLE and an absolute Pearson Correlation improvement from 0.9978 to 0.998 in comparison with the state-of-the-art models. According to the authors, the combination of graph-based deep learning approaches can be very useful to aid to understand the spread and evolution of Covid-19.

de Lima et al. [21] proposed a real-time surveillance, forecast, and spatial visualization of Covid-19, named COVID-SGIS. As a case study, the forecasting system was applied to monitor Brazil. The system captures routinely reported Covid-19 information from 27 federative units from the Brazil.io database. It uses Covid-19 confirmed case data notified through Brazil's National Notification System, SINAN, from March to May 2020. Time series ARIMA models were integrated to forecast the cumulative number of Covid-19 cases and deaths. These include 6-days forecasts as graphical outputs for each federal state in Brazil, separately, with its corresponding 95% confidence interval. The worst and the best scenarios are both presented. The overall percentage error between the forecasted values and the actual values varied between 2.56% and 6.50%. For the days when the forecasts fell outside the forecast interval, the percentage errors in relation to the worst case scenario were below 5%. Considering the good results obtained with the proposed tool, the authors claimed that the proposed method for dynamic forecasting may be used to guide social policies and plan direct interventions in a cost-effective, concise, and robust manner.

4   Conclusions

COVID-19 is a disease that was discovered and soon assumed pandemic status as it spread to several countries around the world. It drew attention for its ease of transmission and for exposing the vulnerabilities of health systems around the world. The individuals who were infected and their families were left with the pain and suffering and the after-effects of the disease. Although there are vaccines, there is still no proven effective drug against the disease, so following safety protocols and social isolation are indispensable. In addition to hygiene practices such as the use of masks and hand-washing, the use of models to understand the behavior of the disease and even to predict it helps to shed light on the next steps to be taken in this pandemic.

The representation of disease through mathematical models facilitates monitoring and can help analyze and understand disease dynamics through key characteristics. Through a system of equations, it is possible to model a disease and contribute to a quantitative understanding. These characteristics become useful information on how the spread of the disease works. They also make it possible to understand how to build time prediction and thus help to create measures to control and prevent COVID-19. However, for this type of modeling, some assumptions are made, such as assuming that disease transmission occurs homogeneously, or selecting only one among several climatic factors. Therefore, this limits the model's ability to predict. And if more features were added, the model would lose robustness.

With the large amount of data available and thanks to speed and storage technologies, Artificial Intelligence is increasingly strong and present in several areas. Then, the use of machine learning techniques grew in order to obtain insights from this data. These models are applied in several areas, such as economics, for the performance of a stock in the stock market, in banking, in e-commerce, determining whether a customer will like the product or not, in health, as is done in the present work with digital epidemiology. But models with this kind of approach are black boxes, that is, they are not intelligible to experts, because their goal is to correctly map inputs to outputs. Another type of approach using Artificial Intelligence is the one that uses hybrid models, i.e. it combines machine learning models with statistical models. By doing this they combine the advantages of each of these types of models in order to obtain a more robust prediction model. Another type of hybrid model is one that combines compartmental models and machine learning. With this approach the model does not have as good a prediction quality as machine learning based models; however, it can aid in human understanding of epidemiological aspects as phenomena, while machine learning based models can return accurate predictions, thus combining intelligent systems for accurate human learning emergent predictions. The use of all these approaches is very important to support us in temporal and spatio-temporal prediction of cases and deaths. For these solutions can shed light on strategies to assist decision-making by health managers.

Finally, COVID-19 brings with it all the challenges of a new disease with only 1 year of existence, in facing this unknown, science makes use of all its arsenal. At this time when there is no extensive background to teach how the disease behaves, daily experience determines adjustments and creation of clinical protocols. Predicting the temporal and spatial behavior of COVID-19 through machine learning becomes a valuable tool to guide strategies, policies, and hope.

References

1. Abou-Ismail, A. (2020). Compartmental models of the COVID-19 pandemic for physicians and physician-scientists. *SN Comprehensive Clinical Medicine*, 2, 852–858.
2. Ahmar, A. S., & Del Val, E. B. (2020). SutteARIMA: Short-term forecasting method, a case: Covid-19 and stock market in Spain. *Science of The Total Environment*, 729, 138883.
3. Ahmed, A., Salam, B., Mohammad, M., Akgul, A., & Khoshnaw, S. (2020). Analysis coronavirus disease (covid-19) model using numerical approaches and logistic model. *Aims Bioengineering*, 7(3), 130–146.
4. Almeshal, A. M., Almazrouee, A. I., Alenizi, M. R., & Alhajeri, S. N. (2020). Forecasting the spread of COVID-19 in Kuwait using compartmental and logistic regression models. *Applied Sciences*, 10(10), 3402.
5. Alzahrani, S. I., Aljamaan, I. A., & Al-Fakih, E. A. (2020). Forecasting the spread of the COVID-19 pandemic in Saudi Arabia using ARIMA prediction model under current public health interventions. *Journal of Infection and Public Health*, 13(7), 914–919.
6. Ambikapathy, B., & Krishnamurthy, K. (2020). Mathematical modelling to assess the impact of lockdown on covid-19 transmission in India: Model development and validation. *JMIR Public Health and Surveillance*, 6(2), e19368.

7. Apostolopoulos, I., Aznaouridis, S., & Tzani, M. (2020). Extracting possibly representative covid-19 biomarkers from X-ray images with deep learning approach and image data related to pulmonary diseases. Preprint. arXiv:2004.00338.

8. Apostolopoulos, I. D., & Mpesiana, T. A. (2020). Covid-19: automatic detection from X-ray images utilizing transfer learning with convolutional neural networks. *Physical and Engineering Sciences in Medicine*, 43(2), 635–640.

9. Ayyoubzadeh, S. M., Ayyoubzadeh, S. M., Zahedi, H., Ahmadi, M., & Kalhori, S. R. N. (2020). Predicting COVID-19 incidence through analysis of google trends data in Iran: data mining and deep learning pilot study. *JMIR Public Health and Surveillance*, 6(2), e18828.

10. Barbosa, V. A. d. F., Gomes, J. C., de Santana, M. A., Jeniffer, E. d. A., de Souza, R. G., de, Souza, R. E., & dos Santos, W. P. (2021). Heg.IA: An intelligent system to support diagnosis of Covid-19 based on blood tests. *Research on Biomedical Engineering*, 2021, 1–18.

11. Bastos, S. B., & Cajueiro, D. O. (2020). Modeling and forecasting the early evolution of the covid-19 pandemic in Brazil. *Scientific Reports*, 10(1), 1–10.

12. Basu, S., Mitra, S., & Saha, N. (2020). Deep learning for screening covid-19 using chest X-ray images. In *2020 IEEE Symposium Series on Computational Intelligence (SSCI)* (pp. 2521–2527).

13. Bengio, Y., Courville, A., & Vincent, P. (2013). Representation learning: A review and new perspectives. *IEEE Transactions on Pattern Analysis and Machine Intelligence*, 35(8), 1798–1828.

14. Capasso, V., & Serio, G. (1978). A generalization of the Kermack-McKendrick deterministic epidemic model. *Mathematical Biosciences*, 42(1–2), 43–61.

15. Castilho, C., Gondim, J. A. M., Marchesin, M., & Sabeti, M. (2020). Assessing the Efficiency of Different Control Strategies for the Coronavirus (COVID-19) Epidemic. Preprint. arXiv:2004.03539. Retrieved from http://arxiv.org/abs/2004.03539

16. Chaudhry, R. M., Hanif, A., Chaudhary, M., & Minhas, S. (2020). Coronavirus Disease 2019 (COVID-19): Forecast of an emerging urgency in Pakistan. *Cureus*, 12(5).

17. Chen, D.-G., Chen, X., & Chen, J. K. (2020). Reconstructing and forecasting the COVID-19 epidemic in the United States using a 5-parameter logistic growth model. *Global Health Research and Policy*, 5, 1–7.

18. Chimmula, V. K. R., & Zhang, L. (2020). Time series forecasting of COVID-19 transmission in Canada using LSTM networks. *Chaos, Solitons & Fractals*, 135, 109864.

19. Civit-Masot, J., Luna-Perejón, F., Domínguez Morales, M., & Civit, A. (2020). Deep learning system for covid-19 diagnosis aid using X-ray pulmonary images. *Applied Sciences*, 10(13), 4640.

20. Coronavirus disease (covid-19) pandemic [Computer software manual]. (2020). Retrieved from www.who.int/emergencies/diseases/novel-coronavirus-2019. Last accessed: 22 April 2020.

21. de Lima, C. L., da Silva, C. C., da Silva, A. C. G., Silva, E. L., Marques, G. S., de Araújo, L. J. B., . . . da Silva-Filho, A. G. (2020). COVID-SGIS: A smart tool for dynamic monitoring and temporal forecasting of Covid-19. *Frontiers in Public Health*, 8, 761.

22. Din, A., Shah, K., Seadawy, A., Alrabaiah, H., & Baleanu, D. (2020). On a new conceptual mathematical model dealing the current novel coronavirus-19 infectious disease. *Results in Physics*, 19, 103510.

23. Distante, C., Pereira, I. G., Goncalves, L. M. G., Piscitelli, P., & Miani, A. (2020). Forecasting Covid-19 outbreak progression in Italian regions: A model based on neural network training from Chinese data. MedRxiv.

24. Döhla, M., Boesecke, C., Schulte, B., Diegmann, C., Sib, E., Richter, E., . . . et al. (2020). Rapid point-of-care testing for SARS-CoV-2 in a community screening setting shows low sensitivity. *Public Health*, 182, 170–172.

25. Dong, E., Du, H., & Gardner, L. (2020). An interactive web-based dashboard to track COVID-19 in real time. *The Lancet Infectious Diseases*, *20*(5), 533–534. Retrieved from https://doi.org/10.1016/S1473-3099(20)30120-1

26. Feng, C., Huang, Z., Wang, L., Chen, X., Zhai, Y., Zhu, F., ... Li, T. (2020). A novel triage tool of artificial intelligence assisted diagnosis aid system for suspected COVID-19 pneumonia in fever clinics. *medRxiv*, 1–68. Retrieved from https://ssrn.com/abstract=3551355

27. Fong, S. J., Li, G., Dey, N., Crespo, R. G., & Herrera-Viedma, E. (2020). Finding an accurate early forecasting model from small dataset: A case of 2019-nCoV novel coronavirus outbreak. Preprint. arXiv:2003.10776, *2020*.

28. Gomes, J. C., de Freitas Barbosa, V. A., de Santana, M. A., Bandeira, J., Valenca, M. J. S., de Souza, R. E., ... dos Santos, W. P. (2020). Ikonos: An intelligent tool to support diagnosis of covid-19 by texture analysis of X-ray images. *Research on Biomedical Engineering*, *2020*, 1–14.

29. Gondim, J. A. M., & Machado, L. (2020). Optimal quarantine strategies for the COVID-19 pandemic in a population with a discrete age structure. Preprint. arXiv: 2005.09786.

30. Hajrajabi, A., & Maleki, M. (2019). Nonlinear semiparametric autoregressive model with finite mixtures of scale mixtures of skew normal innovations. *Journal of Applied Statistics*, *2019*, 2010–2029.

31. Hamzah, F. B., Lau, C., Nazri, H., Ligot, D. V., Lee, G., Tan, C. L., ... et al. (2020). Coronatracker: worldwide covid-19 outbreak data analysis and prediction. *Bull World Health Organ*, *1*(32), 1–32.

32. Huang, C.-J., Chen, Y.-H., Ma, Y., & Kuo, P.-H. (2020). Multiple-input deep convolutional neural network model for Covid-19 forecasting in China. *MedRxiv*, *2020*, 1–16.

33. Ismael, A. M., & Şengür, A. (2021). Deep learning approaches for covid-19 detection based on chest X-ray images. *Expert Systems with Applications*, *164*, 114054.

34. Ivorra, B., Ferrández, M. R., Vela-Pérez, M., & Ramos, A. (2020). Mathematical modeling of the spread of the coronavirus disease 2019 (covid-19) taking into account the undetected infections. The case of China. *Communications in Nonlinear Science and Numerical Simulation*, *88*, 105303.

35. Jain, G., Mittal, D., Thakur, D., & Mittal, M. K. (2020). A deep learning approach to detect covid-19 coronavirus with X-ray images. *Biocybernetics and Biomedical Engineering*, *40*(4), 1391–1405.

36. Ji, D., Zhang, D., Xu, J., Chen, Z., Yang, T., Zhao, P., ... Qin, E. (2020). Prediction for progression risk in patients with COVID-19 pneumonia: the CALL score. *Clinical Infectious Diseases*, *71*(6), 1393–1399.

37. Jiang, N., Liu, Y., Yang, B., Li, Z., Si, D., Ma, P., ... & Yu, Q.(2020). Analysis of the factors associated with negative conversion of severe acute respiratory syndrome coronavirus 2 rna of coronavirus disease 2019. *Open Access Macedonian Journal of Medical Sciences*, *8*(1), 436–442.

38. Jin, C., Chen, W., Cao, Y., Xu, Z., Zhang, X., Deng, L., ... Feng, J. (2020). Development and evaluation of an AI system for COVID-19 diagnosis. *medRxiv*. Retrieved from http://medrxiv.org/content/early/2020/03/27/2020.03.20.20039834.abstract

39. Kapoor, A., Ben, X., Liu, L., Perozzi, B., Barnes, M., Blais, M., & O'Banion, S. (2020a). Examining COVID-19 forecasting using spatio-temporal graph neural networks. ArXiv preprint. Retrieved from http://arxiv.org/abs/2007.03113

40. Kapoor, A., Ben, X., Liu, L., Perozzi, B., Barnes, M., Blais, M., & O'Banion, S. (2020b). Examining covid-19 forecasting using spatio-temporal graph neural networks. Preprint. arXiv:2007.03113, *2020*.

41. Kermack, W. O., & McKendrick, A. G. (1932). Contributions to the mathematical theory of epidemics. II.—The problem of endemicity. *Proceedings of the Royal Society of London. Series A, Containing Papers of a Mathematical and Physical Character*, *138*(834), 55–83.

42. Khajanchi, S., Bera, S., & Roy, T. K. (2021). Mathematical analysis of the global dynamics of a HTLV-I infection model, considering the role of cytotoxic t-lymphocytes. *Mathematics and Computers in Simulation*, *180*, 354–378.

43. Khajanchi, S., & Sarkar, K. (2020). Forecasting the daily and cumulative number of cases for the covid-19 pandemic in India. *Chaos: An Interdisciplinary Journal of Nonlinear Science, 30*(7), 071101.
44. Khajanchi, S., Sarkar, K., Mondal, J., & Perc, M. (2020). Dynamics of the covid-19 pandemic in India. Preprint. arXiv:2005.06286.
45. Khan, A. I., Shah, J. L., & Bhat, M. M. (2020). Coronet: A deep neural network for detection and diagnosis of covid-19 from chest X-ray images. *Computer Methods and Programs in Biomedicine, 196*, 105581.
46. Khoshnaw, S. H., Salih, R. H., & Sulaimany, S. (2020). Mathematical modelling for coronavirus disease (covid-19) in predicting future behaviours and sensitivity analysis. *Mathematical Modelling of Natural Phenomena, 15*, 33.
47. Khoshnaw, S. H., Shahzad, M., Ali, M., & Sultan, F. (2020). A quantitative and qualitative analysis of the covid-19 pandemic model. *Chaos, Solitons & Fractals, 138*, 109932.
48. Khrapov, P., & Loginova, A. (2020). Comparative analysis of the mathematical models of the dynamics of the coronavirus covid-19 epidemic development in the different countries. *International Journal of Open Information Technologies, 8*(5), 17–22.
49. Kırbaş, İ., Sözen, A., Tuncer, A. D., & Kazancıoğlu, F. Ş. (2020). Comparative analysis and forecasting of COVID-19 cases in various European countries with ARIMA, NARNN and LSTM approaches. *Chaos, Solitons & Fractals, 138*, 110015.
50. Kufel, T. (2020). ARIMA-based forecasting of the dynamics of confirmed Covid-19 cases for selected European countries. *Equilibrium. Quarterly Journal of Economics and Economic Policy, 15*(2), 181–204.
51. LeCun, Y., Bengio, Y., & Hinton, G. (2015). Deep learning. *Nature, 521*(7553), 436–444.
52. Li, Q., Feng, W., & Quan, Y.-H. (2020). Trend and forecasting of the COVID-19 outbreak in China. *Journal of Infection, 80*(4), 469–496.
53. Luz, E., Silva, P. L., Silva, R., & Moreira, G. (2020). Towards an efficient deep learning model for covid-19 patterns detection in X-ray images. Preprint. arXiv:2004.05717.
54. Maghdid, H. S., Asaad, A. T., Ghafoor, K. Z., Sadiq, A. S., & Khan, M. K. (2020). Diagnosing covid-19 pneumonia from x-ray and ct images using deep learning and transfer learning algorithms. Preprint. arXiv:2004.00038.
55. Maleki, M., & Arellano-Valle, R. B. (2017). Maximum a-posteriori estimation of autoregressive processes based on finite mixtures of scale-mixtures of skew-normal distributions. *Journal of Statistical Computation and Simulation, 87*(6), 1061–1083.
56. Maleki, M., Arellano-Valle, R. B., Dey, D. K., Mahmoudi, M. R., & Jalali, S. M. J. (2017). A Bayesian approach to robust skewed autoregressive processes. *Calcutta Statistical Association Bulletin, 69*(2), 165–182.
57. Maleki, M., Mahmoudi, M. R., Wraith, D., & Pho, K.-H. (2020). Time series modelling to forecast the confirmed and recovered cases of COVID-19. *Travel Medicine and Infectious Disease, 37*, 101742.
58. Maleki, M., & Nematollahi, A. (2017). Autoregressive models with mixture of scale mixtures of gaussian innovations. *Iranian Journal of Science and Technology, Transactions A: Science, 41*(4), 1099–1107.
59. Mandal, M., Jana, S., Nandi, S. K., Khatua, A., Adak, S., & Kar, T. (2020). A model based study on the dynamics of covid-19: Prediction and control. *Chaos, Solitons & Fractals, 136*, 109889.
60. Massonis, G., Banga, J. R., & Villaverde, A. F. (2020). Structural identifiability and observability of compartmental models of the covid-19 pandemic. *Annual Reviews in Control*. Volume 51, 2021, Pages 441–459
61. Mbuvha, R. R., & Marwala, T. (2020). On data-driven management of the Covid-19 outbreak in South Africa. *medRxiv, 2020*.
62. Meng, Z., Wang, M., Song, H., Guo, S., Zhou, Y., Li, W., ... Ying, B. (2020). Development and utilization of an intelligent application for aiding COVID-19 diagnosis. *medRxiv* (37), Volume 2020, Pages 1–21.

63. Minaee, S., Kafieh, R., Sonka, M., Yazdani, S., & Soufi, G. J. (2020). Deep-covid: Predicting covid-19 from chest X-ray images using deep transfer learning. *Medical Image Analysis, 65*, 101794.
64. Moftakhar, L., Mozhgan, S., & Safe, M. S. (2020). Exponentially increasing trend of infected patients with COVID-19 in Iran: a comparison of neural network and ARIMA forecasting models. *Iranian Journal of Public Health, 2020*.
65. Nabi, K. N., Abboubakar, H., & Kumar, P. (2020). Forecasting of covid-19 pandemic: from integer derivatives to fractional derivatives. *Chaos, Solitons & Fractals, 141*, 110283.
66. Narin, A., Kaya, C., & Pamuk, Z. (2020). Automatic detection of coronavirus disease (covid-19) using X-ray images and deep convolutional neural networks. Preprint. arXiv:2003.10849.
67. Ndaïrou, F., Area, I., Nieto, J. J., & Torres, D. F. (2020). Mathematical modeling of covid-19 transmission dynamics with a case study of Wuhan. *Chaos, Solitons & Fractals, 135*, 109846.
68. Pal, R., Sekh, A. A., Kar, S., & Prasad, D. K. (2020). Neural network based country wise risk prediction of COVID-19. *Applied Sciences, 10*(18), 6448.
69. Peng, L., Yang, W., Zhang, D., Zhuge, C., & Hong, L. (2020). Epidemic analysis of COVID-19 in China by dynamical modeling. Preprint. arXiv:2002.06563, *2020*.
70. Pourghasemi, H. R., Pouyan, S., Heidari, B., Farajzadeh, Z., Fallah Shamsi, S. R., Babaei, S., ... Sadeghian, F. (2020). Spatial modeling, risk mapping, change detection, and outbreak trend analysis of coronavirus (COVID-19) in Iran (days between February 19 and June 14, 2020). *International Journal of Infectious Diseases, 98*, 90–108. Retrieved from https://doi.org/10.1016/j.ijid.2020.06.058
71. Putra, S., & Khozin Mu'tamar, Z. (2019). Estimation of parameters in the SIR epidemic model using particle swarm optimization. *American Journal of Mathematical and Computer Modelling, 4*(4), 83–93.
72. Qeadan, F., Honda, T., Gren, L. H., Dailey-Provost, J., Benson, L. S., VanDerslice, J. A., ... Shoaf, K. (2020). Naive forecast for COVID-19 in Utah based on the South Korea and Italy models-the fluctuation between two extremes. *International Journal of Environmental Research and Public Health, 17*(8), 2750.
73. Qi, H., Xiao, S., Shi, R., Ward, M. P., Chen, Y., Tu, W., ... Zhang, Z. (2020). COVID-19 transmission in Mainland China is associated with temperature and humidity: a time-series analysis. *Science of the Total Environment, 728*, 138778.
74. Rahimi, I., Chen, F., & Gandomi, A. H. (2021). A review on COVID-19 forecasting models. *Neural Computing and Applications, 2020*, 1–11.
75. Rajagopal, K., Hasanzadeh, N., Parastesh, F., Hamarash, I. I., Jafari, S., & Hussain, I. (2020). A fractional-order model for the novel coronavirus (covid-19) outbreak. *Nonlinear Dynamics, 101*(1), 711–718.
76. Ren, H., Zhao, L., Zhang, A., Song, L., Liao, Y., Lu, W., & Cui, C. (2020). Early forecasting of the potential risk zones of COVID-19 in China's megacities. *Science of the Total Environment, 729*, 138995. Retrieved from https://doi.org/10.1016/j.scitotenv.2020.138995
77. Ribeiro, M. H. D. M., da Silva, R. G. da, Mariani, V. C., & dos Santos Coelho, L. (2020). Short-term forecasting COVID-19 cumulative confirmed cases: Perspectives for Brazil. *Chaos, Solitons & Fractals, 135*, 109853.
78. Roda, W. C., Varughese, M. B., Han, D., & Li, M. Y. (2020). Why is it difficult to accurately predict the covid-19 epidemic? *Infectious Disease Modelling, 5*, 271–281.
79. Roy, S., Bhunia, G. S., & Shit, P. K. (2020a). Spatial prediction of COVID-19 epidemic using ARIMA techniques in India. *Modeling Earth Systems and Environment, 2019*(0123456789). Retrieved from https://doi.org/10.1007/s40808-020-00890-y
80. Roy, S., Bhunia, G. S., & Shit, P. K. (2020b). Spatial prediction of covid-19 epidemic using arima techniques in India. *Modeling Earth Systems and Environment, 2020*, 1–7.
81. Sadun, L. (2020). Effects of latency on estimates of the covid-19 replication number. *Bulletin of Mathematical Biology, 82*(9), 1–14.
82. Salgotra, R., Gandomi, M., & Gandomi, A. H. (2020a). Evolutionary modelling of the COVID-19 pandemic in fifteen most affected countries. *Chaos, Solitons & Fractals, 140*, 110118.

83. Salgotra, R., Gandomi, M., & Gandomi, A. H. (2020b). Time series analysis and forecast of the covid-19 pandemic in India using genetic programming. *Chaos, Solitons & Fractals, 138*, 109945.

84. Samui, P., Mondal, J., & Khajanchi, S. (2020). A mathematical model for covid-19 transmission dynamics with a case study of India. *Chaos, Solitons & Fractals, 140*, 110173.

85. Sarkar, K., Khajanchi, S., & Nieto, J. J. (2020). Modeling and forecasting the covid-19 pandemic in India. *Chaos, Solitons & Fractals, 139*, 110049.

86. Schmidhuber, J. (2015). Deep learning in neural networks: An overview. *Neural Networks, 61*, 85–117.

87. Shao, N., Zhong, M., Yan, Y., Pan, H., Cheng, J., & Chen, W. (2020). Dynamic models for coronavirus disease 2019 and data analysis. *Mathematical Methods in the Applied Sciences, 43*(7), 4943–4949.

88. Singh, S., Parmar, K. S., Kumar, J., & Makkhan, S. J. S. (2020). Development of new hybrid model of discrete wavelet decomposition and autoregressive integrated moving average (ARIMA) models in application to one month forecast the casualties cases of COVID-19. *Chaos, Solitons & Fractals, 135*, 109866.

89. Suba, M., Shanmugapriya, R., Balamuralitharan, S., & Joseph, G. A. (n.d.). Current mathematical models and numerical simulation of sir model for coronavirus disease-2019 (covid-19). *European Journal of Molecular & Clinical Medicine, 7*(05), 2020.

90. Sujath, R., Chatterjee, J. M., & Hassanien, A. E. (2020). A machine learning forecasting model for COVID-19 pandemic in India. *Stochastic Environmental Research and Risk Assessment, 34*, 959–972.

91. Tamang, S., Singh, P., & Datta, B. (2020). Forecasting of Covid-19 cases based on prediction using artificial neural network curve fitting technique. *Global Journal of Environmental Science and Management, 6*(Special Issue (Covid-19)), 53–64.

92. Tomar, A., & Gupta, N. (2020). Prediction for the spread of COVID-19 in India and effectiveness of preventive measures. *Science of the Total Environment, 728*, 138762. Retrieved from https://doi.org/10.1016/j.scitotenv.2020.138762

93. Varotsos, C. A., & Krapivin, V. F. (2020). A new model for the spread of covid-19 and the improvement of safety. *Safety Science, 132*, 104962.

94. Velásquez, R. M. A., & Lara, J. V. M. (2020). Forecast and evaluation of COVID-19 spreading in USA with reduced-space Gaussian process regression. *Chaos, Solitons & Fractals, 136*, 109924.

95. Viguerie, A., Veneziani, A., Lorenzo, G., Baroli, D., Aretz-Nellesen, N., Patton, A., . . . Auricchio, F. (2020). Diffusion–reaction compartmental models formulated in a continuum mechanics framework: application to covid-19, mathematical analysis, and numerical study. *Computational Mechanics, 66*(5), 1131–1152.

96. Wang, L., Lin, Z. Q., & Wong, A. (2020). Covid-net: A tailored deep convolutional neural network design for detection of covid-19 cases from chest X-ray images. *Scientific Reports, 10*(1), 1–12.

97. Wang, L., Wang, G., Gao, L., Li, X., Yu, S., Kim, M., . . . Gu, Z. (2020). Spatiotemporal dynamics, nowcasting and forecasting of COVID-19 in the United States. ArXiv, 1–26. Retrieved from http://arxiv.org/abs/2004.14103

98. WHO. (2021). WHO Coronavirus (COVID-19) Dashboard [Computer software manual]. Retrieved from https://covid19.who.int/. Last accessed: 06 April 2021.

99. Wieczorek, M., Siłka, J., & Woźniak, M. (2020). Neural network powered COVID-19 spread forecasting model. *Chaos, Solitons & Fractals, 140*, 110203.

100. Xie, J., Hungerford, D., Chen, H., Abrams, S. T., Li, S., Wang, G., . . . Toh, C.-H. (2020). Development and external validation of a prognostic multivariable model on admission for hospitalized patients with COVID-19. *The Lancet, 2020*, 1–29.

101. Yang, Z., Zeng, Z., Wang, K., Wong, S.-S., Liang, W., Zanin, M., . . . et al. (2020). Modified SEIR and AI prediction of the epidemics trend of COVID-19 in China under public health interventions. *Journal of Thoracic Disease, 12*(3), 165.

102. Yesilkanat, C. M. (2020). Spatio-temporal estimation of the daily cases of COVID-19 in worldwide using random forest machine learning algorithm. *Chaos, Solitons and Fractals, 140*, 110210.
103. Zarrin, P., Maleki, M., Khodadai, Z., & Arellano-Valle, R. B. (2019). Time series models based on the unrestricted skew-normal process. *Journal of Statistical Computation and Simulation, 89*(1), 38–51.
104. Zeroual, A., Harrou, F., Dairi, A., & Sun, Y. (2020). Deep learning methods for forecasting covid-19 time-series data: A comparative study. *Chaos, Solitons & Fractals, 140*, 110121.
105. Zhong, L., Mu, L., Li, J., Wang, J., Yin, Z., & Liu, D. (2020). Early prediction of the 2019 novel coronavirus outbreak in the mainland China based on simple mathematical model. *IEEE Access, 8*, 51761–51769.

Image Reconstruction for COVID-19 Using Multifrequency Electrical Impedance Tomography

Julia Grasiela Busarello Wolff, David William Cordeiro Marcondes, Wellington P. dos Santos ⓘ, and Pedro Bertemes-Filho

1 Introduction

Electrical impedance tomography (EIT) is a technique for visualizing the passive electromagnetic properties of a biological tissue, human or animal organ. The electromagnetic properties that can be visualized in an EIT system are an electrical conductivity $(\int)$, an electrical permittivity $(\sum)$, a magnetic permeability $(\int)$, or an electrical impedance [1].

When the tissue is excited by an alternating current in a frequency range, an electrical potential is generated in the conductive volume under study. This current injection occurs between two neighboring (adjacent) electrodes, and the voltage measurement is performed by two other electrodes, adjacent or not.

In the last 40 years, there has been a great academic production on electrical impedance tomography – EIT [2–22].

The interest in this tomography modality arose due to its characteristics, such as, the rapid production of images [23–25], the low cost of implementing electronic circuits [26, 27], the nonemission of ionizing radiation [1, 28], obtaining images in real time [29], besides the technique being noninvasive [1, 9]. While the spatial resolution of these images is relatively low, the temporal resolution of data from the EIT may be high [28].

To improve the resolution of the generated images, the number of electrodes used in the tomograph is increased. This increases the number of elements in the

J. G. B. Wolff · D. W. C. Marcondes · P. Bertemes-Filho (✉)
Department of Electrical Engineering, Universidade do Estado de Santa Catarina, Joinville, Santa Catarina, Brazil

W. P. dos Santos
Department of Biomedical Engineering, Federal University of Pernambuco, Recife, Pernambuco, Brazil

© The Author(s), under exclusive license to Springer Nature Switzerland AG 2022
S. K. Pani et al. (eds.), *Assessing COVID-19 and Other Pandemics and Epidemics using Computational Modelling and Data Analysis*,
https://doi.org/10.1007/978-3-030-79753-9_19

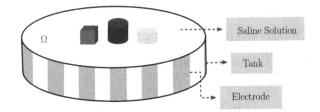

Fig. 1 Electrode tray for potential measurement and visualization of object characteristics

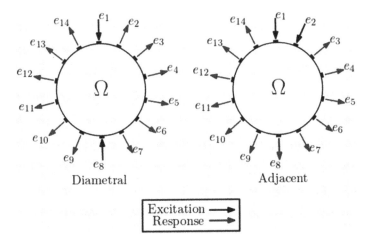

Fig. 2 Measurement standards. (Source: The authors, 2021)

sensitivity matrix. The measurements of potential drop in each pair of electrodes must be carried out in a given frequency range. It was observed that, when using the potential difference measurement method between neighboring interval electrodes, the crosstalk is reduced [30].

The algorithms used to generate tomographic images have a strong influence on the resolution of the images. The electronic instrumentation itself also interferes with the quality of the images generated [1, 9, 31].

Figure 1 shows a circular vat with stainless steel electrodes located around it, filled with a saline solution. The objects arranged inside it represent an organ, tissue, or bone.

The objective is to measure the properties of the objects or to visualize them through a sliced image of the vat.

The EIT does not use ionizing radiation, which offers safety to the patient, as it can be used in the treatment of neoplasms and mechanical ventilation, in ICU beds, through an electrode belt placed on the patient's body. Through this belt with electrodes, an alternating electric current of low intensity and high frequencies is injected, in order to measure the resulting electrical voltages in the other electrodes. Figure 2 shows two current injection and voltage measurement schemes.

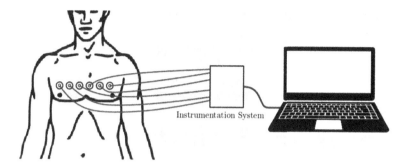

Fig. 3 Electrode belt with computer interface for analysis of reconstructed images. (Source: The authors, 2021)

Figure 3 shows the electrodes attached to the human chest. The data collected from the potential difference generated in a given number of electrodes are sent to a computer, in which an image reconstruction and optimization algorithm shows the characteristics of the organ on the screen.

A complete scan cycle around the chest generates an image of the chest. This cycle maps the variation of the impedance ($\otimes Z$), the change in permittivity ($\otimes \sum$), or the change in conductivity ($\otimes \int$) in the organ under analysis.

A standard EIT system consists of a Howland current source, which injects standardized values of sinusoidal electrical current into the electrodes, a data acquisition unit that collects measurements of potential differences between the electrodes, an analog to digital converter (A/D), several transimpedance amplifiers, switching circuits, filters, and an even number of electrodes. A waveform is created by an oscillator, which modulates a variable source in frequency and amplitude, with a high impedance output. In most EIT systems, the excitation/injection signal is an alternating current, in which case it is multiplexed to the electrodes arranged around the organ under analysis. The measurement electrodes are connected to an instrumentation amplifier, whose amplified output is demodulated and converted into digital values for further processing and image reconstruction, via software ([32], p. 23).

Current injection patterns can be performed in different ways. The most common are the adjacent and diametrical patterns ([28], p. 56).

The adjacent pattern consists of applying the current to two adjacent electrodes, where one electrode is next to the other. In the diametrical pattern, the current is injected into two diametrically opposed electrodes. The potential difference is measured in the other electrodes that make up the EIT system. There are other patterns known as multiport and multiterminal, in which a pair of electrodes is called a port and a single electrode is called a terminal. Thus, in the multiport pattern, a current is applied to a pair of electrodes (port) and the resulting potential differences are measured in other pairs of electrodes along the object. In the multiterminal pattern, the current is injected into one electrode (terminal) and the transfer currents are measured in the other electrodes in relation to a fixed reference that, generally, is the "ground" of the system ([32], p. 24).

There are other less-used patterns which use the sinusoidal, triangular, trigonometric functions [33], and the Dirac delta ([28], p. 56). On the other hand, some EIT systems, instead of injecting sinusoidal electrical currents into the electrodes attached to the object, inject alternating electrical potentials and measure the resulting currents. ([28], pp. 54–55).

Voltage measurements can use two or four electrodes, which is a four-wire measurement. The four-wire measurement technique is used with the current injected by the external electrodes and the difference in voltage measured between the internal electrodes. Some systems apply a potential difference to the object, instead of current, which is technically simpler; however, in these cases, the potential difference is more sensitive to noise. An alternating current is injected between electrodes 2 and 3 and a measurement scan of the resulting potentials is carried out from electrode 4, that is, between electrodes 4 and 5, then 5 and 6, and so on.

2 Hardwares EIT and MfEIT

EIT systems have continued to evolve since their conception in the 1970s and 1980s. Modern digital systems today include several programmable logic devices, such as digital signal processors (DSPs) and programmable port arrays (FPGAs). There are several types of EIT systems. Regarding the structure, they can be categorized as serial, parallel, distributed, or adaptive systems [34]. EIT systems can also be categorized according to the frequency of operation, which can be single or multiple frequencies [1]. A system that operates with a single frequency is said to be monofrequency and can be used in applications such as tumors and bones. Multifrequency systems use a range with different frequency values and their purpose is to monitor brain functions [15, 35–39] and pulmonary functions [40–43] and perform breast imaging [6, 44].

In the EIT, 1 mA of electrical current is generally applied [45]. The application of excitation signals over a frequency range is called electrical impedance spectroscopy. In this case, the variation of the impedance often allows to generate more reliable and significant images than those generated with a mono-frequency tomograph.

In the 1980s and 1990s, the first EIT systems used frequencies in the range of 10 to 20 kHz. Over time, the latest tomographs use frequencies from 1 to 10 MHz [9].

Many modern CT scanner designs have multiple frequencies, but there are certain technical problems that are still reported in the literature. They can, however, also be classified according to the number of electrodes, which can vary between 8, 16, 32, 64, and 128 electrodes [34, 46]. And yet, the systems can be differentiated in terms of dimensionality, that is, two-dimensional or three-dimensional. In these tomographs, the electrodes are positioned on one or more tomographic planes.

Table 1 EIT hardwares

EIT systems	References
Sheffield Mark 1	Brown et al. [47]
Sheffield Mark 2	Smith et al. [19]
Sheffield Mark 3	Brown et al. [30]
Sheffield Mk 3.5	Wilson [48]
OXPACT II	Lidgey et al. [49], Zhu et al. [50]
OXPACT III	Zhu et al. [51], McLeod [52]
OXPACT V	Bayford [1]
UCLH Mark 1b	Yerworth et al. [53]
ACT3	Edic et al. [29], Cook [26], Newell et al. [54], Gisser et al. [55]
ACT 4	Liu et al. [56]
Dartmouth system	Hartov et al. [57]
EIT-4 sys	Casas et al. [58]
KHU Mark 1	Oh et al. [59, 60]
KHU Mark 2.5	Mcewan et al. [39]
IMPETOM	Santos et al. [61]
Timpel Enlight 1800	Lima et al. [62], Timpel [63]
PulmoVista 500	Dräger [64]
KIT4 system (University of Eastern Finland)	Hauptmann et al. [65]

Source: The authors (2021)

Some EIT systems that have been developed over the past 40 years are shown in Table 1.

Changes in the electrical properties of the object whose image is to be reconstructed may represent a tumor in an organ [6, 15], pH changes during electrolysis [13] can assist in artificial mechanical ventilation in the ICU beds [1, 22, 66], in addition to having potential applications in the industry [67, 68], in geophysics [69], and damage in tubulations [70].

Images of the interior of the object are formed based on the distribution of passive electrical properties of the object [71, 72]. For this reason, the EIT technique has been potentially used in several medical applications, as shown in Table 2.

Different clinical applications are only possible given that there are typical differences between healthy tissues and tumor tissues, since tumor tissue has a higher permissiveness and conductivity value compared to normal tissues. This reason is due to the fact that tumor cells have a higher water content and sodium concentration than normal cells, as well as, they have different electrochemical properties in their cell membranes. Biological tissues, in the frequency range of 100 kHz, which is the range of interest of MfEIT, have electrical conductivity and permittivity values as shown in Table 3, adapted from Andreuccetti et al. [104].

Visualization of pulmonary activity via EIT is only possible given that the empty lung has twice as much resistance as the lung filled with air. Anomalies in the tissues also modify the region in which they are present; the impedance is lower in the case

Table 2 EIT for medical diagnosis

Medical application	References
Lung function	Henderson and Webster [169], Hahn [73], Mellenthin et al. [42], Kobylianskii et al. [74], Wals et al. [75], Rosa et al. [66], Teschner and Imhoff [22]
Prostate tumors	Borsic et al. [76], Wan et al. [77], Borsic et al. [78]
Breast cancer	Cherepenin, [6], Zou and Guo [79], Hong et al. [80]
Skin cancer	Barber and Brown, [3], Aberg et al. [81], Braun et al. [82]
Cerebral anomalies	Abboud et al. [83], Holder [84], Dowrick et al. [36]
Gastric function	Trokhanova et al. [85], Mangnall et al. [86]
Oral tumors	Murdoch et al. [87], Richter et al. [88], Sun et al. [89]
Monitoring of cardiac function	Vonk-Noordegraaf et al. [90], Deibele et al. [91], Vonk-Noordegraaf et al. [92], Maisch et al. [93]
Cervical cancer	Trokhanova et al. [94], Barrow and Wu [95], Trokhanova et al. [85]
Location of epileptic outbreaks	Holder [9]
Images of brain activity	Carpenter [96], Romsauerova [15]
Changes in pH during electrolysis	Meier and Rubinsky [13]
Thoracic textiles for the measurement of lung characteristics in neonates	Sophocleous et al. [97]
Tools for assistance on COVID-19 in U.T.I.	Chen et al. [98], Diehl et al. [99], Tomasino et al. [100], Zhao et al. [101, 102], Shono et al. [103]

Source: The authors (2021)

of a tumor, and the resistance increases due to the air in the case of a pneumothorax (air sac between the lung and the pleura) [105].

3 Types of Images Generated

Multifrequency electrical impedance tomography (MfEIT) is a relatively new imaging method that has evolved over the past 40 years. It has been successfully applied in the biomedical area, in situations where it is desired to obtain images of the impedance or conductivity distribution, in a specific organ, for the clinical diagnosis of neoplasms. However, MfEIT uses algorithms for data processing and image reconstruction of biological tissues under analysis. This systematic review aims to analyze the evolution of the technique, from its beginning in 1983 to the present day, both in terms of hardware and software. Through this review, deficiencies in the techniques and computer programs for image reconstruction used so far will be evidenced, suggesting alternative solutions for the improvement of these technologies.

Images generated by EIT systems can be of the monofrequency type (MnEIT) and multifrequencial (MfEIT). MnEIT uses a single frequency value for the

Table 3 Conductivities and biological permissivities at 100 kHz

Biological tissue	Electrical conductivity	Electrical permittivity
Aorta	0.3187	929.93
Bladder	0.2189	1231.1
Blood	0.70292	5120
Blood vessel	0.3187	929.93
Body fluid	1.5	97.99
Cancellous bone	0.083892	471.71
Cortical bone	0.020791	227.64
Bone marrow	0.0033172	110.72
Gray brain matter	0.13366	3221.8
White brain matter	0.081845	2107.6
Breast fat tissue	0.025048	70.61
Cartilage	0.17854	2572.2
Cerebellum	0.1537	3515.3
Cerebrospinal fluid	2	109
Cervix	0.5476	1750.8
Colon	0.24778	3722
Cornea	0.49934	10,567
Duodenum	0.53605	2860.9
Tough	0.50187	326.33
Sclera of the eye	0.51848	4745.3
Fat	0.024414	92.885
Gallbladder	0.90014	107.26
Gallbladder bile	1.4	120
Gland	0.53697	3301.2
Heart	0.21511	9845.8
Kidney	0.17134	7651.6
Lens	0.34012	2067.6
Liver	0.084568	7498.9
Deflated lung	0.27161	5145.3
Inflated lung	0.10735	2581.3
Lymph	0.53697	3301.2
Mucous membrane	0.065836	15357
Muscle	0.36185	8089.2
Nail	0.020791	227.64
Nerve	0.080776	5133
Esophagus	0.53605	2860.9
Ovary	0.33939	1941.7
Pancreas	0.53697	3301.2
Prostate	0.43861	5717
Retina	0.51848	4745.3
Dry skin	0.00045128	1119.2

(continued)

Table 3 (continued)

Biological tissue	Electrical conductivity	Electrical permittivity
Moist skin	0.065836	15357
Small intestine	0.5942	13847
Spinal cord	0.080776	5133
Spleen	0.12218	4222.3
Stomach	0.53605	2860.9
Tendon	0.38853	472.42
Thymus	0.53697	3301.2
Thyroid	0.53697	3301.2
Tongue	0.28795	4745.6
Teeth	0.020791	227.64
Trachea	0.33801	3734.8
Uterus	0.53144	3411.3
Vitreous humor	1.5	97.99

Source: Adapted from Andreuccetti et al. [104]

reconstruction of images of the distribution of passive properties within the organ under analysis.

MfEIT injects current into an object and measures the potential difference at various sinusoidal frequencies, generating a Bode diagram for analyzing the frequency spectrum of the object to be visualized, as shown in Figs. 4 and 5.

The EIT is based on the measurement of electrical impedance, that is, the module of the total complex resistance of the human organ or tissue, as shown in Eq. (1):

$$|Z| = \sqrt{R^2 + X^2}. \tag{1}$$

When an alternating current is applied to an object, the energy is dissipated as it travels through it, thus producing a voltage that can be measured. In addition, the time between the measured voltages is out of phase with the current values. Equation (2) shows the impedance phase:

$$\theta = \tan^{-1}\left(\frac{jX}{R}\right). \tag{2}$$

With Eqs. (1) and (2) distributed over a frequency spectrum, two Bode diagrams are obtained.

Differences between two EIT techniques are evident in the reconstruction of images. MfEIT generates static or absolute images with low precision, so, for example, it does not detect water or blood in the lungs. MfEIT generates dynamic or differential images of organs with more visible contours, which allows a more realistic image. Differential images are constructed using the difference between the data collected at different times of frequency.

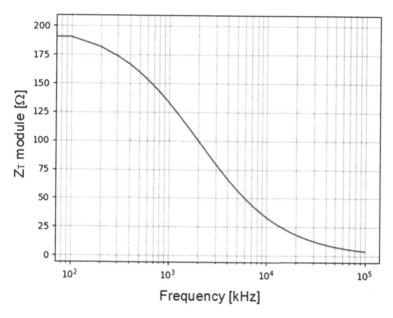

Fig. 4 Electrical impedance module – Bode diagram. (Source: The authors, 2021)

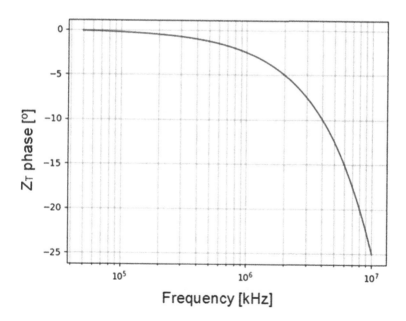

Fig. 5 Electrical impedance phase – Bode diagram. (Source: The authors, 2021)

Therefore, the justification for this proposal is due to the fact that the images in MfEIT still have low resolution and slow reconstruction process, when compared to other available tomography techniques. This work will use an MfEIT system and the two-dimensional D-bar method for reconstructing the images of tissues and organs in patients with COVID-19.

4 Forward Problem and Inverse Problem in MfEIT

In MfEIT there are two mathematical problems to be solved, namely, the forward problem [46, 106–108] and the inverse problem [109–113]. Most methods to solve the forward problem use finite elements [108, 114, 115].

Inverse problem in MfEIT is nonlinear, poorly placed, and ill-conditioned, that is, extremely sensitive to modeling errors and measurement noise. A typical case of an ill-conditioned system is one whose coefficients of the conductivity or resistivity matrix are very close to that of a single matrix. It is well known in the literature that the matrices that characterize linear systems can behave in three different ways :

- Matrices admit inverse, that is, given a square matrix $[A]$, of "n" order, its inverse $[A^{-1}]$ is calculated, which is unique and not singular.
- Matrices do not admit inverse, that is, given a square matrix $[A]$, of "n" order, you cannot find your inverse matrix $[A^{-1}]$.

The singularity is related to the fact that the determinant of the coefficient matrix $[A]$ is equal to zero.

From a mathematical point of view, tomographic reconstruction can be considered as a common class of inverse problems. Inverse problems are generally poorly placed and poorly conditioned. A problem is poorly conditioned or sensitive, if small changes in excitation cause major errors in the final result.

For a linear system, given the generic equations in compact form $[A].[X] = [b]$, conditioning number, Cond (A), by definition, is given by calculating mathematical expression shown in Eq. (3):

$$\text{Cond}(A) = \|A\| . \left\|A^{-1}\right\| . \tag{3}$$

where $\|A\|$ is the Euclidean norm of sensitivity matrix and $\|A^{-1}\|$ is the Euclidean norm of inverse matrix of $[A]$.

Sensitivity can be defined in relation to the amplitude or phase variation of the electrical impedance. Measurement of small phase variations is less susceptible to errors and interference than the measurement of small variations in amplitude [72]. For this reason, there are EIT systems that measure the differences in amplitude and phase of the impedance of the object whose image will be reconstructed [1, 116].

Thus, it can be concluded that the greater the number of conditions in the matrix $[A]$, the greater the sensitivity of the system, that is, matrices with a high number of conditions are called quasi-singular matrices. They differ from singular matrices because they admit to having inverses, but as the coefficients are very close to zero or the lines are linearly dependent on each other, therefore, the system solution is strongly influenced by small modeling or measurement errors. This includes numerical truncation due to finite precision in storing and processing values on a computer [72].

Small variations in the coefficients of the matrices make the solutions very distinct, that is, small variations in the input data resulted in large variations in the system solution. When this happens, we say that the problem is ill-conditioned.

The Tikhonov regularization method was developed by David Phillips and Andrey Tychonoff and aims to obtain approximate solutions for ill-conditioned systems, adding additional information to the desired solution, in order to reduce the number of conditioning. The conditioning number of a square matrix of "n" order represents the capacity that an almost singular matrix has to be treated numerically, that is, regularized or optimized in order not to generate reconstructed images that do not correspond to the original image.

Thus, Tikhonov's regularization method leads us to minimize the following problem shown in Eq. (4):

$$\min \|Ax - b\|_2^2 + \lambda^2 \|L\,(x - x_0)\|_2^2. \tag{4}$$

minimizing the regularization term L related to the minimization of the residual Euclidean norm. If we make the regularization matrix L equal to the identity matrix of the same order, this matrix regularization is equivalent to minimizing the solution norm.

The most appropriate value of the regularization parameter $\lfloor$ is obtained from the L curve. This graph is shown in Fig. 6.

5   Usual Software and Methods for Image Reconstruction

Several methods for solving the inverse problem in tomography systems have already been suggested in the literature, as shown in Table 4.

Most used and available software is EIDORS – *Electrical Impedance and Diffuse Optical Tomography Reconstruction Software* [116], DICOM – *Digital Imaging and Communications in Medicine* [143], and GREIT – *Graz consensus Reconstruction algorithm for EIT* [144].

Figure 7 shows a reconstructed image from the EIDORS software. The small reddish circles in the center of the object represent conductivities of different values.

The number of independent measurements in an MfEIT system varies depending on the number of electrodes in domain, the current injection pattern, and the potential difference measurement pattern. It indicates the quality of the spatial resolution and the noise level in the images. By increasing the number of electrodes, resolution and quality of image reconstructed can be improved. Instrumentation

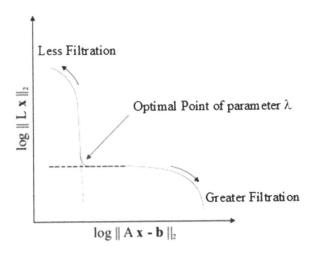

Fig. 6 L curve for determining optimal parameter $\lfloor$. (Source: [72], p. 54)

Table 4 Software used in EIT systems

Numerical method	References
Backprojection	Barber and Brown [117]
Filtered Backprojection	Gonçalves [118], Barber et al. [119]
Tikhonov Regularization	Sun et al. [89], Cohen-Bacrie et al. [120], Dusek et al. [121], Liu et al. [122]
Classical Variational Methods	Kohn and McKenney [123], Kohn et al. [124], Knowles [125]
Total Variation Smoothing Method	Borsic et al. [109]
Gauss-Newton Method	Polydorides et al. [126], Brandstatter [127], Islam and Kiber [128]
Regularized Gauss-Newton Method	Bakushinskii [129], Brandstätter [127]
Newton-Raphson Method	Rao et al. [130]
D-bar Method	Alsaker [131], Dangelo and Mueller [132], Mueller and Siltanen [133], Isaacson et al. [134], Hamilton et al. [135, 136]
Neural Networks	Hrabuska et al. [137], Hamilton and Hauptmann [138], Agnelli et al. [139]
Deep Learning	Gomes et al. [140], Siltanen and Ide [141]
Genetic Algorithms	Feitosa et al. [142]

Source: The authors (2021)

used in the construction of tomograph, as well as the appropriate choice of forward and inverse methods for image reconstructing, also influences the reliability and accuracy of images generated [146, 147].

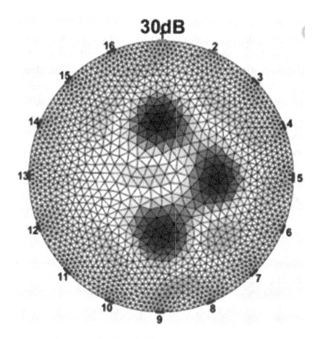

Fig. 7 Image reconstruction using EIDORS. (Source: EIDORS [145])

6 EIT Applications with Respect to COVID-19

Coronavirus disease appeared in the world in December 2019 in the city of Wuhan and it has caused panic in the population. The first clinical features of the disease characterized it as pneumonia, preceded by fever, sore throat, nasal congestion, fatigue, shortness of breath, headache, and other less frequent symptoms, such as diarrhea, myalgia, nausea, or vomit. In view of this clinical picture, some scientific communities around the world have proposed the use of EIT as an adjunct in the individualized treatment of patients with severe symptoms that need to be intubated and monitored in the ICU [148].

EIT can improve regional lung parameters and provide useful data on the patient's health status, time, and intensity that are unique to each individual, as well as assessing an autoimmune response to the drugs administered. It assists in volume, current and positive expiratory pressure (PEEP); in the choice of the withdrawal mode; in the removal of the mechanical ventilator; in the titration and recruitment; in perioperative monitoring; and in the monitoring of pulmonary perfusion [102, 149, 150].

These studies have shown that EIT provides monitoring at the ICU bedside of the patient's response to various drug interventions and helps guide treatments, as it closely monitors lung function. Based on this, we proposed an MfEIT for use at the bedside in patients with severe symptoms of COVID-19. The system can subsequently be used in several units of the Unified Health System (SUS) in Brazil as an adjunct in the individual monitoring of pulmonary ventilation. The D-bar method performs better than traditional methods, with Tikhonov regularization, backprojection, and finite element method, for example. With this, it is intended to expand the possibilities of using the multifrequency EIT tomography and to contribute to the advancement of regional science and technology.

Although some experimental and clinical results have been obtained in Europe and Asia, according to citations in the last line of Table 2, portable electronic instrumentation, safe, secure and accurate images are still aspects to be researched in MfEIT for monitoring the pulmonary activity during the treatment of COVID-19-associated acute respiratory distress syndrome (ARDS).

The accumulation of fluids and secretions in the airways compromise the severity of the lungs during concentration and an ARDS associated with COVID-19 can lead to death. In this context, an EIT is effective in managing the effects of different levels of PEEP, pulmonary mechanics, aeration asymmetry between the right and left lungs, lungs that are poorly recruitable or with increased recruitability, and alternation of body positions between supine and supine prone.

Therefore, the objective of this work is to optimize the open-source 2D D-bar algorithm for imaging in electrical impedance multifrequency tomography systems for use in patients with COVID-19.

7 D-Bar Method

Several numerical methods have already been employed to solve direct and inverse problems; however, after extensive literary research in the EIT area, we chose to use two-dimensional D-bar method for obtaining medical images. It is a method that has evolved over the years and has very promising results compared to other reconstruction methods. D-bar method is a noniterative method, that is, a direct method that promotes the regular reconstruction of images, based on a low-pass filter of a nonlinear Fourier transform. Conductivity data of the organ under analysis is filtered and then the Fourier transform is inverted. Over the years of research, D-bar method has been improved so that there is no need to solve the forward problem in MfEIT, so this method is more attractive than traditional methods [151]. They consist of a family of computational inversion methods, which involve Fourier transforms, scattering transforms, CGO solutions, which are complex geometric optical solutions. Several modifications to the method are necessary to deal with inherent problems, such as noise and artifacts in the data, and to deal with the finite nature of the measurements obtained, in practice, with the MfEIT system [152].

Regularization is provided by low-pass filtering in the domain of nonlinear frequency. This method is characterized by finding the solution of the inverse problem through exponentially increasing solutions (Complex Geometrical Optics Solutions – CGO) and the inverse scattering theory. The method allows nonlinear reconstruction in one step and without having to estimate intermediate conductivities.

In a multifrequency EIT system, data is obtained by measuring voltages that arise in the surface electrodes after the injection of a low amplitude electrical current. For L electrodes, linearly independent (L-1) current patterns are applied to understand a frame in the data set, and the Dirichlet-Neumann map will be calculated. There are several formulations for MfEIT using the D-bar method in the literature [33, 153, 154].

8   Future of MfEIT

Several authors believe that the future of electrical impedance tomography is in hybrid systems. Hybrid systems are composed of two electrical and/or radiological techniques. These systems have been reported in the literature since the 1990s [46, 155–158].

One way to improve the detection and classification of tumors and other clinical applications is to combine the EIT with other imaging modalities, such as ultrasound, magnetic resonance, and computed tomography.

According to Bayford [1], there is interest from the scientific community in the development of new systems, which use another image modality combined with EIT. The research excels in noncontact methods that use solenoids to induce electric current or measure the generated field [1].

The combination of magnetic resonance imaging (MRI) and electrical impedance tomography is a technique known as electrical impedance tomography by magnetic resonance (MREIT). Magnetic resonance electrical impedance tomography (MREIT) is a method of producing medical images of the electrical conductivity of a conductive object. It aims to replace EIT and MIT because it offers reconstructed images with greater accuracy. It was developed from 1990 and is also called a hybrid system. The systems of electrical impedance tomography by magnetic resonance were improved after the year 2001, with the objective of solving the inverse problems badly placed, in image reconstruction in the conventional EIT systems [46, 159–163]. There is a strong tendency to mix electrical impedance tomography with magnetic resonance.

According to Seo and Woo [164] "several experiences show that data sets measurable results in an EIT system are insufficient for a robust reconstruction of a high-resolution static conductivity image, due to its poorly laid out nature and influences of errors in direct modeling." . MREIT uses internal density data of magnetic flux in three dimensions, which is induced by an injected current externally to the system. MREIT uses an MRI scanner as a tool to measure the magnetic induction component in the z direction. That is, z is the axial magnetization direction of the magnetic resonance scanner [164].

The relationship between current and voltage limit data and the distribution of internal conductivity has a nonlinearity and low sensitivity. Hence, the inverse

Table 5 Systematic review of hybrid systems

Hybrid systems	Researchers group
EIT with induced currents imaging achieved with a multiple coil system	Freeston and Tozer [155]
MIT with magnetic excitation with coils and measurement of surface potentials with electrodes	Gençer et al. [156]
MIT with current injection via electrodes and sensing of the external magnetic field with coils	Tozer et al. [157]
MREIT (KHU)	Woo et al. [46], Kwon et al. [159], Seo et al. [166], Woo and Seo [165], Seo and Woo [164]
Injected current and induced current MREIT	Liu et al. [56]
ETS (electromagnetic tomography system)	Wang et al. [167]

Source: The authors (2021)

problem that recovers the conductivity distribution is ill-posed and misplaced. Another aspect of MfEIT is that it is difficult to obtain accurate information about the limit of the geometry of the electrodes, as well as their positions around the object. In practice, they will change places with each measurement rotation. With that, the inverse problem is sensitive to these modeling errors and to the artifacts inherent in measurements and noise. All of these factors result in EIT images with a poor resolution space. For these reasons, tomography by electrical impedance resonance (MREIT) was developed, with the objective of producing images of high resolution. It generates conductivity images with a spatial resolution of some millimeters or less. This makes it a very useful tool in medical practice [165]. In 2008, the MREIT technique evolved into human experiments and it is currently well known in South Korea, England, the United States of America, and Finland. Suggestions for future research applications in MREIT are given in the areas of biomedicine, biology, chemistry, and materials science. Table 5 shows some hybrid EIT systems described in the literature up to the present.

In imaging of organs and biological tissues, there is a long way to go, such as use of deep neural networks to accelerate numerical methods, as well as genetic algorithms and clusters.

9   Methods and Materials

In this work, we use the two-dimensional D-bar method scripts developed by Siltanen et al. [168] to generate some test images. Later, we intend to apply these images in a low-cost EIT system for monitoring respiratory diseases and COVID-19, as well as their variants, at the edge of ICU beds.

The computer used was a Centrium with an Intel Xeon processor, CPU E3-1231 v3, 3.4 GHz, and 16 GB of RAM. The images were reconstructed from original images simulated in MalLab 2013b. Several sets of conductivity have been defined for the lungs and the heart, in order to simulate the most unusual images of respiratory and cancerous pathologies.

We will show how to simulate the voltage data for current distribution and how to recover the internal conductivity of the contours, using two-dimensional D-bar method. The algorithms were obtained from the following website: https://blog.fips.fi/author/samu/. The codes were compiled in the following sequence:

(i)   heartNlungs.m
(ii) DbarEIT01_heartNlungs_plot.m
(iii) DbarEIT02_mesh_comp.m
(iv) FEMconductivity.m
(v)   DbarEIT03_ND_comp.m
(vi) DbarEIT04_Kvec_comp.m
(vii) DbarEIT05_psi_BIE_comp.m
(viii) solveBIE.m
(ix) DbarEIT06_tBIE_comp.m
(x)   DbarEIT07_tBIErecon_comp.m

The objective of the two-dimensional D-bar method is to obtain internal images of cross sections of patients whose body is surrounded by a 16 electrode band. A certain voltage potential was maintained at each electrode, and measurements of the resulting electric current through the electrodes were collected. This measurement is repeated for several voltage patterns. Numerical phantoms of radial sections were used with the presence of the lungs and the heart.

Assuming that there are 16 electrodes in total, we can use a maximum of 14 linearly independent voltage patterns, since one of the electrodes is considered the potential of the earth and the other potentials are compared to that of the earth. In the tested algorithm, trigonometric voltage patterns approximated by continuous sinusoidal curves in the contour were used. The conductivities of the heart and lungs were simulated in the first two files that are rotated, and then the maps of Dirichlet-Neumann and Neumann-Dirichlet are generated. According to Siltanen [154], the solution to the conductivity equation is given as a linear combination of basic functions that are linear by parts in a triangular mesh. The mesh is constructed by the mesh_comp.m routine containing the Nrefine parameter. Here we use Nrefine = 6.

10 Discussions and Results

The two-dimensional D-bar method has shown promising results in 2D reconstruction of electrical impedance tomography images. The colors of the simulated and reconstructed original images represent the numerical values of conductivity in the organs and, therefore, are directly comparable. Note that reconstruction underestimates the magnitude of electrical conductivity in the heart. This is a characteristic of the truncated Fourier nonlinear transform, calculated in the D-bar method, already predicted by Siltanen [154]. He approached the 3D situation from the D-bar method to a 2D computational model. In practice, this model generates relevant and quick results.

The experimental results show that the method can be used to monitor PEEP and other pulmonary activities in patients positive for COVID-19 as a support device. Three different scenarios were simulated in Matlab to provide the effectiveness and robustness of the algorithm. Each image took about 11 minutes to be reconstructed.

We performed three sets of simulated tests to obtain reconstruction of medical images of the heart and lungs::

- Conductivity of the heart remains constant at 2 S/m, while the conductivity of the lungs varies from 0.1 to 2.5 S/m.
- Conductivity of the lungs remains constant at 0.4 S/m (inspiration), while the conductivity of the heart varies from 1 to 2.5 S/m.
- Conductivity of the lungs remains constant at 0.7 S/m (expiration), while the conductivity of the heart varies from 1 to 2.5 S/m.

Figure 8 shows the lungs with electrical conductivity of 0.1 S/m and the heart with 2 S/m.

Since the electrical conductivity of the heart is greater than the conductivity of the lungs, the background conductivity has changed.

Figure 9 shows the lungs with electrical conductivity of 0.2 S/m and the heart with 2 S/m.

Figure 10 shows the lungs with electrical conductivity of 0.3 S/m and the heart with 2 S/m.

Figure 11 shows the lungs with 0.4 S/m electrical conductivity and the heart with 2 S/m.

Figure 12 shows the lungs with 0.5 S/m electrical conductivity and the heart with 2 S/m.

Fig. 8 Lungs with 0.1 S/m and heart with 2 S/m

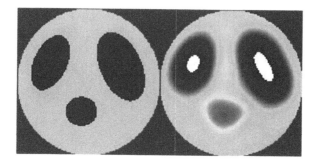

Fig. 9 Lungs with 0.2 S/m and heart with 2 S/m

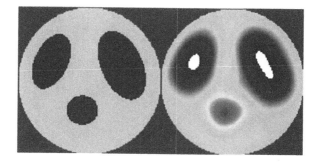

Fig. 10 Lungs with 0.3 S/m and heart with 2 S/m

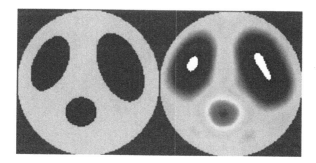

Fig. 11 Lungs with 0.4 S/m and heart with 2 S/m

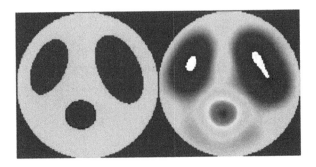

Fig. 12 Lungs with 0.5 S/m and heart with 2 S/m

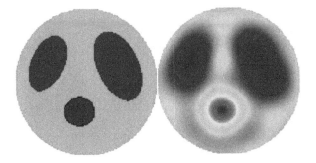

Fig. 13 Lungs with 0.6 S/m and heart with 2 S/m

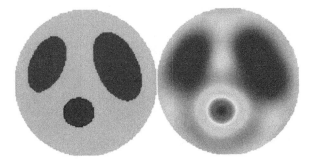

Fig. 14 Lungs with 0.7 S/m and heart with 2 S/m

Figure 13 shows the lungs with 0.6 S/m electrical conductivity and the heart with 2 S/m.

Figure 14 shows a phantom of the lungs with an electrical conductivity of 0.7 S/m and the heart with 2 S/m.

Figure 15 shows a phantom of the lungs with an electrical conductivity of 0.8 S/m and the heart with 2 S/m.

Figure 16 shows a phantom of the lungs with an electrical conductivity of 0.9 S/m and the heart with 2 S/m.

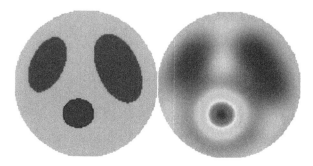

Fig. 15 Lungs with 0.8 S/m and heart with 2 S/m

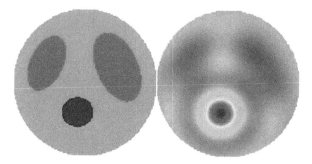

Fig. 16 Lungs with 0.9 S/m and heart with 2 S/m

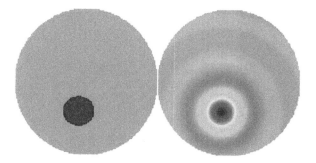

Fig. 17 Lungs with 1.0 S/m and heart with 2 S/m

Figure 17 shows a phantom of the lungs with an electrical conductivity of 1.0 S/m and the heart with 2 S/m.

Figure 18 shows a phantom of the lungs with an electrical conductivity of 1.1 S/m and the heart with 2 S/m.

Figure 19 shows a phantom of the lungs with an electrical conductivity of 1.2 S/m and the heart with 2 S/m.

Figure 20 shows a phantom of the lungs with electrical conductivity of 1.3 S/m and the heart with 2 S/m.

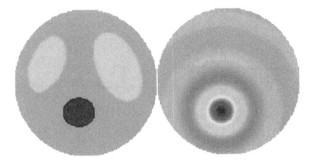

Fig. 18 Lungs with 1.1 S/m and heart with 2 S/m

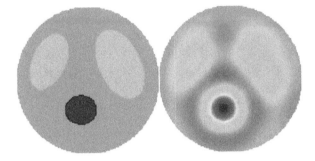

Fig. 19 Lungs with 1.2 S/m and heart with 2 S/m

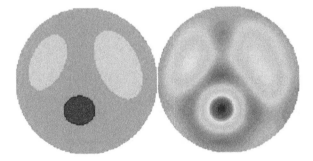

Fig. 20 Lungs with 1.3 S/m and heart with 2 S/m

Figure 21 shows a phantom of the lungs with electrical conductivity of 1.4 S/m and the heart with 2 S/m.

Figure 22 shows a phantom of the lungs with electrical conductivity of 1.5 S/m and the heart with 2 S/m.

Figure 23 shows a phantom of the lungs with electrical conductivity of 1.6 S/m and the heart with 2 S/m.

Figure 24 shows a phantom of the lungs with an electrical conductivity of 1.7 S/m and the heart with 2 S/m.

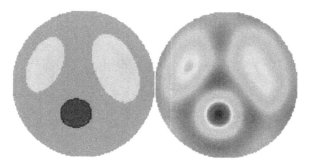

Fig. 21 Lungs with 1.4 S/m and heart with 2 S/m

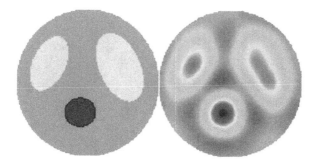

Fig. 22 Lungs with 1.5 S/m and heart with 2 S/m

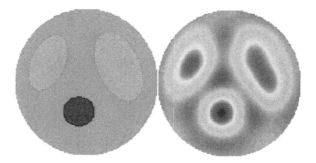

Fig. 23 Lungs with 1.6 S/m and heart with 2 S/m

Figure 25 shows a phantom of the lungs with an electrical conductivity of 1.8 S/m and the heart with 2 S/m.

Figure 26 shows a phantom of the lungs with an electrical conductivity of 1.9 S/m and the heart with 2 S/m.

Figure 27 shows a phantom of the lungs with electrical conductivity of 2 S/m and the heart with 2 S/m.

Figure 28 shows a phantom of the lungs with electrical conductivity of 2.1 S/m and the heart with 2 S/m.

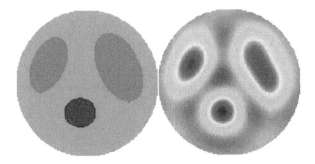

Fig. 24 Lungs with 1.7 S/m and heart with 2 S/m

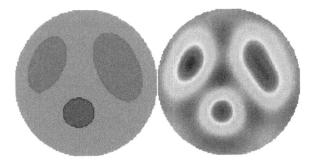

Fig. 25 Lungs with 1.8 S/m and heart with 2 S/m

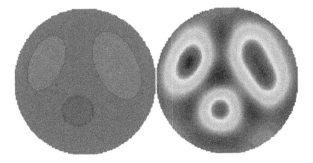

Fig. 26 Lungs with 1.9 S/m and heart with 2 S/m

Figure 29 shows a phantom of the lungs with an electrical conductivity of 2.2 S/m and the heart with 2 S/m.

Figure 30 shows a phantom of the lungs with electrical conductivity of 2.3 S/m and the heart with 2 S/m.

Figure 31 shows a phantom of the lungs with electrical conductivity of 2.4 S/m and the heart with 2 S/m.

Figure 32 shows a phantom of the lungs with an electrical conductivity of 2.5 S/m and the heart with 2 S/m.

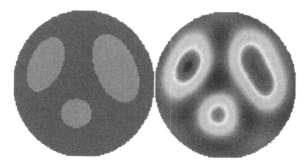

Fig. 27 Lungs with 2 S/m and heart with 2 S/m

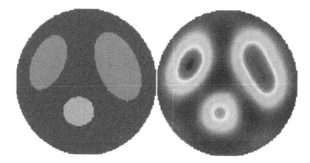

Fig. 28 Lungs with 2.1 S/m and heart with 2 S/m

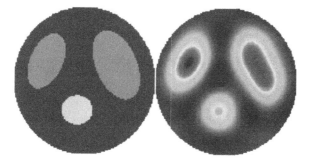

Fig. 29 Lungs with 2.2 S/m and heart with 2 S/m

We selected another set of measures, a title of proof of concept, making the lung with constant conductivity of inspiration 0.4 S/m and varying the electrical conductivity of the heart, we obtain the following original images simulated on the left and reconstructed using the D-bar method on the right. These conductivities of lungs refer to frequency 100 kHz.

Figure 33 shows a phantom of the lungs with an electrical conductivity of 0.4 S/m and the heart with 1 S/m.

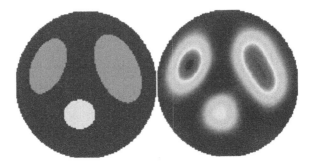

Fig. 30 Lungs with 2.3 S/m and heart with 2 S/m

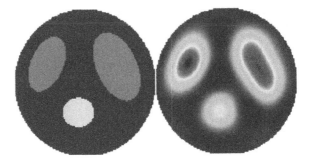

Fig. 31 Lungs with 2.4 S/m and heart with 2 S/m

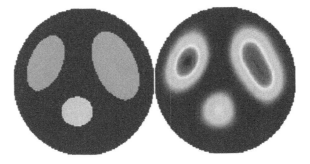

Fig. 32 Lungs with 2.5 S/m and heart with 2 S/m

Figure 34 shows a phantom of the lungs with an electrical conductivity of 0.4 S/m and the heart with 1.1 S/m.

Figure 35 shows a phantom of the lungs with an electrical conductivity of 0.4 S/m and the heart with 1.2 S/m.

Figure 36 shows a phantom of the lungs with an electrical conductivity of 0.4 S/m and the heart with 1.3 S/m.

Figure 37 shows a phantom of the lungs with an electrical conductivity of 0.4 S/m and the heart with 1.4 S/m.

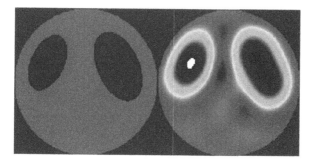

Fig. 33 Heart with 1.0 S/m and lungs with 0.4 S/m

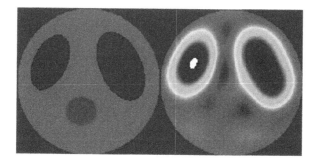

Fig. 34 Heart with 1.1 S/m and lungs with 0.4 S/m

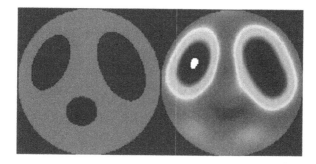

Fig. 35 Heart with 1.2 S/m and lungs with 0.4 S/m

Figure 38 shows a phantom of the lungs with an electrical conductivity of 0.4 S/m and the heart with 1.5 S/m.

Figure 39 shows a phantom of the lungs with an electrical conductivity of 0.4 S/m and the heart with 1.6 S/m.

Figure 40 shows a phantom of the lungs with an electrical conductivity of 0.4 S/m and the heart with 1.7 S/m.

Figure 41 shows a phantom of the lungs with an electrical conductivity of 0.4 S/m and the heart with 1.8 S/m.

Fig. 36 Heart with 1.3 S/m and lungs with 0.4 S/m

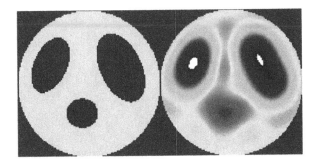

Fig. 37 Heart with 1.4 S/m and lungs with 0.4 S/m

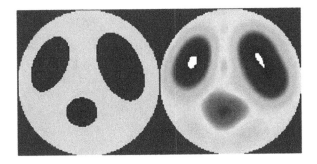

Fig. 38 Heart with 1.5 S/m and lungs with 0.4 S/m

Figure 42 shows a phantom of the lungs with an electrical conductivity of 0.4 S/m and the heart with 1.9 S/m.

Figure 43 shows a phantom of the lungs with an electrical conductivity of 0.4 S/m and the heart with 2.0 S/m.

Figure 44 shows a phantom of the lungs with an electrical conductivity of 0.4 S/m and the heart with 2.1 S/m.

Figure 45 shows a phantom of the lungs with an electrical conductivity of 0.4 S/m and the heart with 2.2 S/m.

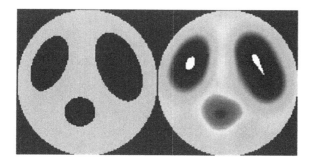

Fig. 39 Heart with 1.6 S/m and lungs with 0.4 S/m

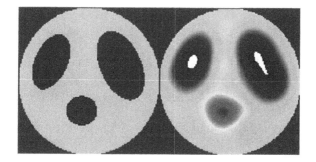

Fig. 40 Heart with 1.7 S/m and lungs with 0.4 S/m

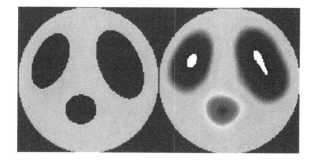

Fig. 41 Heart with 1.8 S/m and lungs with 0.4 S/m

Figure 46 shows a phantom of the lungs with an electrical conductivity of 0.4 S/m and the heart with 2.3 S/m.

Figure 47 shows a phantom of the lungs with an electrical conductivity of 0.4 S/m and the heart with 2.4 S/m.

Figure 48 shows a phantom of the lungs with an electrical conductivity of 0.4 S/m and the heart with 2.5 S/m.

MfEIT combined with the D-bar method can provide images of internal organs which have nonuniform electrical properties, such as the heart and lungs. Biological

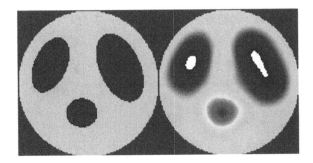

Fig. 42 Heart with 1.9 S/m and lungs with 0.4 S/m

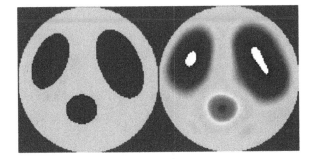

Fig. 43 Heart with 2.0 S/m and lungs with 0.4 S/m

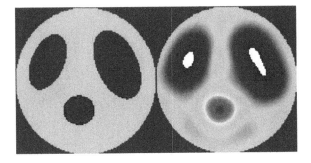

Fig. 44 Heart with 2.1 S/m and lungs with 0.4 S/m

tissues differ considerably in the values of their electrical properties, especially the organs affected by tumors. This can be explained due to the fact that cancerous tissues have a higher concentration of water and sodium compared to normal tissues.

Now, let us simulate a set of data with the conductivity of the lungs constant at 0.7 S/m (expiration) and vary the conductivity of the heart.

Figure 49 shows a phantom of the lungs with an electrical conductivity of 0.7 S/m and the heart with 1 S/m.

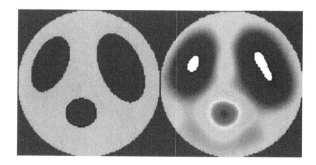

Fig. 45 Heart with 2.2 S/m and lungs with 0.4 S/m

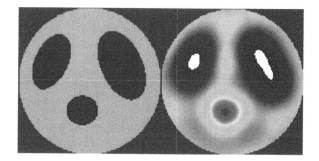

Fig. 46 Heart with 2.3 S/m and lungs with 0.4 S/m

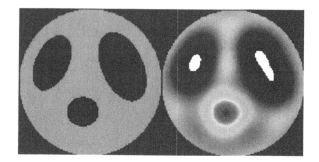

Fig. 47 Heart with 2.4 S/m and lungs with 0.4 S/m

Figure 50 shows a phantom of the lungs with an electrical conductivity of 0.7 S/m and the heart with 1.1 S/m.

Figure 51 shows a phantom of the lungs with an electrical conductivity of 0.7 S/m and the heart with 1.2 S/m.

Figure 52 shows a phantom of the lungs with an electrical conductivity of 0.7 S/m and the heart with 1.3 S/m.

Figure 53 shows a phantom of the lungs with an electrical conductivity of 0.7 S/m and the heart with 1.4 S/m.

Fig. 48 Heart with 2.5 S/m and lungs with 0.4 S/m

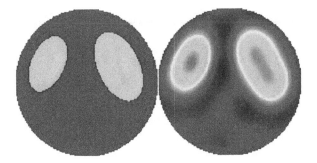

Fig. 49 Heart with 1.0 S/m and lungs with 0.7 S/m

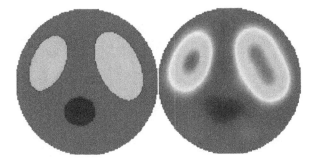

Fig. 50 Heart with 1.1 S/m and lungs with 0.7 S/m

Figure 54 shows a phantom of the lungs with an electrical conductivity of 0.7 S/m and the heart with 1.5 S/m.

Figure 55 shows a phantom of the lungs with an electrical conductivity of 0.7 S/m and the heart with 1.6 S/m.

Figure 56 shows a phantom of the lungs with an electrical conductivity of 0.7 S/m and the heart with 1.7 S/m.

Figure 57 shows a phantom of the lungs with an electrical conductivity of 0.7 S/m and the heart with 1.8 S/m.

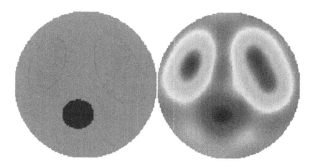

Fig. 51 Heart with 1.2 S/m and lungs with 0.7 S/m

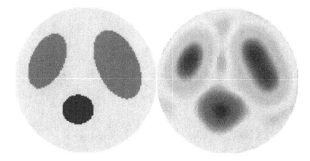

Fig. 52 Heart with 1.3 S/m and lungs with 0.7 S/m

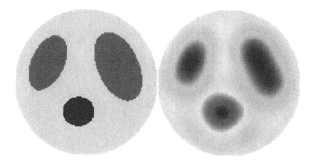

Fig. 53 Heart with 1.4 S/m and lungs with 0.7 S/m

Figure 58 shows a phantom of the lungs with an electrical conductivity of 0.7 S/m and the heart with 1.9 S/m.

Figure 59 shows a phantom of the lungs with an electrical conductivity of 0.7 S/m and the heart with 2.0 S/m.

Figure 60 shows a phantom of the lungs with an electrical conductivity of 0.7 S/m and the heart with 2.1 S/m.

Figure 61 shows a phantom of the lungs with an electrical conductivity of 0.7 S/m and the heart with 2.2 S/m.

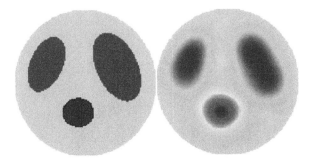

Fig. 54 Heart with 1.5 S/m and lungs with 0.7 S/m

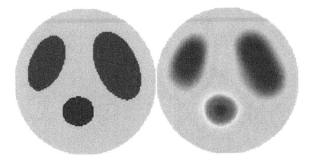

Fig. 55 Heart with 1.6 S/m and lungs with 0.7 S/m

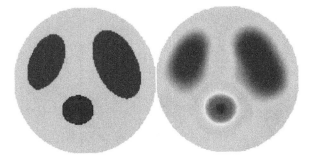

Fig. 56 Heart with 1.7 S/m and lungs with 0.7 S/m

Figure 62 shows a phantom of the lungs with an electrical conductivity of 0.7 S/m and the heart with 2.3 S/m.

Figure 63 shows a phantom of the lungs with an electrical conductivity of 0.7 S/m and the heart with 2.4 S/m.

Figure 64 shows a phantom of the lungs with an electrical conductivity of 0.7 S/m and the heart with 2.5 S/m.

Siltanen [154] points out that the reconstructed images blurry due to the use of the low-pass filter, but this blur can be avoided using deep learning.

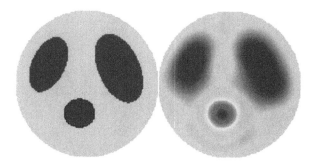

Fig. 57 Heart with 1.8 S/m and lungs with 0.7 S/m

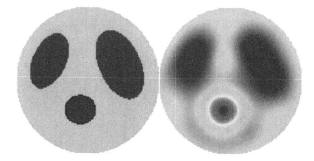

Fig. 58 Heart with 1.9 S/m and lungs with 0.7 S/m

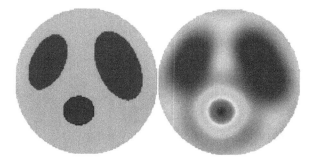

Fig. 59 Heart with 2.0 S/m and lungs with 0.7 S/m

D-bar method applied in an MfEIT system reconstructs the conductivity images with accuracy, for the detection of tumors and for the monitoring of regional pulmonary ventilation and aeration, predominantly applied in mechanically ventilated patients. The numerical results are encouraging and open possibility to be used in conjunction with a low-cost, wireless MfEIT system, at the bedside for managing respiratory activities in the COVID-19 pandemic.

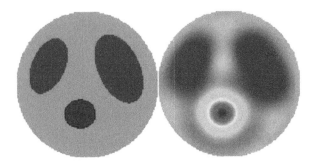

Fig. 60 Heart with 2.1 S/m and lungs with 0.7 S/m

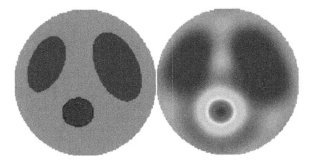

Fig. 61 Heart with 2.2 S/m and lungs with 0.7 S/m

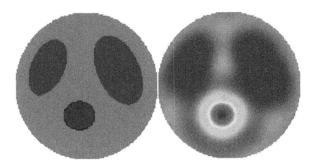

Fig. 62 Heart with 2.3 S/m and lungs with 0.7 S/m

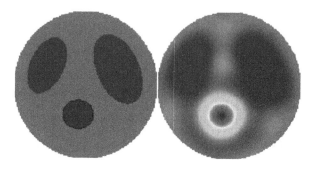

Fig. 63 Heart with 2.4 S/m and lungs with 0.7 S/m

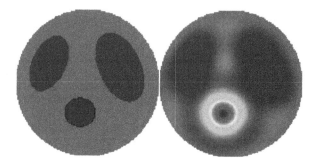

Fig. 64 Heart with 2.5 S/m and lungs with 0.7 S/m

11   Conclusion

D-bar method is a direct method, based on the low-pass filtering of a nonlinear Fourier transform, very promising for use in multifrequency electrical impedance systems for biomedical applications. The original images were simulated with the conductivity of organs: human heart and lung organs at 100 kHz, and the numerical results obtained with the method showed a good agreement with the original image. However, it was noted that in some cases the conductivity of the heart was overestimated in the reconstructed images and that some images of the lungs were somewhat blurred. These aspects can be improved by increasing the number of sources and refining the problem's discretization grid. The advantage of the D-bar method in relation to the others is that it is a noniterative method, therefore, faster and more robust. The images took about 11 minutes to compile.

As our future work, we intend to optimize the processing time of images in MfEIT using deep learning. We suggest that the codes be developed in Python and be applied in the analysis of experimental results obtained in a multifrequency electrical impedance tomography.

Acknowledgments The authors would like to thank UDESC, FAPESC, and UFPE for the institutional support of this project and especially grateful to Professor Dr. Samuli Siltanen and his collaborators for sharing the Matlab scripts we used in this research.

References

1. Bayford, R. H. (2006). Biompedance tomography. *Annual Review of Biomedical Engineering, 8*, 63–91.
2. Ammari, H., Triki, F., & Tsou, C.-H. (2017). Numerical determination of anomalies in multifrequency electrical impedance tomography. *Numerical Analysis (math.NA)*. arXiv:1704.04878 [math.NA].
3. Barber, D. C., & Brown, B. H. (1984). Applied potential tomography. *Journal of Physics E: Scientific Instruments, 17*, 723–733.
4. Brown, B. H. (2003). Electrical impedance tomography (EIT): A review. *Journal of Medical Engineering & Technology, 27*(3), 97–108. https://doi.org/10.1080/0309190021000059687.
5. Cheney, M., Isaacson, D., & Newell, J. C. (1999). Electrical impedance tomography. *Society for Industrial and Applied Mathematics, 41*(1), 85–101.
6. Cherepenin, V. A., Karpov, A. Y., Korjenevsky, A. V., Kornienko, V. N., Kultiasov, Y. S., et al. (2002). Three-dimensional EIT imaging breast tissues: System design and clinical testing. *IEEE Transactions on Medical Imaging, 21*(6), 662–667.
7. Cheng, J., Choulli, M., & Lu, S. (2018, October 12–14). An inverse conductivity problem in multifrequency electrical impedance tomography. In *Inverse problems and related topics*. Shanghai, China.
8. Darnajou, M., Dupré, A., Dang, C., Ricciardi, G., Bourennani, S., & Bellis, C. (2019). On the implementation of simultaneous multi-frequency excitations and measurements for electrical impedance tomography. *Sensors, 19*(17), 3679. https://doi.org/10.3390/s19173679.
9. Holder, D. S. (2004). *Electrical impedance tomography: Methods, history and applications.* Institute of Physics: CRC Press, 456 p. ISBN: 9780750309523. https://doi.org/10.1201/9781420034462.
10. Jiang, Y. D., & Soleimani, D. (2019). Capacitively coupled electrical impedance tomography for brain imaging. *IEEE Transactions on Medical Imaging, 38*(9), 2104–2113.
11. Liston, A., Bayford, R., & Holder, D. (2012). A cable theory based biophysical model of resistance change in crab peripheral nerve and human cerebral cortex during neuronal depolarisation: Implications for electrical impedance tomography of fast neural activity in the brain. *Medical & Biological Engineering & Computing, 50*(5), 425–437.
12. McDermott, B., Avery, J., O'Halloran, M., Aristovich, K., & Porter, E. (2019). Bi-frequency symmetry difference electrical impedance tomography—a novel technique for perturbation detection in static scenes. *Physiological Measurement, 40*(4), 044005.
13. Meier, A., & Rubinsky, B. (2015). Electrical impedance tomography of electrolysis. *PLoS One, 10*(6), e0126332. https://doi.org/10.1371/journal.pone.0126332.
14. Oh, T. I., et al. (2008). Validation of a multi-frequency electrical impedance tomography (mfEIT) system KHU Mark1: Impedance spectroscopy and time-difference imaging. *Physiological Measurement, 29*, 295–307.
15. Romsauerova, A., McEwan, A., Horesh, L., Yerwort, R., Bayford, R. H., & Holder, D. S. (2006). Multi-frequency electrical impedance tomography (EIT) of the adult human head: Initial findings in brain tumours, arteriovenous malformations and chronic stroke, development of an analysis method and calibration. *Physiological Measurement, 27*, S147–S161. https://doi.org/10.1088/0967-3334/27/5/S13.
16. Rymarczyk, T., & Vejar, A. (2019). Multi frequency electrical tomography with re-configurable excitation waveforms. In *2019 applications of electromagnetics in modern engineering and medicine (PTZE)*.
17. Saulnier, G. J., Blue, S. R., & Isaacson, D. (2001). Electrical impedance tomography. *IEEE Signal Processing Magazine, 18*(6), 31–43.
18. Singh, G., Anand, S., Lall, B., SRIVASTAVA, A., & Singh, V. (2019). A low-cost portable wireless multi-frequency electrical impedance tomography system. *Arabian Journal for Science and Engineering, 44*, 2305–2320.

19. Smith, R. W. M., Freeston, I. L., & Brown, B. H. (1995). A real-time electrical impedance tomography system for clinical use-design and preliminary results. *IEEE Transactions on Biomedical Engineering, 42*(2), 133–140. https://doi.org/10.1109/10.341825.
20. Soni, N. K., Paulsen, K. D., Dehghani, H., & Hartov, A. (2005). A 3-D reconstruction algorithm for EIT planner electrode arrays. In *6th conference on Biomedical Applications of Electrical Impedance Tomography.* London, UK. Disponível em: http://www.eit.org.uk. Acesso em: 18/04/2019.
21. Tarek, N. A., Alam, F., Jalal, A. H., & Ahad, M. A. (2019). Multi frequency assessment of the electrical impedance myography parameters on 3D malignant breast. *Biomaterials and Soft Materials, 4*(22), 1285–1291.
22. Teschner, E., & Imhoff, M. (2011). *Electrical impedance tomography: The realization of regional ventilation monitoring.* Lübeck, Alemanha: Dräger Medical GmbH.
23. Adler, A., Dai, T., & Lionheart, W. R. B. (2007). Temporal image reconstruction in electrical impedance tomography. *Physiological Measurement, 28*(7), S1.
24. Filipowicz, S. F., & Rymarczyk, T. (2012). Measurement methods and image reconstruction in electrical impedance tomography. *Przegląd Elektrotechniczny* (electrical review), ISSN 0033-2097, R. 88 NR 6/2012.
25. Kim, K. Y., Kang, S. I., Kim, M. C., Kim, S., Lee, Y. J., & Vauhkonen, M. (2002). Dynamic image reconstruction in electrical impedance tomography with known internal structures. *IEEE Transactions on Magnetics, 38*(2), 1301–1304.
26. Cook, R. D., et al. (1994). ACT3-a high-speed, high-precision electrical-impedance tomograph. *IEEE Transactions on Biomedical Engineering, 41*, 713–722.
27. Kim, D. Y., Wi, H., Yoo, P. J., Oh, T. I., & Woo, E. J. (2010). Performance evaluation of KHU Mark2 parallel multi-frequency EIT system. *Journal of Physics: Conference Series, 224*, 012013.
28. Menin, O. H., Martinez, A. S., & Rolnick, V. (2016). *Tomografia de impedância elétrica: métodos computacionais.* Editora Livraria da Física: São Paulo.
29. Edic, P. M., Saulnier, G. J., Newell, J. C., & Isaacson, D. (1995). A real-time electrical impedance tomograph. *IEEE Transactions on Biomedical Engineering, 42*, 9. [PubMed: 7558059].
30. Brown, B. H., et al. (1994). Multi-frequency imaging and modelling of respiratory related electrical impedance changes. *Physiological Measurement, 15*, 1–12.
31. Bera, T. K. (2018). Applications of electrical impedance tomography (EIT): A short review. In *International conference on Communication Systems (ICCS-2017). IOP Publishing. IOP Conference Series: Materials Science and Engineering* (Vol. 331, p. 012004). https://doi.org/10.1088/1757-899X/331/1/012004.
32. Rocha, M. D. L. (2019). *Um modelo de pesos para a reconstrução de imagens em tomografia por impedância elétrica via mínimos quadrados ponderados aplicável ao monitoramento de fluxo em dutos.* 107 f.: il.; 30 cm. Dissertação (Mestrado). Universidade de Caxias do Sul, Programa de Pós-Graduação em Engenharia Mecânica.
33. Mueller, J. L., & Siltanen, S. (2012). *Linear and nonlinear inverse problems with practical applications.* SIAM.
34. Xu, G., Wang, R., Zhang, S., Yang, S., Justin, G. A., Sun, M., & Yan, W. (2007, August). A 128-electrode three dimensional electrical impedance tomography system. In *Proceedings of the 29th annual international conference of the IEEE EMBS* (pp. 23–26). Cité Internationale, Lyon, France.
35. Aristovich, K. Y., Packham, B. C., Koo, H., dos Santos, G. S., McEvob, A., & Holder, D. S. (2016). Imaging fast electrical activity in the brain with electrical impedance tomography. *NeuroImage, 124*, 204–213. https://doi.org/10.1016/j.neuroimage.2015.08.071.
36. Dowrick, T., Santos, G. S., Vongerichten, A., & Holder, D. (2015). Parallel, multi frequency EIT measurement, suitable for recording impedance changes during epilepsy. *Journal of Electrical Bioimpedance, 6*(1), 37–43.
37. Faulkner, M., Hannan, S., Aristovich, K., Avery, J., & Holder, D. (2018). Feasibility of imaging evoked activity throughout the rat brain using electrical impedance tomography. *NeuroImage, 178*, 1–10. https://doi.org/10.1016/j.neuroimage.2018.05.022.

38. Goren, N., Avery, J., Dowrick, T., Mackle, E., Witkowska-Wrobel, A., Werring, D., & Holder, D. (2018). Multi-frequency electrical impedance tomography and neuroimaging data in stroke patients. *Scientific Data, 5*, 180112. https://doi.org/10.1038/sdata.2018.112.

39. McEwan, A., Romsauerova, A., Yerworth, R., Horesh, L., Bayford, R., & Holder, D. (2006). Design and calibration of a compact multi-frequency EIT system for acute stroke imaging. *Physiological Measurement, 27*, S199–S210.

40. Leite, E. V. C., Gomes, S., Beraldo, M. A., Volpe, M. S., Schettino, I. A. L., Tucci, M. R., Bohm, S. H., Tanaka, H., Lima, R. G., Carvalho, C. R. R., & Amato, M. B. P. (2008). Real-time detection of pneumothorax using electrical impedance tomography. *Critical Care Medicine, 36*, 1–10.

41. Masner, A., Blasina, F., & Simini, F. (2019). Electrical impedance tomography for neonatal ventilation assessment: A narrative review. In *3rd Latin-American conference on Bioimpedance IOP Conf. Series: Journal of Physics: Conf. Series 1272.* 012008 IOP Publishing. https://doi.org/10.1088/1742-6596/1272/1/012008.

42. Mellenthin, M. M., et al. (2019). The ACE1 electrical impedance tomography system for thoracic imaging. *IEEE Transactions on Instrumentation and Measurement, 68*, 3137–3150.

43. Yoshida, T., et al. (2016). Spontaneous effort during mechanical ventilation. *Critical Care Medicine, 44*, 1–e688.

44. Cherepenin, V., Karpov, A., Korjenevsky, A., Kornienko, V., Kultiasov, Y., Mazaletskaya, A., & Mazourov, D. (2001). Preliminary static EIT images of the thorax in health and disease. *Physiological Measurement, 23*(1), 33.

45. Rosell, J., Colominas, J., Riu, P., Pallas-Areny, R., & Webster, J. G. (1988). Skin impedance from 1 Hz to 1 MHz. *IEEE Transactions on Biomedical Engineering, 35*(8), 649–651.

46. Woo, E. J., Hua, P., Webster, J. G., & Tompkins, W. J. (1994). Finite-element method in electrical impedance tomography. *Medical and Biological Engineering and Computing, 32*, 530–536.

47. Brown, B. H., & Seagar, A. D. (1987). The Sheffield data collection system. *Clinical Physics and Physiological Measurement, 8*, 91–97.

48. Wilson, A. J., Milnes, P., Waterworth, A. R., Smallwood, R. H., & Brown, B. H. (2001). Mk3.5: A modular, multi-frequency successor to the Mk3a EIS/EIT system. *Physiological Measurement, 22*, 49–54.

49. Lidgey, F. J., et al. (1992). Electrode current determination from programmable voltage sources. *Clinical Physics and Physiological Measurement, 13*, 43–46.

50. Zhu, Q., Lionheart, W. R. B., Lidgey, F. J., McLeod, C. N., Paulson, K. S., & Pidcock, M. K. (1993). An adaptive current tomograph using voltage sources. *IEEE Transactions on Biomedical Engineering, 40*(2), 63–68.

51. Zhu, Q. S., et al. (1994). Electrical impedance tomography electrode current determination from programmable voltage sources. *Clinical Physics and Physiological Measurement, 15*, 37–43.

52. McLeod, C. N., Denyer, C. W., Lidgey, F. J., Lionheart, W. R. B., Paulson, K. S., Pidcock, M. K., & Shi, Y. (1996). High speed in vivo chest imaging with OXBACT III. In *Proceedings of 18th annual international conference of the IEEE Engineering in Medicine and Biology Society.* Amsterdam.

53. Yerworth, R. J., Bayford, R. H., Cusick, G., Conway, M., & Holder, D. S. (2002). Design and performance of the UCLH Mark 1b 64 channel electrical impedance tomography (EIT) system, optimized for imaging brain function. *Physiological Measurement, 23*, 149–158.

54. Newell, J. C., Gisser, D. G., & Isaacson, D. (1988). An electric current tomography. *IEEE Transactions on Biomedical Engineering, n., 35*, 828–833.

55. Gisser, D. G., et al. (1991). Analog electronics for a high-speed high-precision electrical impedance tomography. In *Proceedings of the annual international conference of the IEEE EMBS (Engineering in Medicine and Biology Society)* (Vol. 13, pp. 23–24).

56. Liu, Z., He, M., & Xiong, H. (2005). Simulation study of the sensing field in electromagnetic tomography for two-phase flow measurement. *IEEE Transactions on Medical Imaging, 16*, 199–204.

57. Hartov, A., Mazzarese, R. A., Reiss, F. R., Kerner, T. E., Osterman, K. S., Williams, D. B., & Paulsen, K. D. A. (2000). Multichannel continuously selectable multifrequency electrical impedance spectroscopy measurement system. *IEEE Transactions on Biomedical Engineering, 47*(1), 49–58.

58. Casas, O., Rosell, J., Bragós, R., Lozano, A., & Riu, P. J. (1996). A parallel broadband real-time system for electrical impedance tomography. *Physiological Measurement, 17*, A1–A6.

59. Oh, T. I., Lee, K. H., Kim, S. M., Koo, H., Woo, E. J., & Holder, D. (2007). Calibration methods for a multi-channel multi-frequency EIT system. *Physiological Measurement, 28*, 1175–1188.

60. Oh, T. I., Lee, J., Seo, J. K., Kim, S. W., & Woo, E. J. (2007). Feasibility of breast cancer lesion detection using a multi-frequency trans-admittance scanner (TAS) with 10 Hz to 500 kHz bandwidth. *Physiological Measurement, 28*(7), S71.

61. Santos, E., Oliveira, W., Hurtado, J., & Simini, F. (2011). Tomografia de impedância elétrica para el seguimento del edema de pulmón: estado del arte y propuesta del proyecto IMPETOM. In *XVIII Congreso Argentino de Bioingeniería SABI 2011*. Mar del Plata.

62. Lima, C. R., et al. (2007). Electrical impedance tomography through constrained sequential linear programming: a topology optimization approach. *Measurement Science and Technology, 18*(9), 2847.

63. TIMPEL. (2019). ENLIGHT 1800. Disponível em: http://www.timpel.com.br/pt/. Acesso em: 12 abr. de 2019.

64. Dräger. (2020). Dräger PulmoVista® 500. Disponível em: https://static.draeger.com/trainer/pulmovista_500_trainer_en1/index.html#id=A1100. Acesso em: 25 jul. de 2020.

65. Hauptmann, A., Kolehmainen, V., Mach, N. M., Savolainen, T., Seppänen, A., & Siltanen, S. (2017). Open 2D electrical impedance tomography data archive. *arXiv:1704.01178v1 [physics.med-ph]*.

66. Rosa, R. G., Rutzen, W., Madeira, L., Ascoli, A. M., Dexheimer Neto, F. L., Maccari, J. G., de Oliveira, R. P., & Teixeira, C. (2015). Uso da tomografia por impedância elétrica torácica como ferramenta de auxílio às manobras de recrutamento alveolar na síndrome do desconforto respiratório agudo: relato de caso e breve revisão da literatura. *Revista Brasileira de Terapia Intensiva, 27*(4), 406–411.

67. Feliu, S., Galvan, J. C., & Morcillo, M. (1989). Interpretation of electrical impedance diagram for painted galvanized steel. *Progress in Organic Coatings, 17*(2), 143–153.

68. Webster, J. G. (1990). *Electrical impedance tomography*. Adam Hilger Pub.

69. Dines, K. A., & Lytle, R. J. (1981). Analysis of electrical conductivity imaging. *Geophysics, 46*(7), 1025–1036.

70. Thomas, A. J., Kim, J. J., Tallman, T. N., & Bakis, C. E. (2019). Damage detection in self-sensing composite tubes via electrical impedance tomography. *Composites Part B: Engineering, 177*(15), 107276, ISSN 1359-8368. https://doi.org/10.1016/j.compositesb.2019.107276.

71. Barber, D. C., & Brown, B. C. (1983). Imaging spatial distributions of resistivity using applied potential tomography. *Electronics Letters, 19*(22), 933.

72. Wolff, J. G. B. (2011). *Análise Computacional de Campos e Correntes em um sistema de tomografia de indução magnética* (Dissertação de Mestrado). Joinville: Universidade do Estado de Santa Catarina – UDESC. Orientador: Dr. Airton Ramos.

73. Hahn, G., et al. (1995). Changes in the thoracic impedance distribution under different ventilatory conditions. *Physiological Measurement, 16*, 161–173.

74. Kobylianskii, J., et al. (2016). Electrical impedance tomography in adult patients undergoing mechanical ventilation: A systematic review. *Journal of Critical Care, 35*, 33–50. https://doi.org/10.1016/j.jcrc.2016.04.028.

75. Wals, B. K., & Smallwood, C. D. (2016). Electrical impedance tomography during mechanical ventilation. *Respiratory Care, 61*(10), 1417–1424. https://doi.org/10.4187/respcare.04914.

76. Borsic, A., Halter, R., Wan, Y., Hartov, A., & Paulsen, K. D. (2009). Sensitivity study and optimization of a 3D electric impedance tomography prostate probe. *Physiological Measurement, 30*(6), S1–S18. https://doi.org/10.1088/0967-3334/30/6/S01.

77. Wan, Y., Halter, R., Borsic, A., Manwarin, P., Hartov, A., & Paulsen, K. D. (2010). Sensitivity study of an ultrasound coupled transrectal electrical impedance tomography system for prostate imaging. In *In: International conference on Electrical Bioimpedance. IOP Publishing. Journal of Physics: Conference Series, 224* (p. 012067). https://doi.org/10.1088/1742-6596/224/1/012067.

78. Borsic, A., Halter, R., Wan, Y., Hartov, A., & Paulsen, K. D. (2010). Electrical impedance tomography reconstruction for three dimensional imaging of the prostate. *Physiological Measurement, 31*(8), S1–S16. https://doi.org/10.1088/0967-3334/31/8/S01.

79. Zou, Y., & Guo, Z. (2003). A review of electrical impedance techniques for breast cancer detection. *Medical Engineering & Physics, 25*(2), 79–90. https://doi.org/10.1016/S1350-4533(02)00194-7.

80. Hong, S., Lee, K., Há, U., Kim, H., Lee, Y., Kim, Y., & Yoo, H.-J. (2015). A 4.9 mΩ-sensitivity mobile electrical impedance tomography IC for early breast-cancer detection system. *IEEE Journal of Solid-State Circuits, 50*(1), 245–257.

81. Aberg, P., Nicander, I., Hansson, J., Geladi, P., Holmgren, U., & Ollmar, S. (2004). Skin cancer identification using multifrequency electrical impedance – a potential screening tool. *IEEE Transactions on Biomedical Engineering, 51*(12), 2097–2102.

82. Braun, R. P., Mangana, J., Goldinger, S., French, L., Dummer, R., & Marghoob, A. A. (2017). Electrical impedance spectroscopy in skin cancer diagnosis. (Elsevier). *Dermatologic Clinics, 35*(4), 489–493.

83. Abboud, M. et al. (1995). Monitoring of peripheral edema using electrical bioimpedance measurements. In *Conf IEEE EMBS*, pp. 641–642.

84. Holder, D. S. (1992). Detection of cerebral ischaemia in the anaesthetised rat by impedance measurement with scalp electrodes: implications for non-invasive imaging of stroke by electrical impedance tomography. *Clinical Physics and Physiological Measurement, 13*(1), 36. https://doi.org/10.1088/0143-0815/13/1/006.

85. Trokhanova, O. V., Chijova, Y. A., Okhapkin, M. B., et al. (2013). Possibilities of electrical impedance tomography in gynecology. *Journal of Physics: Conference Series, 434*, 012038.

86. Mangnall, Y. F., Kerrigan, D. D., Johnson, A. G., & Read, N. W. (1991). Applied potential tomography noninvasive method for measuring gastric emptying of a solid test meal. *Digestive Diseases and Sciences, 36*(12), 1680–1684.

87. Murdoch, C., Brown, B. H., Hearnden, V., Speight, P. M., D'Apice, K., Hegarty, A. M., Tidy, J. A., Healey, T. J., Highfield, P. E., & Thornhill, M. H. (2014). Use of electrical impedance spectroscopy to detect malignant and potentially malignant oral lesions. *International Journal of Nanomedicine, 9*, 4521–4532. https://doi.org/10.2147/IJN.S64087.

88. Richter, I., Alajbeg, I., Boras, V. V., Rogulj, A. A., & Brailo, V. (2015). Mapping electrical impedance spectra of the healthy oral mucosa: A pilot study. *Acta Stomatologica Croatica, 49*(4), 331–339. https://doi.org/10.15644/asc49/4/9.

89. Sun, T.-P., et al. (2010). The use of bioimpedance in the detection/screening of tongue cáncer. *Cancer Epidemiology, 34*, 207–211.

90. Vonk-Noordegraaf, A., et al. (1997, May). *Noninvasive assessment of right ventricular diastolic function by electrical impedance tomography, Chest, 111*(5), 1222–1228.

91. Deibele, J. M., Luepschen, H., & Leonhardt, S. (2008). Dynamic separation of pulmonary and cardiac changes in electrical impedance tomography. *Physiological Measurement, 29*(6). https://doi.org/10.1088/0967-3334/29/6/S01.

92. Vonk-Noordegraaf, A., et al. (2000). Determination of stroke volume by means of electrical impedance tomography. *Physiological Measurement, 21*, 285–293.

93. Maisch, S., Bohm, S. H., Sola, J., Goepfert, M. S., Kubitz, J. C., Richter, H. P., Ridder, J., Goetz, A. E., & Reuter, D. A. (2011). Heart–lung interactions measured by electrical impedance tomography. Brief Communications. *Critical Care Medicine, 39*(9), 2173–2176.

94. Trokhanova, O. V., Chijova, Y. A., Okhapkin, M. B., Korjenevsky, A. V., & Tuykin, T. S. (2010). Using of electrical impedance tomography for diagnostics of the cervix uteri diseases. *Journal of Physics: Conference Series, 224*, 012068. https://doi.org/10.1088/1742-6596/224/1/012068.

95. Barrow, A. J., & Wu, S. M. (2007, October). Impedance measurements for cervical cancer diagnosis. *Gynecologic Oncology, 107*(1, Supplement), S40–S43. https://doi.org/10.1016/j.ygyno.2007.07.030.
96. Carpenter, J. (2013). Images capture moment brain goes unconscious. *BBC News: Science & Environment.* UK: BBC.
97. Sophocleous, L., Frerichs, I., Miedema, M., Kallio, M., Papadouri, T., Karaoli, C., Bechter, T., Tingay, D. G., Van Kaam, A. H., Bayford, R., & Waldmann, A. D. (2018). Clinical performance of a novel textile interface for neonatal chest electrical impedance tomography. *Physiological Measurement, 39*, 044004, 11 pp.
98. Chen, R., Lovas, A., Krüger-Ziolek, S., Benyó, B., & Möller, K. (2021). EIT based time constant analysis to determine different types of patients in COVID-19 pneumonia. In T. Jarm, A. Cvetkoska, S. Mahnič-Kalamiza, & D. Miklavcic (Eds.), *8th European Medical and Biological Engineering conference. EMBEC 2020. IFMBE proceedings* (Vol. 80). Cham, Springer. https://doi.org/10.1007/978-3-030-64610-3_52.
99. Diehl, J.-L., Peron, N., Chocron, R., Debuc, B., Guerot, E., Hauw-Berlemont, C., Hermann, B., Augy, J. L., Younan, R., Novara, A., Langlais, J., Khider, L., Gendron, N., Goudot, G., Fagon, J.-F., Mirault, T., & Smadja, D. M. (2020). Respiratory mechanics and gas exchanges in the early course of COVID-19 ARDS: A hypothesis-generating study. *Annals of Intensive Care, 10*(95). https://doi.org/10.1186/s13613-020-00716-1.
100. Tomasino, S., Sassanelli, R., Marescalco, C., Meroi, F., Vetrugno, L., & Bove, T. (2020). Electrical impedance tomography and prone position during ventilation in COVID-19 pneumonia: Case reports and a brief literature review. *Seminars in Cardiothoracic and Vascular Anesthesia, 24*(4), 287–292. https://doi.org/10.1177/1089253220958912.
101. Zhao, Z., Kung, W.-H., Chang, H.-T., Hsu, Y.-L., & Frerichs, I. (2020). COVID-19 pneumonia: Phenotype assessment requires bedside tools. *Critical Care, 24*(272). https://doi.org/10.1186/s13054-020-02973-9.
102. Zhao, Z., Zhang, J. S., Chen, Y. T., Chang, H. T., Hsu, Y. L., Frerichs, I., & Adler, A. (2021). The use of electrical impedance tomography for individualized ventilation strategy in COVID-19: A case report. *BMC Pulmonary Medicine, 21*, 38.
103. Shono, A., Kotani, T., & Frerichs, I. (2021). Personalisation of therapies in COVID-19 associated acute respiratory distress syndrome, using electrical impedance tomography. *The Journal of Critical Care Medicine, 7*(1), 62–66.
104. Andreuccetti, D., Fossi, R., & Petruccii, C. (2017). *Calculation of the dielectric properties of body tissues in the frequency range 10 Hz–100 GHz.* Disponível em: http://niremf.ifac.cnr.it/tissprop/document/tissprop.pdf. Acesso em: 10 mar. de 2021.
105. Yojo, A. Y. (2008). *Construção do Hardware de um Tomógrafo por Impedância Elétrica.* São Paulo: USP.
106. Pursiainen, S., & Hakula, H. (2006, March 26–29). A high-order finite element method for electrical impedance tomography. In *Progress in electromagnetics research symposium.* Cambridge, USA, pp. 260–264.
107. Xu, Y., & He, B. (2005). Magnetoacoustic tomography with magnetic induction (MAT-MI). *Physics in Medicine & Biology, 50*, 5175–5187. https://doi.org/10.1088/0031-9155/50/21/015.
108. Bagshaw, A. P., Liston, A. D., Bayford, R. H., Tizzard, A., Gibson, A. P., Tidswell, A. T., Sparkes, M. K., Dehghani, H., Binnie, C. D., & Holder, D. S. (2003). Electrical impedance tomography of human brain function using reconstruction algorithms based on the finite element method. *Neuroimage, 20*, 752–764.
109. Borsic, A., McLeod, C. N., & Lionheart, W. R. B. (2001). Total variation regularization in EIT reconstruction. In *Proceedings of the 2nd world congress on Industrial Process Tomography* (p. 433).
110. Woo, E. J., Webster, J. G., & Tompkins, W. J. (1990). The improved Newton-Raphson method and its parallel implementation for static impedance imaging. *Annual International Conference of the IEEE Engineering in Medicine and Biology Society, 12*(1), 102–103.

111. Aster, R. C., Borchers, B., & Thurber, C. H. (2018). *Parameter estimation and inverse problems* (3rd ed., p. 404). Elesevier.
112. Hansen, P. C. (1998). Analysis of discrete ill-posed problems by means of the L-curve. *SIAM Rev, 34*, 561–580.
113. Jin, B., & Maass, P. (2010). *An analysis of electrical impedance tomography with applications to Tikhonov regularization*. Available in: http://citeseerx.ist.psu.edu/viewdoc/download;jsessionid=8B900A3C238DA0361A6177B9C7100460?doi=10.1.1.660.1512&rep=rep1&type=pdf. Accessed 26 July 2021.
114. Murai, T., & Kagawa, Y. (1985). Electrical impedance computed tomography based on a finite element model. *IEEE Transactions on Biomedical Engineering, 32*(3), 177–184.
115. Hua, et al. (1993). Finite element modeling of electrode-skin contact impedance in electrical impedance tomography. *IEEE Transactions on Biomedical Engineering, 40*(4), 335–343.
116. Adler, A., & Lionheart, W. R. B. (2006). *Uses and abuses of EIDORS: An extensible software base for EIT*. Disponível em: http://sce.carleton.ca/faculty/adler/publications/2005/adler-lionheart-2005-EIDORS.pdf. Acesso em: 11/03/2019.
117. Barber, D. C., & Seagar, A. D. (1987). Fast reconstruction of resistance images. *Clinical Physics and Physiological Measurement, 8*(Suppl. A), 47–54.
118. Gonçalves, R. F. L. (2014). *Electrical impedance tomography – thorax*. Instituto Superior Técnico, Lisboa, Portugal. Disponível em: https://pdfs.semanticscholar.org/4a7e/1adecf3841cde5ebf996a03bb505662d14f4.pdf. Acesso em: 11 abr. de 2020.
119. Barber, D. C., Brown, B. H., & Avis, N. J. (1992). Image reconstruction in electrical impedance tomography using filtered back-projection. In *14th annual international conference of the IEEE Engineering in Medicine and Biology Society*.
120. Cohen-Bacrie, C., Goussard, Y., & Guardo, R. (1997). Regularized reconstruction in electrical impedance tomography using a variance uniformization constraint. *IEEE Transactions on Medical Imaging, 16*(5), 562–571.
121. Dusek, J., Hladky, D., & Mikulka, J. (2017). Electrical impedance tomography methods and algorithms processed with a GPU. In *2017 Progress In Electromagnetics Research Symposium – Spring (PIERS)* (pp. 22–25).
122. Liu, J., Lin, L., Zhang, W., & Li, G. (2013). A novel combined regularization algorithm of total variation and Tikhonov regularization for open electrical impedance tomography. *Physiological Measurement, 34*(7), 823–838.
123. Kohn, R. V., & McKenney, A. (1987). Numerical implementation of a variational method for electrical impedance tomography. *Inverse Problems, 6*(3), 389.
124. Kohn, R. V., & Vogelius, M. (1987). Relaxation of a variational method for impedance computed tomography. *Communication on Pure and Applied Mathematics, 40*(6), 745–777. https://doi.org/10.1002/cpa.3160400605.
125. Knowles, I., & Renka, R. J. (2014). Methods for numerical differentiation of noisy data. Variational and topological methods: Theory, applications, numerical simulations, and open problems (2012). *Electronic Journal of Differential Equations, Conference, 21*, 235–246. ISSN: 1072-6691. http://ejde.math.txstate.edu, http://ejde.math.unt.edu, http://ejde.math.txstate.edu
126. Polydorides, N., & Lionheart, W. R. B. (2002). A Matlab toolkit for three-dimensional electrical impedance tomography: A contribution to the electrical impedance and diffuse optical reconstruction software project. *Measurement Science and Technology, 13*, 1871–1883.
127. Brandstätter, B., Hollaus, K., Hutten, H., Mayer, M., Merwa, R., & Scharfetter, H. (2003). Direct estimation of Cole parameters in multifrequency EIT using a regularized Gauss–Newton method. *Physiological Measurement, 24*(2), 437.
128. Islam, M. R., & Kiber, M. A. (2014). Electrical impedance tomography imaging using Gauss–Newton algorithm. In *International conference on Informatics, Electronics Vision (ICIEV)* (pp. 1–4).
129. Bakushinskii, A. B. (1992). The problem of the convergence of the iteratively regularized Gauss-Newton method. *Computational Mathematics and Mathematical Physics, 32*, 1353–1359.

130. Rao, L., He, R., Wang, Y., Yan, W., Bai, J., & Ye, D. (1999). An efficient improvement of modified Newton-Raphson algorithm for electrical impedance tomography. *IEEE Transactions on Magnetics, 35*(3), 1562–1565.
131. Alsaker, M. (2016). Computational advancements. In *The D-bar reconstruction method for 2-D electrical impedance tomography* (Dissertation). Colorado, Colorado State University. Advisor: Jennifer L. Mueller.
132. Dangelo, M., & Mueller, J. L. (2010). 2D D-bar reconstructions of human chest and tank data using an improved approximation to the scattering transform. *Physiological Measurement, 31*(2), 221.
133. Mueller, J. L., & Siltanen, S. (2009). The D-bar method for electrical impedance tomography—Demystified. *Inverse Problems, 36*, 3001, 28 pp. https://doi.org/10.1088/1361-6420/aba2f5.
134. Isaacson, D., Mueller, J. L., Newell, J. C., & Siltanen, S. (2006). Imaging cardiac activity by the D-bar method for electrical impedance tomography. *Physiological Measurement, 27*, S43–S50. https://doi.org/10.1088/0967-3334/27/5/S04.
135. Hamilton, S. J.; Herrera, C. N. L.; Mueller, J. L.; Von Herrmann A. (2012). A direct D-bar reconstruction algorithm for recovering a complex conductivity in 2D. Inverse Problems, vol. 28, 095005 24 pp. DOI:https://doi.org/10.1088/0266-5611/28/9/095005.
136. Hamilton, S. J., Isaacson, D., Kolehmainen, V., Muller, P. A., Toivanen, J., & Bray, P. F. (2020). *3D EIT reconstructions from electrode data using direct inversion D-bar and Calderon methods.* https://arxiv.org/pdf/2007.03018.pdf
137. Hrabuska, R., Prauzek, M., Venclikova, M., & Konecny, J. (2018). Image reconstruction for electrical impedance tomography: Experimental comparison of radial basis neural network and Gauss – Newton method. *IFAC-PapersOnLine, 51*(6), 438–443.
138. Hamilton, S. J., Hänninen, A., Hauptmann, A., & Kolehmainen, V. (2019). Beltrami-net: domain-independent deep D-bar learning for absolute imaging with electrical impedance tomography (a-EIT). *Physiological Measurement, 40*(7), 074002.
139. Agnelli, J. P., Öl, A. C., Lassas, M., Murthy, R., Santacesaria, M., & Siltanen, S. (2020). Classification of stroke using neural networks in electrical impedance tomography. *Inverse Problems, 36*, 115008, 26 pp. https://doi.org/10.1088/1361-6420/abbdcd.
140. Gomes, J. C., Barbosa, V. A. F., Ribeiro, D. E., de Souza, R. E., & dos Santos, W. P. (2020). *Electrical impedance tomography image reconstruction based on backprojection and extreme learning machines* (Manuscript). Recife, Universidade Federal de Pernambuco.
141. Siltanen, S., & Ide, T. (2020, September 21–24). Electrical impedance tomography, enclosure method and machine learning. In *2020 IEEE international workshop on Machine Learning for Signal Processing*. Espoo, Finlândia.
142. Feitosa, A. R. S., Ribeiro, R. R., Barbosa, V. A. F., de Souza, R. E., & dos Santos, W. P. (2014). Reconstruction of electrical impedance tomography images using particle swarm optimization, genetic algorithms and non-blind search. In *5th ISSNIP-IEEE biosignals and biorobotics conference: Biosignals and robotics for better and safer living (BRC)*. Salvador, Bahia. https://doi.org/10.1109/BRC.2014.6880996.
143. Yerworth, & Bayford. (2017). DICOM for EIT. In *Proceedings of the 18th international conference on biomedical applications of electrical impedance tomography. 21-24 June 2017.* Hanover, New Hampshire, USA. [Conference or Workshop Item]. https://doi.org/10.5281/zenodo.557093.
144. Adler, et al. (2009). GREIT: a unified approach to 2D linear EIT reconstruction of lung images. *Physiological Measurement, 30*(6), S35.
145. EIDORS. (2021). EIDORS: electrical impedance tomography and diffuse optical tomography reconstruction software. Available in: http://eidors3d.sourceforge.net/. Accessed 26 July 2021.
146. Dalvi-Garcia, F. (2012). *Algoritmo para Reconstrução de Imagens Bidimensionais para Sistema de Tomografia por Impedância Elétrica Baseado em Configuração Multiterminais.* Rio de Janeiro: UFRJ/COPPE, 2012. XX, 112 p. Orientadores: Marcio Nogueira de Souza e Alexandre Visintainer Pino.

147. Tang, M., Wang, W., Wheeler, J., McCormick, M., & Dong, X. (2002). The number of electrodes and basis functions in EIT image reconstruction. *Physiological Measurement, 23,* 129–140.

148. Guan, W. J., Ni, Z. Y., Hu, Y., et al. (2020). Clinical characteristics of coronavirus disease 2019 in China. *The New England Journal of Medicine, 382,* 1708–1720. https://doi.org/10.1056/NEJMoa2002032.

149. Fu, et al. (2020). Deep learning in medical image registration: a review. *Physics in Medicine and Biology, 65,* 20TR01.

150. Wolf, et al. (2012). Reversal of dependent lung collapse predicts response to lung recruitment in children with early acute lung injury. *Pediatric Critical Care Medicine, 13*(5), 509–515. https://doi.org/10.1097/PCC.0b013e318245579c.

151. Siltanen, S., Mueller, J., & Isaacson, D. (2000). An implementation of the reconstruction algorithm of A Nachman for the 2-D inverse conductivity problem. *Inverse Problems, 16,* 681–699.

152. Siltanen, S., & Mueller, J. (2012). *Linear and nonlinear inverse problems with practical applications.* USA: SIAM.

153. Dodd, M., & Mueller, J. L. (2014). A real-time D-bar algorithm for 2-D electrical impedance tomography data. *Inverse Problems and Imaging, 8*(4), 1013–1031. https://doi.org/10.3934/ipi.2014.8.1013.

154. Siltanen, S. (2017). *The D-bar method for electrical impedance tomography – Simulated data.* Available in: https://blog.fips.fi/tomography/eit/the-d-bar-method-forelectrical-impedance-tomography-simulated-data/. Access in: 28 dez. de 2020.

155. Freeston, I. L., & Tozer, R. C. (1995). Impedance imaging using induced currents. *Physiological Measurement, 16,* A257–AZ66.

156. Gençer, N. G., Ider, Y. Z., & Williamsom, S. J. (1996). Electrical impedance tomography: Inducedcurrent imaging achieved with multiple coil system. *IEEE Transactions on Biomedical Engineering, 43*(2), 139–149.

157. Tozer, J. C., Ireland, R. H., Barber, D. C., & Baker, A. T. (1998, April 5–9). Magnetic impedance tomography. In *Proceedings of the 10th international conference on Electrical Bioimpedance* Barcelona, Spain, pp. 369–372.

158. Lee, et al. (2005). Electrical conductivity images of biological tissue phantoms in MREIT. *Physiological Measurement, 26*(2005), S279–S288.

159. Kwon, O., Woo, E. J., Yoon, J. R., & Seo, J. K. (2002). Magnetic resonance electrical impedance tomography (MREIT): Simulation study of J-substitution algorithm. *IEEE Transactions on Biomedical Engineering, 48,* 160–167.

160. Lee, et al. (2003). Three-dimensional forward solver and its performance analysis for magnetic resonance electrical impedance tomography (MREIT) using recessed electrodes. *Physics in Medicine and Biology, 48*(13), 1971–1986.

161. Birgul, et al. (2003). Experimental results for 2D magnetic resonance electrical impedance tomography (MR-EIT) using magnetic flux density in one direction. *Physics in Medicine and Biology, 48,* 3485–3504.

162. Hasanov, et al. (2004). A new approach to current density impedance imaging. In *The 26th annual international conference of the IEEE engineering in medicine and biology society* (pp. 1321–1324).

163. Joy, M. L. G. (2004). MR current density and conductivity imaging: the state of the Aart. In *The 26th annual international conference of the IEEE engineering in medicine and biology society* (pp. 5315–5319).

164. Seo, J. K., & Woo, E. J. (2011). Magnetic resonance electrical impedance tomography (MREIT). *SIAM Review, 53*(1), 40–68.

165. Woo, E. J., & Seo, J. K. (2008). Magnetic resonance electrical impedance tomography (MREIT) for high-resolution conductivity imaging. *Physiological Measurement, 29*(10), 49.

166. Seo, J. K., Yoon, J. R., Woo, E. J., & Kwon, O. (2003). Reconstruction of conductivity and current density images using only one component of magnetic field measurements. *IEEE Transactions on Biomedical Engineering, 50*(9), 1121–1124.

167. Wang, C., Zhang, J., Li, F., Cui, Z., & Xu, C. (2011). Design of a non-magnetic shielded and integrated electromagnetic tomography system. *Measurement Science and Technology, 22*(10), 104007.
168. Siltanen, et al. (2017b). The D-bar method for electrical impedance tomography – simulated data. 2017. Available in: https://blog.fips.fi/tomography/eit/the-d-bar-method-for-electrical-impedance-tomography-simulated-data/. Accessed 28 Dec 2020.
169. Henderson, R. P., & Webster, J. G. (1978). An impedance camera for spatially specific measurements of the thorax. *IEEE Transactions on Biomedical Engineering, BME-25*(3), 250–254.

Printed in the United States
by Baker & Taylor Publisher Services